Adsorbents in Treatment of Pollutants

Adsorbents in Treatment of Pollutants

Guest Editor
Wei Wang

MDPI
Basel • Beijing • Wuhan • Barcelona • Belgrade • Novi Sad • Cluj • Manchester

Guest Editor
Wei Wang
Beijing Key Laboratory of
Environmentally Harmful
Chemical Analysis
Beijing University of
Chemical Technology
Beijing
China

Editorial Office
MDPI AG
Grosspeteranlage 5
4052 Basel, Switzerland

This is a reprint of the Special Issue, published open access by the journal *Molecules* (ISSN 1420-3049), freely accessible at: https://www.mdpi.com/journal/molecules/special_issues/N75LB53J38.

For citation purposes, cite each article independently as indicated on the article page online and as indicated below:

Lastname, A.A.; Lastname, B.B. Article Title. *Journal Name* **Year**, *Volume Number*, Page Range.

ISBN 978-3-7258-6368-6 (Hbk)
ISBN 978-3-7258-6369-3 (PDF)
https://doi.org/10.3390/books978-3-7258-6369-3

Contents

molecules

MDPI

Article

Simple Co-Precipitation of Iron Minerals for the Removal of Phenylarsonic Acid: Insights into the Adsorption Performance and Mechanism

Lili Wang [1], Changchao Hu [1,*], Ze Yang [2], Songding Guo [2], Tingting Zhang [2] and Shangyi Li [2,*]

1 Petroleum Exploration and Production Research Institute, SINOPEC, Beijing 100083, China; wangll.syky@sinopec.com
2 Department of Environmental Science and Engineering, Research Centre for Resource and Environment, Beijing University of Chemical Technology, Beijing 100029, China; yzzb_0118@163.com (Z.Y.); guosongding123@163.com (S.G.); zhangtt@mail.buct.edu.cn (T.Z.)
* Correspondence: hucc2049@163.com (C.H.); lishangyi@buct.edu.cn (S.L.); Tel.: +86-15810171131 (S.L.)

Abstract: In this study, three kinds of iron minerals, ferrihydrite, hematite, and goethite, were prepared by a simple coprecipitation method for the adsorption and removal of phenylarsonic acid (PAA). The adsorption of PAA was explored, and the influences of ambient temperature, pH, and co-existing anions on adsorption were evaluated. The experimental results show that rapid adsorption of PAA occurs within 180 min in the presence of iron minerals, and the adsorption process conforms to a pseudo-second-order kinetic model. The isothermal adsorption of PAA by ferrihydrite, goethite, and hematite agrees with the Redlich–Peterson model. The maximum adsorption capacities of PAA are 63.44 mg/g, 19.03 mg/g, and 26.27 mg/g for ferrihydrite, goethite, and hematite, respectively. Environmental factor experiments illustrated that an alkaline environment will significantly inhibit the adsorption of PAA by iron minerals. CO_3^{2-}, SiO_3^{2-}, and PO_4^{3-} in the environment will also significantly reduce the adsorption performance of the three iron minerals. The adsorption mechanism was analyzed by FTIR and XPS, which indicated that ligand exchange between the surface hydroxyl group and the arsine group leads to the formation of an Fe-O-As bond, and electrostatic attraction between the iron minerals and PAA played an important role in the adsorption.

Keywords: iron minerals; adsorption; phenylarsonic acid; ligand exchange interaction

1. Introduction

Arsenic (As) is ubiquitous and humans are certainly exposed to arsenic via atmospheric air, groundwater, and food sources [1,2]. Most environmental arsenic contamination originates from anthropogenic activities which leads to millions of people experiencing life-threatening complications by drinking water poisoned and food produced from arsenic-contaminated soils irrigated with arsenic-contaminated water. Arsenic contamination has been noted as both a twentieth and twenty-first century catastrophe by researchers and authorities [3,4]. Unlike inorganic arsenic, the toxicity of organic arsenic compounds depends on the associated organic functional groups [5]. Chronic exposure of humans to organic arsenic contamination can lead to central nervous system disorders and optic nerve atrophy [6]. Although benzene organic arsenic compounds are generally considered less toxic than inorganic arsenic [7–9], organic arsenic substances may still be converted to inorganic arsenic through biotic and abiotic reactions in the natural environment, thereby increasing the risks of inorganic arsenic contamination [10]. However, Kroening et al., after biological toxicity research on the degradation products of arsenic-containing agents, found that the toxicity sequence, from high to low, was phenylarsine oxide, triphenylarsonic acid, phenylarsonic acid, diphenylarsonic acid, and pentavalent inorganic arsenic [11]. Therefore, the removal of organic arsenic from the environment needs to be closely monitored.

The removal of arsenic from water bodies can be performed by chemical flocculation, membrane filtration, and adsorption [12–16] to separate arsenic from the contaminated media. Due to the advantages of simple operation, high removal efficiency, low economic cost, and renewable solid and sustainable utilization, adsorption technologies for removing arsenic from sewage stand out among many treatment technologies. The adsorption performance of the adsorbent materials is the key to removing arsenic from water [17]. Compared with other functional materials, such as activated carbon, clay minerals, and biomass materials, the adsorption and removal performance of iron-based materials for arsenic is better [18] and more cost effective. In studies of arsenic adsorption, iron minerals such as ferrihydrite, hematite (α-Fe_2O_3), and goethite (α-FeOOH) are the most widely used. Different iron minerals exhibit different adsorption properties due to the differences in their chemical structures. Ferrihydrite is the largest of the iron minerals in terms of specific surface area, which is up to 700 m^2/g, followed by goethite at 8~200 m^2/g, and hematite, which exhibits the smallest surface area at only 2~115 m^2/g [19]. Xie et al. [20] completed experiments on iron mineral adsorption of arsenic acid and loxarsine to eliminate the difference in specific surface area between different materials, and the adsorption capacity was normalized and found to satisfy the trend ferrihydrite > hematite > goethite.

The behavior and mechanisms occurring at the interface between iron oxides and contaminant water are important for the study of typical contaminant migration and transformation. Donald Sparks [21–23] worked on the interactions between minerals and contaminants, including outer sphere coordination, inner sphere coordination, particle diffusion, cluster formation or deposition, film or surface diffusion, co-precipitation or doping, and desorption or dissolution processes. Similar studies on the surface behavior of iron oxides and pollutants are also available. Based on this, we studied the interfacial reaction between iron minerals and organic arsenic to gain further insight into the immobilization of organic arsenic by adsorption at the molecular or atomic level. Iron minerals mainly adsorb arsenic by forming inner sphere complexes and outer sphere complexes with arsenic contaminants, where inner sphere complexes are chemisorbed and outer sphere complexes are physically adsorbed; thus, the stability of inner sphere complexes is significantly higher than that of outer sphere complexes [24]. Further studies on the structure of inner sphere complexes have revealed that there are three types of inner sphere complexes, namely bidentate binuclear, bidentate mononuclear, and monodentate mononuclear, among which the bidentate binuclear structure is the most stable [25]. Ferrihydrite is a type of adsorbent material with a strong affinity for arsenic contaminants. Li et al. [26] found that 2-line ferrihydrite with a mesoporous structure possesses excellent adsorption characteristics for arsenic (As(III)) contaminants. Experimental results showed that As (III) contamination can be effectively removed within 2 h by this 2-line ferrihydrite, which is attributed to its special mesoporous structure and large surface area (133 $m^2 \cdot g^{-1}$). The As (III) loading capacity of this mesoporous 2-line ferrihydrite was up to 128 mg (As) g^{-1}(Fe). Masato et al. [27] used XAFS to characterize ferrihydrite adsorbed with organic arsenic and found that PAA and diphenylarsonic acid formed bidentate binuclear and monodentate mononuclear inner sphere complexation models with ferrihydrite, respectively. Goethite, a highly crystalline and stable iron mineral, is dehydroxylated to hematite when the temperature exceeds 120 °C. Compared with goethite, hematite has the same high crystallinity but a slightly weaker thermal stability. Goethite and hematite are also widely used for the removal of various arsenic contaminants due to their abundance, easy preparation, and low cost.

There is no doubt that different crystalline surfaces of iron minerals have different adsorption effects on arsenic contaminants, and the reactive sites on the crystalline surfaces are the key factors controlling the adsorption process. Yan et al. [28] concluded that arsenite and arsenate formed single dentate complexes with (001) crystals and double-dentate complexes with (110) and (214) crystals, determined by XAFS characterization and density functional theory calculations. Chen et al. [29] indicated that DFT-calculated adsorption energies and charge density redistributions for various surface complexes on different charged (101) and (210) facets are consistent with the trends observed in batch measurements, suggesting that the observed behavior reflects the primary controlling influence of goethite surface chemistry at the molecular scale. Cao et al. [30] found that organic arsenic adsorbed on two facets (e.g., (001) and (012)) of hematite through the same non-protonated inner sphere coordination but in different molecular configurations. Surface complexation models were set up to analyze the preferred coordination geometries dependent on the facets, and the (012) facet was found to favor PAA adsorption more than the (001) facet.

In this work, three common iron minerals, including ferrihydrite, goethite, and hematite, were prepared by the co-precipitation method as adsorbent materials for the adsorption and removal of PAA. The physicochemical properties of the three iron minerals were characterized by scanning electron microscopy, N_2 adsorption and desorption tests, Fourier infrared spectroscopy, and X-ray energy spectroscopy. The differences in the adsorption capacity of the three iron minerals for PAA were evaluated by isothermal adsorption, adsorption thermodynamics, and adsorption kinetics experiments. The effects of environmental factors such as pH and coexisting ions on the adsorption performance of the iron minerals were investigated, and the possible adsorption mechanisms were inferred by material characterization before and after adsorption.

2. Results

2.1. Characterizations of Ferrihydrite, Hematite, and Goethite

Figure 1 shows the SEM image of ferrihydrite (a), hematite (b), and goethite (c). The ferrihydrite exhibited an irregular particle structure with a rough surface and abundant pores [31]. Fine particles were agglomerated on the surface of hematite, which was very rough. Goethite showed a typical 3D needle-like structure, which had a certain mechanical stability against wear. The XRD characterization results are shown in Figure 2a–c. The XRD pattern of ferrihydrite shows wide diffraction peaks at 36° and 62°, indicating that the preparation is definitely an amorphous iron mineral. In contrast, goethite and hematite exhibited multiple diffraction peaks in the range of 10°~80°, and the peaks are narrow and pointed, indicating that synthetic goethite and synthetic hematite showed high crystallinity [32]. By comparing with the XRD patterns of the standard goethite and hematite, it was found that the two kinds of iron minerals are in good conformity with the standard materials. During the synthesis of iron minerals, it was found that the pH of the solution and the synthesis temperature played an important role in their crystal transformation.

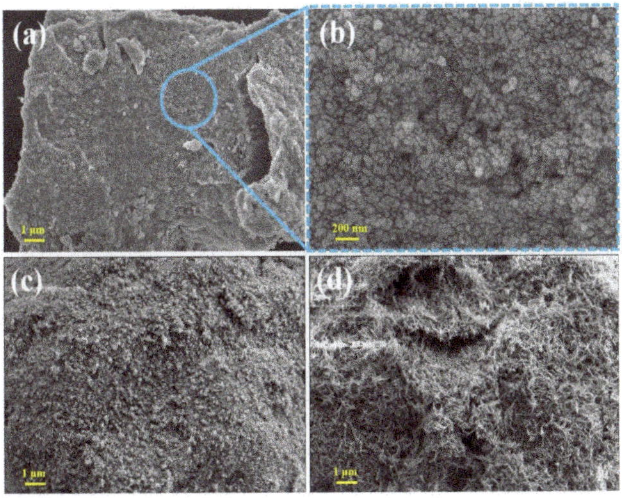

Figure 1. SEM images of (**a**) ferrihydrite, (**b**) ferrihydrite with a higher magnification, (**c**) hematite, and (**d**) goethite.

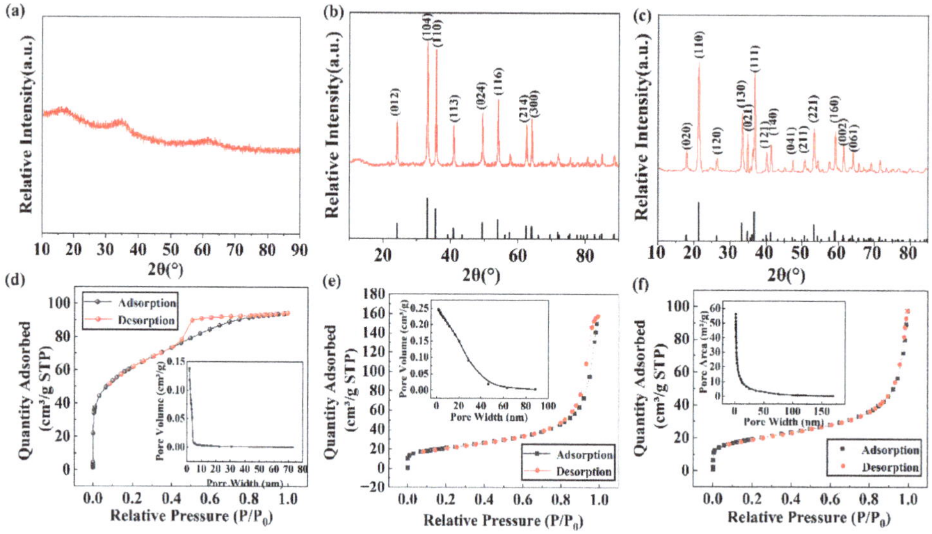

Figure 2. XRD patterns of iron minerals with different crystallinity: (**a**) ferrihydrite, (**b**) hematite, and (**c**) goethite. N_2 adsorption/desorption isotherms and pore size distribution of (**d**) ferrihydrite, (**e**) hematite, and (**f**) goethite. a.u.—arbitrary units.

The pore structure analysis of the three iron minerals is shown in Figure 2d–e. The adsorption isotherm of ferrihydrite is more in line with a type I adsorption isotherm, which may theoretically provide space for physical adsorption via pore filling. Both goethite and hematite belong to the type IV adsorption isotherm with H1 hysteresis loops, indicating that the pore size of the adsorbent material distribution was more uniform. The specific surface areas of ferrihydrite, goethite, and hematite were 214 m^2/g, 65 m^2/g, and 73 m^2/g, respectively. The pore volumes were 0.146 cm^3/g, 0.142 cm^3/g, and 0.246 cm^3/g, respectively. Additionally, the tested pore diameters were 2.719 nm, 8.502 nm, and 13.271 nm, respectively.

2.2. PAA Adsorption Performance of Ferrihydrite, Hematite, and Goethite

Freundlich, Langmuir, and Redlich–Peterson isotherms were used to fit the experimental data in Figure 3a. The specific fitting parameters are shown in Table 1. The Langmuir isothermal adsorption model is used to describe a monolayer adsorption process, assuming that the active sites on the adsorbent surface are uniformly distributed. The Freundlich isothermal adsorption model is an empirical model, which can describe multilayer adsorption on the heterogeneous surface of an adsorbent [33]. The Redlich–Peterson isothermal adsorption model is a three-parameter model combining the features of Langmuir and Freundlich isothermal adsorption models for describing the adsorption process in homogeneous and non-homogeneous systems, while also considering the interaction between the adsorbent active site and the adsorbed material. The results showed that in contrast with the results fitted by the Langmuir or Freundlich models, the Redlich–Peterson isothermal model can achieve a higher R^2 value (0.999) and it was suitable for describing the adsorption process of PAA in iron minerals. In the fitting parameter, β deviates from 1, indicating that the adsorption process is mainly controlled by chemisorption [34]. According to the fitting results, the maximum adsorption capacities of ferrihydrite, goethite, and hematite are 62.21 mg/g, 22.99 mg/g, and 35.58 mg/g, respectively.

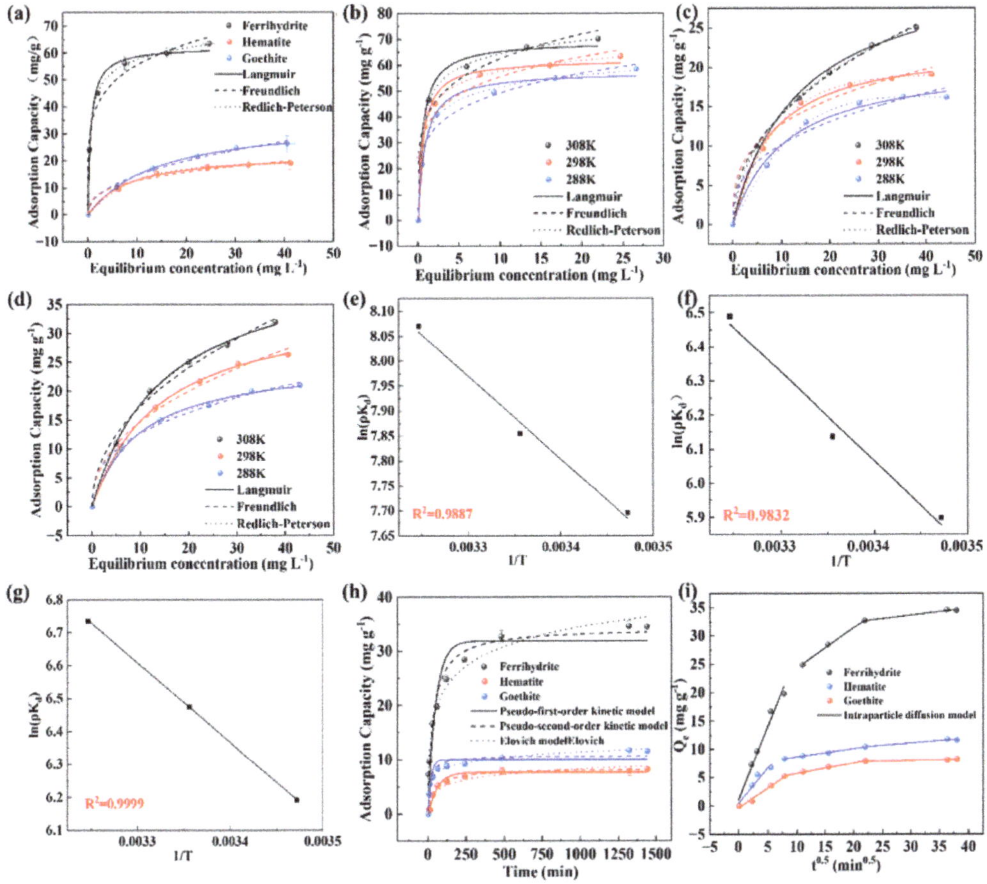

Figure 3. (**a**) Isothermal adsorption model of iron minerals, isothermal adsorption models of (**b**) ferrihydrite, (**c**) goethite, and (**d**) hematite at different temperatures. Van 't Hoff curve of adsorption of PAA by (**e**) ferrihydrite, (**f**) goethite, and (**g**) hematite. (**h**) Kinetic model of iron minerals. (**i**) Intragranular diffusion model.

Table 1. Isothermal adsorption model fitting results of PAA adsorption by iron minerals.

Adsorbent	Langmuir			Freundlich			Redlich–Peterson			
	Q_{max} (mg/g)	k_L	R^2	k_F	n	R^2	k_R	α	β	R^2
Ferrihydrite	62.21	1.63	0.992	35.84	0.19	0.981	155	3.11	0.92	0.999
Goethite	22.99	0.13	0.996	6.21	0.314	0.982	2.14	0.036	1.24	0.999
Hematite	35.58	0.072	0.999	5.36	0.440	0.992	2.47	0.061	1.031	0.999

In the adsorption process, the temperature is an important factor affecting the adsorption effect. At three different reaction temperatures of 288K, 298K, and 308K, ferrihydrite, goethite, and hematite were used to absorb PAA to explore the effect of temperature on the adsorption process. The experimental results are shown in Figure 3b–d. The adsorption capacity of three iron minerals for PAA increases with the increase in adsorption temperature, and the isothermal adsorption models are consistent with the Redlich–Peterson model, which agrees with the results of the isothermal adsorption experiments. The experimental data were fitted and analyzed using the Van 't Hoff equation, and the feasibility of adsorption was determined by analyzing ΔG, ΔH, and ΔS (Figure 3e–g). The fitting results of thermodynamic experimental data are shown in Tables S1 and S2. Under different temperature conditions, the ΔG in the adsorption process of PAA for all three iron minerals is all less than 0, and ΔG decreases with the increase in temperature [35], indicating that the degree of thermodynamic responsivity increases with the increase in temperature. A $\Delta S > 0$ indicates that the adsorption process of the three iron minerals on PAA is a spontaneous entropy increasing process. A $\Delta H > 0$ indicates that the adsorption process is a endothermic process and a high temperature is favorable for the adsorption reaction [36].

The experimental results of adsorption kinetics are shown in Figure 3h. In the case of the three iron minerals, the adsorption capacity of PAA increases rapidly within 3 h after the adsorption begins, and then the adsorption rate slows down gradually and finally reaches the adsorption equilibrium. The adsorption equilibrium of goethite and hematite was reached very quickly, while the adsorption equilibrium of ferrihydrite was slightly delayed in contrast with the former materials. To further accurately describe the adsorption behavior of iron minerals in PAA, pseudo-first-order kinetic, pseudo-second-order kinetic, and Elovich models were fitted to the adsorption results. Studies have shown that the pseudo-first-order kinetic model tends to describe the initial concentration of pollutants with a relatively high number of active sites on the adsorbent, which is usually used to describe the physical adsorption process [37], while the pseudo-second-order kinetic model is more suitable to describe the initial concentration of pollutants with a relatively low number of active sites on the adsorbent, which is usually used to describe the chemisorption process. It is easy to see from Table 2 that the fitting results of the pseudo-second-order kinetic model for the adsorption of PAA by ferrihydrite, goethite, and hematite (R^2: 0.98, 0.99, and 0.96, respectively) are better than those of the pseudo-first-order kinetic model (R^2: 0.93, 0.98, and 0.89, respectively), indicating that chemisorption is likely to occur in the adsorption of PAA by the three iron minerals [38]. Meanwhile the fitted results of the adsorption rates of ferrihydrite, goethite, and hematite were 0.00085 $mg \cdot g^{-1} \cdot min^{-1}$, 0.00267 $mg \cdot g^{-1} \cdot min^{-1}$, and 0.00725 $mg \cdot g^{-1} \cdot min^{-1}$, respectively, indicating that hematite had the fastest adsorption rate of PAA, followed by goethite, and ferrihydrite had the slowest adsorption rate. The fitting results of the Elovich model also verified that chemisorption occurred between the iron minerals and PAA.

Table 2. Fitting results of the adsorption kinetics model of iron minerals.

Adsorbent	Pseudo First Order			Pseudo Second Order			Elovich		
	Q_e/(mg/g)	k_1	R^2	Q_e/(mg/g)	k_2	R^2	α	β	R^2
Ferrihydrite	31.91	0.020	0.93	34.26	0.00085	0.98	3.87	0.19	0.99
Goethite	7.70	0.017	0.98	8.39	0.00267	0.99	0.48	0.70	0.95
Hematite	10.05	0.057	0.89	10.75	0.00725	0.96	6.78	0.75	0.99

To explore whether the velocity control step of the adsorption process is external diffusion or internal diffusion, the intraparticle diffusion model was chosen to fit the kinetic data again. The fitting results of each adsorbent in Figure 3i are divided into three linear parts, which proves the complexity of the adsorption process and also indicates that the adsorption process includes three diffusion steps [39]: (1) the external diffusion process of PAA from the liquid PAAe to the surface of iron minerals; (2) the internal diffusion process of PAA from the adsorbent surface into the pores after adsorption to the internal active site; and (3) the process of adsorption gradually reaching equilibrium. As shown in Table 3, it can be observed that the adsorption rate of ferrihydrite is much larger than that of goethite in stage I. As the adsorption process continues, the adsorption rate of ferrihydrite in stages II and III has decreased to a very low level, indicating that ferrihydrite quickly reaches saturation due to rapid adsorption, leading to a decrease in the mass transfer driving force. According to the theory of Weber and Morris, intraparticle diffusion can be considered the rate-controlling step when the solute adsorption, Q_e, versus $t^{0.5}$ is a straight line through the origin [40]. According to the fitting results in Table 3, it was found that the fitting of the adsorption results for all iron minerals had a strong linear relationship. It is noteworthy that none of them passed through the origin, proving that the adsorption process was influenced by other mechanisms besides the intraparticle diffusion process. It is inferred that a thick boundary layer may be present on the surface of ferrihydrite, where PAA needs to approach the surface of the adsorbent from the main body of the liquid PAA by membrane diffusion and then enter the adsorbent by intraparticle diffusion internally.

Table 3. Fitting results of intragranular diffusion models of iron minerals.

Adsorbent		Stage I			Stage II			Stage III	
	k_{1d}	C_1	R^2	k_{2d}	C_2	R^2	k_{3d}	C_3	R^2
Ferrihydrite	2.60	1.03	0.98	0.71	17.23	0.99	0.12	30.10	0.98
Goethite	0.70	−0.28	0.98	0.19	3.88	0.99	0.02	7.52	0.97
Hematite	1.26	0.57	0.93	0.14	7.23	0.99	0.08	8.70	0.96-

2.3. Effect of Main Parameters on the Adsorption of PAA

The adsorption effects of iron minerals under different pH conditions were investigated. As can be seen from Figure 4a, all iron minerals exhibit remarkable adsorption properties, and the adsorption properties change little in the range of pH = 3–7. However, in the range of pH = 7–11, the adsorption capacity decreases sharply with the increase in pH. Under acidic conditions, there is little difference in the adsorption performance of ferrihydrite. When pH = 5, the adsorption capacity of ferrihydrite reaches the maximum, at 65 mg/g. Similarly, hematite and goethite also obtained the maximum adsorption capacity at pH = 5, which were 22 mg/g and 17 mg/g, respectively. The effect of pH on the adsorption properties is usually closely related to the electrostatic attraction between adsorbed materials and pollutants. Hence, we measured the zeta potential of three iron minerals to analyze the change in surface charge. From Figure 4b, we can see that the isoelectric points of ferrihydrite, hematite, and goethite are very close to each other at pH = 8. In this case, when the environmental pH is less than 8, the surface of the three iron minerals is positively charged [41], and when the environmental pH is greater than 8, the surface of the three iron minerals is negatively charged [42]. In addition, the pH affects the dissociation equilibrium and morphological changes of PAA molecules in an aqueous solution. Visual MINTEQ was used to simulate the dissociation of PAA in water. According to Figure 4c, it is not difficult to see that PAA is present in molecular form at pHs less than 6. In the range of pH = 2–11, $C_6H_6AsO_3^-$ is generated due to the loss of one proton, and in the range of pH = 5–8, the content of $C_6H_6AsO_3^-$ would be close to 100% [43]. $C_6H_5AsO_3^{2-}$ is formed by the loss of two protons when the pH greater is than 6, which may mean that when the pH is higher than the isoelectric point of ferrihydrite, deprotonation will occur on its surface and it will have a negative charge. Meanwhile, according to the morphological distribution of PAA under alkaline conditions, PAA mainly exists in the forms of $C_6H_6AsO_3^-$ and $C_6H_5AsO_3^{2-}$, and the electrostatic repulsion between them will prevent a certain amount of PAA from being adsorbed to ferrihydrite, thus reducing the adsorption capacity. Under the conditions of strong acidity, ferrihydrite has a positive charge due to protonation; the adsorption capacity of PAA can be improved theoretically through electrostatic attraction. However, under these conditions, PAA exists in the form of molecules, and the electrostatic attraction between them will be relatively weakened, resulting in a decrease in the adsorption capacity.

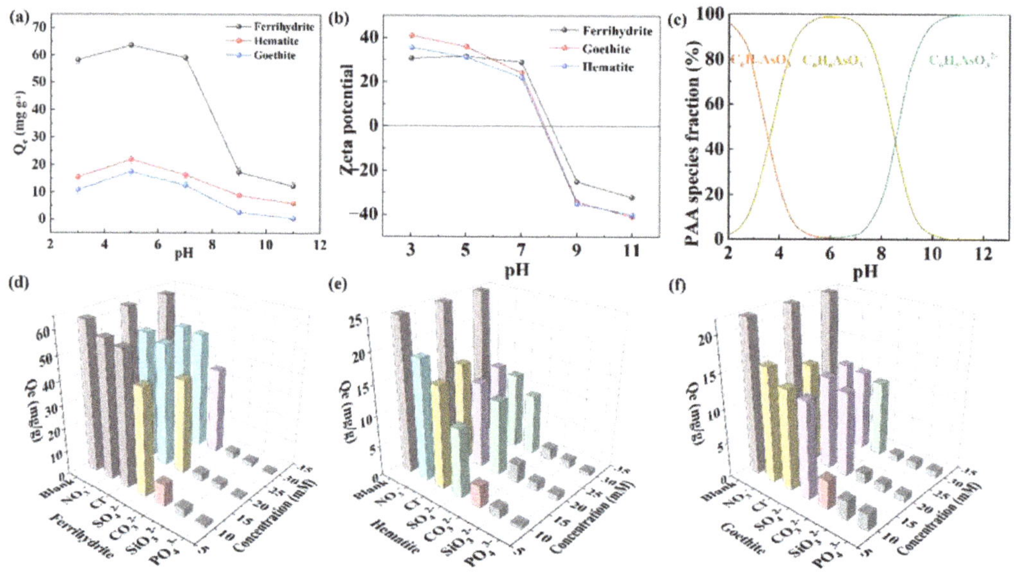

Figure 4. (**a**) The effect of different pH conditions on the adsorption effect of iron minerals on PAA. (**b**) Zeta potential diagram of iron minerals. (**c**) Relationship between the presence of PAA and pH. Effect of coexisting anions on adsorption of PAA by (**d**) ferrihydrite, (**e**) hematite, and (**f**) goethite.

The effect of coexisting ions in water on the adsorption of PAA by iron minerals was further considered. The adsorption of PAA by iron minerals in the presence of NO_3^-, Cl^-, SO_4^{2-}, CO_3^{2-}, SiO_3^{2-}, and PO_4^{3-} was investigated in Figure 4d–f. The adsorption effect of the three iron minerals on PAA decreases with the increase in anion concentration. From observing the data in ferrihydrite, we can see that the removal of PAA decreased by 8.3% in the presence of 20 mM NO_3^- and 10.8% in the presence of 20 mM Cl^-. In hematite, the removal of PAA decreased by 8.9% in the presence of 20 mM NO_3^- and 11% in the presence of 20 mM Cl^-. The adsorption of PAA by goethite decreased by 7.7% in the presence of 20 mM NO_3^- and 8.6% in the presence of 20 mM Cl^-. CO_3^{2-}, SiO_3^{2-}, and PO_4^{3-} all have strong inhibitory effects on the three iron minerals, as their adsorption capacity almost decreased to 0 at the concentration of 10 mM. On the one hand, CO_3^{2-} hydrolyzes in water to produce OH^-, which increases the environmental pH value and is not conducive to the adsorption of PAA by iron minerals. On the other hand, SiO_3^{2-} and PO_4^{3-} both have a similar tetrahedral spatial structure with AsO_4^{3-}, and also form endosphere complexes with hydroxyl groups on the surface of iron minerals, which leads to the adsorption inhibition of PAA [41]. The experimental results also show that SO_4^{2-} can inhibit the adsorption of phenylarsonic acid on iron minerals to a certain extent. Xie et al., obtained similar results when exploring the performance of metal oxides in adsorbing phenylarsonic acid compounds, which may be due to the SO_4^{2-} competing with PAA for the active site on the iron minerals, thus affecting the iron mineral's adsorption of PAA [20,44]. Figure S1 presents the effect of humic acid on the adsorption process of PAA with ferrihydrite, with the results showing that the adsorption capacity was virtually unchanged. Namely, in the system containing organic matter, ferrihydrite still exhibited a good adsorption performance. Figure S2 presents the material reusability in the adsorption process of PAA by ferrihydrite, with the results showing that the adsorption capacity of ferrihydrite decreased by around 40% after two cycles.

As an in situ solution to pollution problems, a permeable reactive wall (PRB) could remove pollutants at their original position during groundwater remediation. Therefore, a column filled with ferrihydrite was used to simulate the PRB process for PAA adsorption under different flow rates (Figure S3). The value of C/C_0 was 0.1 and 0.95 represented the penetration and depletion point, respectively. It can be seen from the figure that when the flow rates were 5 mL/min and 10 mL/min, the fixed bed of ferrihydrite was penetrated after only 5 pore volumes, and the depletion points appeared via 410 pore volumes and 240 pore volumes, respectively. When the flow rate was 1 mL/min, the fixed bed was penetrated through 35 pore volumes and reached the exhaustion point through 465 pore volumes. Therefore, the slower flow rate (1 mL/min) better facilitated the removal of phenylarsonic acid by an adsorption fixed bed owing to delayed the penetration and depletion point.

2.4. Mechanism of Adsorption

Figure 5a–c shows the FTIR spectra of the three different iron minerals. According to Figure 5a, after the adsorption of PAA by ferrihydrite, relatively strong peaks appeared near 3440 cm^{-1} and 1642 cm^{-1}, which was attributed to the stretching vibration of hydroxyl groups in water [45]. The peaks at 691 cm^{-1} and 419 cm^{-1} belong to the vibration of the Fe-O bond. Before adsorption, the stretching vibration of hydroxyl groups in hydroxyl oxide appeared at 1137 cm^{-1}. After adsorption, the characteristic peak was redshifted to 1096 cm^{-1}. This phenomenon may be caused by hydrogen bonding between the hydroxyl group in the ferrihydrite and PAA, which decreases the bonding force constant of the hydroxyl group. At the same time, the characteristic peak of As-O-Fe was formed at 814 cm^{-1} in the adsorbed ferrihydrite{10}, which further proved that the adsorption of PAA on ferrihydrite occurred. It can be seen from Figure 5b that there were some obvious characteristic peaks in the FTIR spectra of hematite. Before and after adsorption, the stretching vibration peaks of the hydroxyl groups in water appeared near 3160 cm^{-1} and 1671 cm^{-1} in the two samples. The peaks at 556 cm^{-1} and 474 cm^{-1} belong to the vibration of the Fe-O bond [46]. Before adsorption, the stretching vibration belonging to hydroxyl group appeared at 1136 cm^{-1} for hematite. After adsorption, the characteristic peak was redshifted to 1094 cm^{-1}. The reason for this phenomenon may be that the hydroxyl group in hematite is hydrogen-bonded with PAA, which reduces the bonding force constant of the hydroxyl group. At the same time, the peaks of the Fe-O bond moved from 556 cm^{-1} and 474 cm^{-1} to 550 cm^{-1} and 461 cm^{-1}, respectively. However, the characteristic peak of As-O-Fe still did not appear in the spectrum of hematite after adsorption [20]. In addition, compared with before adsorption, after adsorption, hematite has an obvious peak at 1440 cm^{-1}, which represents the characteristic vibration of -CH_2 [47]. Compared with the FTIR spectrum of ferrihydrite, some obvious characteristic peaks appeared in the FTIR spectra of goethite (Figure 5c). Before and after adsorption, the two samples showed strong peaks near 3143 cm^{-1} and 1660 cm^{-1}, which were caused by the stretching vibration of hydroxyl groups in water. Strong peaks of Fe-O bonds appeared at 892 cm^{-1}, 795 cm^{-1}, and 639 cm^{-1}. Unfortunately, the characteristic peak of Fe-O-As did not appear in the spectra after adsorption, but it did not mean that there was no chemical interaction between goethite and PAA, but was probably due to the low adsorption capacity [48]. In addition, compared with before adsorption, goethite has an obvious peak at 1438 cm^{-1} after adsorption, which was attributed to the characteristic vibration of -CH_2 [47].

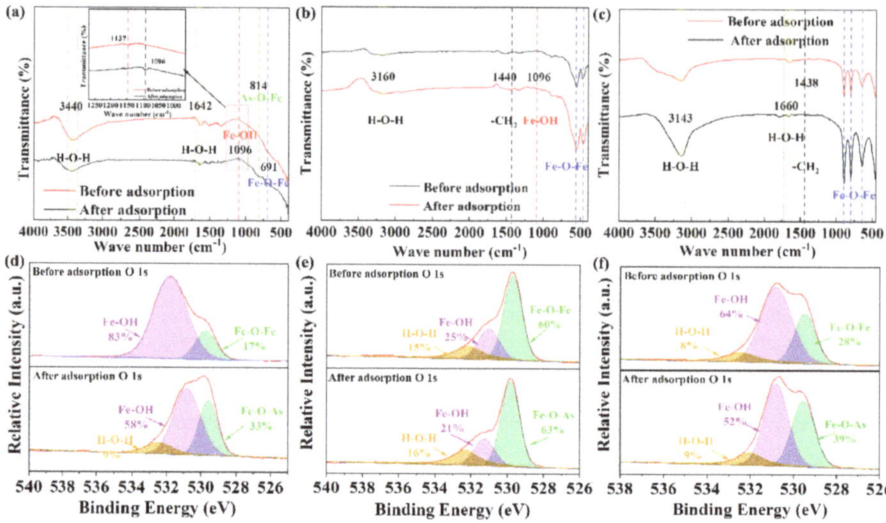

Figure 5. FTIR spectra of (**a**) ferrihydrite, (**b**) hematite, and (**c**) goethite. O 1s XPS spectra of (**d**) ferrihydrite, (**e**) hematite, and (**f**) goethite.

To further explore the PAA adsorption mechanism of the three iron minerals, the XPS analysis results are shown in Figure 5d–e. A comparison of the O 1s orbital XPS spectra of the iron minerals found three peaks at 532, 529, and 531 eV, which correspond to water, Fe-O-Fe, and Fe-OH, respectively [43]. It can be seen from Figure 5d that ferrihydrite has only two oxygen-containing functional groups before adsorption, namely Fe-OH and Fe-O-Fe, with contents of 83% and 17%, respectively. This also shows that ferrihydrite contains a large number of amorphous iron oxide functional groups on its surface, which gives ferrihydrite its excellent adsorption capacity. After adsorption, three absorption peaks appeared in ferrihydrite, showing that ferrihydrite adsorbed about 10% of the water molecules in the adsorption process of PAA, the content of Fe-OH decreased by 25%, and the content of Fe-O-Fe increased by 16%. It can be seen from Figure 5e that before the adsorption by hematite, the percentages of H-O-H, Fe-OH, and Fe-O-Fe were 15%, 25%, and 60%, respectively, while the percentage of Fe-OH was relatively small and the content of Fe-O-Fe was large, which proves the dehydroxylation of iron oxides after the high-temperature synthesis. The percentage of water molecules in hematite after adsorption was 16%, almost unchanged, while the percentage of hydroxyl groups decreased to 31% and the content of Fe-O-As increased to 63%. Figure 5f shows that before the adsorption of goethite, the percentages of H-O-H, Fe-OH, and Fe-O-Fe were 8%, 64%, and 28%, respectively. After adsorption, the percentage of H-O-H in goethite is 9%, which was almost unchanged, and the percentage of Fe-OH decreased to 52%, while the percentage of Fe-O-As was 39%. Furthermore, Figures S4–S6 show the As XPS spectra of the used iron minerals, indicating obvious PAA adsorption. The overall adsorption mechanism diagram is shown in Figure S7.

3. Materials and Methods

3.1. Chemicals

All chemicals were analytically pure and used without further purification. PAA ($C_6H_7AsO_3$) was purchased from the Tokyo Chemical Industry. Ferric chloride hexahydrate ($FeCl_3 \cdot 6H_2O$), ferric nitrate nonahydrate ($Fe(NO_3)_3 \cdot 9H_2O$), sodium chloride (NaCl), sodium carbonate (Na_2CO_3), sodium sulphate (Na_2SO_4), sodium silicate (Na_2SiO_3), and sodium phosphate (Na_3PO_4) were purchased from the Macklin Biochemical Co. Hydrochloric acid (HCl, 36%) and sodium hydroxide (NaOH) were purchased from the Beijing Chemical Factory Co.

3.2. Preparation of Ferrihydrite, Goethite, and Hematite

For ferrihydrite, a certain amount of ferric chloride hexahydrate was dissolved in deionized water to prepare a ferric chloride solution with a concentration of 0.01 mol/L, then it was sonicated for 20 min so that ferric chloride was evenly dispersed. The beaker containing the ferric chloride solution was placed in a mixer with a rotation speed set to 600 r/min. The pH value was adjusted by a sodium hydroxide solution with a concentration of 5 mol/L until the pH was 7. After stirring for 30 min, the solution was aged for 12 h. After washing with deionized water 5 times, centrifugal separation was carried out, and the obtained iron mud was freeze-dried. In the pre-experimental stage, we synthesized ferrihydrite by coprecipitation at room temperature. When the synthesis temperature was raised to 70 °C, the generated iron minerals were still ferrihydrite, and the adsorption effect of PAA was very similar due to the unchanged crystal type. When the reaction pH was adjusted to 12 at the reaction temperature of 70 °C, the iron minerals changed from red-brown to the characteristic earthy yellow of goethite. Similarly, in the process of preparing hematite, we raised the synthesis temperature to 90 °C and controlled the pH of the reaction to obtain hematite. Cudennec et al. studied the conversion conditions of ferrihydrite to goethite and hematite and came to a similar conclusion; high temperature and neutral pH are more conducive to the formation of hematite, while low pH (pH = 2–5) and high pH (pH = 10–14) are more conducive to the formation of goethite.

For goethite, a certain amount of ferric nitrate nonahydrate was dissolved in deionized water to prepare a ferric nitrate solution at a concentration of 0.01 mol/L. The beaker containing the ferric nitrate solution was placed into a water bath and 5 mol/L sodium hydroxide solution was added to the solution until the pH was 12, and it was left to stand for 1 h. Then, the temperature of the water bath was set at 70 °C for aging for 12 h. After the reaction temperature dropped to room temperature, the reaction products were washed with deionized water to neutralize them, and the iron mud was separated by centrifugation and freeze-dried.

For hematite, a 200 mL ferric nitrate solution with a concentration of 0.2 mol/L was prepared in a beaker. Then, 24 mL of 5 mol/L sodium hydroxide solution and 20 mL of 1 mol/L potassium bicarbonate solution was added into the ferric nitrate solution at the same time. The mixture was placed in a water bath at 90 °C and heated to age for 12 h. The thermally treated sample was then cooled to room temperature by air and washed with deionized water 3 times. Finally, the sample was centrifuged to separate the iron mud and then freeze-dried.

3.3. Adsorption Experiments

3.3.1. Isothermal Adsorption Experiment

PAA solutions (50 mL) with concentrations of 10, 20, 30, 40, and 50 mg/L were placed several conical bottles, and 0.02 g of adsorption material was added. The conical mouth of the bottle was sealed and placed in a 150 rpm, 298 K constant temperature air shaker for oscillation. When adsorption equilibrium was reached after 1440 min, a needle tube was used for sampling. The collected liquid was treated through a 0.22 μm filter. The equations for isothermal adsorption are shown as follows (Equations (1)–(3)):

Langmuir isotherm:

$$Q_e = \frac{Q_{max} C_e K_L}{1 + C_e K_L} \quad (1)$$

Freundlich isotherm:

$$Q_e = K_f C_e^{\frac{1}{n}} \quad (2)$$

Redlich–Peterson isotherm:

$$Q_e = \frac{C_e K_R}{1 + \alpha C_e{}^{\beta}} \tag{3}$$

where Q_{max} (mg·g^{-1}) is the saturated monolayer adsorption capacity and K_L (L·mg^{-1}) is the Langmuir constant related to the adsorbent energy. K_F (mg·g^{-1}) and n are Freundlich constants, which are related to adsorption capacity and relative adsorption strength, respectively. K_R (mg·g^{-1}), α, and β are constants of the Redlich–Peterson model.

The experimental process to determine the influence of different temperatures is as follows:

A certain amount of adsorbent was put into conical bottles with a concentration of 0.4 g·L^{-1}, and 50 mL of solution was added to each conical flask with concentrations of 5 mg·L^{-1}, 15 mg·L^{-1}, 20 mg·L^{-1}, 25 mg·L^{-1}, and 55 mg·L^{-1}. The conical mouth of the bottle was sealed and oscillated in a constant temperature air shaker at 150 rpm at temperatures of 288 K, 298 K, and 308 K. When adsorption equilibrium was reached after 1440 min, needle tube sampling was used. The collected liquid was treated through a 0.22 μm filter. The calculation of K_d, ΔG, and $\ln(\rho K_d)$ was performed (Equations (4)–(6)):

$$K_d = \frac{Q_e}{C_e} \tag{4}$$

$$\Delta G = -\mathrm{RT} \cdot \ln(\rho K_d) \tag{5}$$

$$\ln(\rho K_d) = \frac{\Delta S}{\mathrm{R}} - \frac{\Delta H}{\mathrm{RT}} \tag{6}$$

where K_d is the allocation coefficient, which is the ratio of adsorption capacity (Q_e, mg·g^{-1}) of the material at equilibrium to pollutant concentration (C_e, mg·L^{-1}) at equilibrium adsorption. ρ is the density of water (g·L^{-1}), R is the gas mole constant (R = 8.314 J·mol^{-1}·K^{-1}), T is the adsorption temperature (K), ΔH is the enthalpy change (kJ·mol^{-1}), ΔS is the entropy change (J·mol^{-1}·K^{-1}), and ΔG is the Gibbs free energy (kJ·mol^{-1}).

3.3.2. Adsorption Kinetics Experiment

An amount of 50 mL of PAA solution with a certain concentration was placed into a conical bottle, 0.02 g of adsorption material was added, the conical bottle was sealed with tin foil, and the solution was oscillated in a constant temperature air shaker at 150 rpm and 298 K. Samples were taken with 1 mL syringes at 0 min, 10 min, 30 min, 60 min, 120 min, 240 min, 360 min, 720 min, 1320 min, and 1440 min after the reaction began, and the collected samples were stored in a refrigerator after being coated. Equations (7)–(10) exhibit the adsorption kinetics models:

Pseudo-first-order kinetic model:

$$\ln(Q_e - Q_t) = \ln Q_e - k_1 t \tag{7}$$

Pseudo-second-order kinetic model:

$$\frac{t}{Q_t} = \frac{1}{k_2 Q_e{}^2} + \frac{t}{Q_e} \tag{8}$$

Elovich model:

$$Q_t = \frac{1}{\beta} \ln(1 + \alpha \cdot \beta \cdot t) \tag{9}$$

Intragranular diffusion model:

$$Q_t = k_{id} t^{0.5} + \mathrm{C} \tag{10}$$

where Q_e and Q_t are the adsorption capacities ($mg \cdot g^{-1}$) at adsorption equilibrium and at time t (min) after adsorption, respectively. k_1 (min^{-1}) and k_2 ($g \cdot mg^{-1} \cdot min^{-1}$) are the adsorption rate constants of pseudo-first-order kinetics and pseudo-second-order kinetics, respectively. α ($mmol \cdot g^{-1} \cdot h^{-1}$) and β ($g \cdot mmol^{-1}$) represent the initial adsorption rate and desorption constant of the Elovich model. k_{id} ($mg \cdot g^{-1} \cdot min^{-1/2}$) is the rate constant of the intragranular diffusion model and C is the intercept, which can give information about the material boundary layer thickness. In general, the larger the intercept, the greater the contribution of surface adsorption to the rate.

3.3.3. Influencing Factor Testing

The initial concentration of PAA was set to 50 $mg \cdot L^{-1}$, and six anions, including nitric acid, chloride ion, sulfate, carbonate, silicate, and phosphate, were added to the solution at dosages of 10, 20, and 30 mM. The concentration of the adsorbent was 0.4 $g \cdot L^{-1}$. The initial pH was adjusted to 3, 5, 7, 9, and 11 by using 0.1 M of HCl and 0.1 M of NaOH. All reactions were carried out in conical bottles and sealed with tin foil during adsorption. The solution was placed in a 150 rpm, 298 K constant temperature air shaker for 1440 min. After this time, a needle tube was used for sampling, and the collected liquid was passed through a 0.22 μm filter 1 mL at a time.

3.4. Characterization and Analytical Methods

SEM images of samples were taken using a scanning electron microscope (Gemini 300, ZEISS, Oberkochen, Germany). XRD was performed on the sample by an X-ray diffractometer (Ultima IV, Rigku, Japan) using Cu Kα radiation, λ = 0.15406 A. The scanning range was 10°–80° and the scanning speed was 10° min^{-1}. FT-IR spectra of the samples were measured using a Fourier infrared spectrometer (Scientific Nicolet iS10, ThermoFischer, Waltham, MA, USA). The scanning wavenumber ranges from 4000 to 400 cm^{-1}, with 32 scans, and the resolution was 4 cm^{-1}. For the FTIR analyses after adsorption, we carried out filtration and freeze drying to obtain the measurement sample. The Zeta potential was analyzed using the nanometer particle size and a Zeta potential analyzer (Zetasizer Nano ZS90, Malvern, UK). The BET curve of the sample was measured by using a fully automated specific surface and porosity analyzer, BET (ASAP 2460 3.01, Micromeritics, USA). The degassing temperature was 423 K, the gas was N_2, and the degassing time was 6 h. XPS was performed using an X-ray photoelectron spectrometer (ESCALAB 250Xi, ThermoFischer, USA). For the XPS analyses after adsorption, we carried out filtration and freeze drying to obtain the measurement sample. High-performance liquid chromatography (LC-6 Shimadzu, Elsichrom, Knivsta, Sweden) was used to detect the concentration of PAA with a C18 column (ZORBAX Agilent, Agilent Technologies, Santa Clara, MO, USA) at 298 K. The mobile PAA consisted of acetonitrile and 0.2% phosphoric acid solution. The mobile PAA ratio was 1:9 at the wavelength of 264 nm and the injection volume was 20 μL each time.

4. Conclusions

Ferrihydrite, hematite, and goethite were prepared by a simple coprecipitation method, and all exhibited good adsorption performance for PAA. The adsorption isotherm of ferrihydrite is more consistent with a type I isotherm, indicating that micropore filling is the main physical mechanism of ferrihydrite adsorption of PAA. According to the BET characterization results, the specific surface area of ferrihydrite is 214 m^2/g, much higher than that of hematite (73 m^2/g) and goethite (65 m^2/g), and ferrihydrite has the best adsorption effect of PAA, at up to 63.44 mg/g. The adsorption kinetics experiment showed that the adsorption of PAA by the three iron minerals was more in line with a pseudo-second-order kinetic model, indicating that there was chemical adsorption between the three iron minerals and PAA molecules. To explore whether the speed control step of the adsorption process was external diffusion or internal diffusion, the intraparticle diffusion model was used to fit the kinetic data again. In addition to the internal diffusion process, the existence of a boundary layer on the surface of the iron minerals will undoubtedly increase the mass transfer resistance, thus affecting the adsorption process. Furthermore, the influence of the environmental pH and co-existing anions on the adsorption of PAA by iron minerals was considered. This study provides new motivation for the development of simple and efficient adsorption removal technologies for PAA. Finally, the adsorption mechanism was discussed through a characterization of FTIR and XPS spectra. The results showed that after adsorption, the number of hydroxyl groups in the structure of iron minerals decreased, with the hydroxyl groups in ferrihydrite decreasing the most, indicating that ferrihydrite has a strong affinity for PAA. The formation of a Fe-O-As bond is due to ligand exchange between the arsenate in PAA and the hydroxyl in the iron minerals, resulting in complexation.

Supplementary Materials: The following supporting information can be downloaded at: https://www.mdpi.com/article/10.3390/molecules28083448/s1, Figure S1: The effect of humic acid on adsorption process of PAA with ferrihydrite; Figure S2: The ferrihydritereusability on adsorption process of PAA; Figure S3: PRB column experimental results; Figure S4: As 3d XPS spectra of ferrihydrite before and after adsorption of phenylarsonic acid; Figure S5: As 3d XPS spectra of hematite before and after adsorption of phenylarsonic acid; Figure S6: As 3d XPS spectra of goethite before and after adsorption of phenylarsonic acid; Figure S7: The adsorption mechanism diagram of PAA adsorption on iron minerals; Table S1: Fitting results of isothermal adsorption models of iron minerals at different temperatures; Table S2: The value of $\triangle G$, $\triangle H$, and $\triangle S$ obtained by Van't Hoff curve.

Author Contributions: L.W.: data curation; formal analysis; and writing—original draft; C.H.: funding acquisition and project administration; Z.Y.: data curation; formal analysis; and writing—original draft; S.G.: methodology; investigation; and validation; T.Z.: writing—review and editing; S.L.: conceptualization; resources; writing—review and editing; supervision; and funding acquisition. All authors have read and agreed to the published version of the manuscript.

Funding: This research was funded by Sinopec excellent youth technological innovation foundation (No. P21059).

Institutional Review Board Statement: Not applicable.

Informed Consent Statement: Not applicable.

Data Availability Statement: Data will be made available on request.

Conflicts of Interest: The authors declare no conflict of interest.

Sample Availability: Samples of ferrihydrite, goethite, and hematite are available from the authors.

References

1. Abdul, K.S.M.; Jayasinghe, S.S.; Chandana, E.P.S.; Jayasumana, C.; De Silva, P.M.C.S. Arsenic and Human Health Effects: A Review. *Environ. Toxicol. Pharmacol.* **2015**, *40*, 828–846. [CrossRef] [PubMed]
2. Deng, H.; Tu, Y.; Wang, H.; Wang, Z.; Li, Y.; Chai, L.; Zhang, W.; Lin, Z. Environmental Behavior, Human Health Effect, and Pollution Control of Heavy Metal(Loid)s toward Full Life Cycle Processes. *Eco-Environ. Health* **2022**, *1*, 229–243. [CrossRef]
3. de Souza, T.D.; Borges, A.C.; Braga, A.F.; Veloso, R.W.; de Matos, A.T. Phytoremediation of Arsenic-Contaminated Water by Lemna Valdiviana: An Optimization Study. *Chemosphere* **2019**, *234*, 402–408. [CrossRef]
4. Hare, V.; Chowdhary, P.; Kumar, B.; Sharma, D.C.; Baghel, V.S. Arsenic Toxicity and Its Remediation Strategies for Fighting the Environmental Threat. In *Emerging and Eco-Friendly Approaches for Waste Management*; Bharagava, R.N., Chowdhary, P., Eds.; Springer Singapore: Singapore, 2019; pp. 143–170. ISBN 978-981-10-8668-7.
5. Liu, Q.; Lu, X.; Peng, H.; Popowich, A.; Tao, J.; Uppal, J.S.; Yan, X.; Boe, D.; Le, X.C. Speciation of Arsenic—A Review of Phenylarsenicals and Related Arsenic Metabolites. *Trends Analyt. Chem.* **2018**, *104*, 171–182. [CrossRef]
6. Vega, L.; Styblo, M.; Patterson, R.; Cullen, W.; Wang, C.; Germolec, D. Differential Effects of Trivalent and Pentavalent Arsenicals on Cell Proliferation and Cytokine Secretion in Normal Human Epidermal Keratinocytes. *Toxicol. Appl. Pharmacol.* **2001**, *172*, 225–232. [CrossRef] [PubMed]
7. Sun, T.; Zhao, Z.; Liang, Z.; Liu, J.; Shi, W.; Cui, F. Efficient Degradation of P-Arsanilic Acid with Arsenic Adsorption by Magnetic CuO-Fe_3O_4 Nanoparticles under Visible Light Irradiation. *Chem. Eng. J.* **2018**, *334*, 1527–1536. [CrossRef]
8. Li, Z.; Liu, X.; Jin, W.; Hu, Q.; Zhao, Y. Adsorption Behavior of Arsenicals on MIL-101(Fe): The Role of Arsenic Chemical Structures. *J. Colloid Interface Sci.* **2019**, *554*, 692–704. [CrossRef]

9. Oya-Ohta, Y.; Kaise, T.; Ochi, T. Induction of Chromosomal Aberrations in Cultured Human Fibroblasts by Inorganic and Organic Arsenic Compounds and the Different Roles of Glutathione in Such Induction. *Mutat. Res.* **1996**, *357*, 123–129. [CrossRef]
10. Peng, X.; Chen, L.; Liu, S.; Hu, L.; Zhang, J.; Wang, A.; Yu, X.; Yan, Z. Insights into the Interfacial Interaction Mechanisms of P-Arsanilic Acid Adsorption on Ionic Liquid Modified Porous Cellulose. *J. Environ. Chem. Eng.* **2021**, *9*, 105225. [CrossRef]
11. Kroening, K.K.; Solivio, M.J.V.; García-López, M.; Puga, A.; Caruso, J.A. Cytotoxicity of Arsenic-Containing Chemical Warfare Agent Degradation Products with Metallomic Approaches for Metabolite Analysis. *Metallomics* **2009**, *1*, 59–66. [CrossRef]
12. Cai, C.; Zhao, M.; Yu, Z.; Rong, H.; Zhang, C. Utilization of Nanomaterials for In-Situ Remediation of Heavy Metal(Loid) Contaminated Sediments: A Review. *Sci. Total Environ.* **2019**, *662*, 205–217. [CrossRef] [PubMed]
13. Wang, Y.; Gu, Y.; Li, H.; Ye, M.; Qin, W.; Zhang, H.; Wang, G.; Zhang, Y.; Zhao, H. Electrodeposition of Hierarchically Amorphous FeOOH Nanosheets on Carbonized Bamboo as an Efficient Filter Membrane for As(III) Removal. *Chem. Eng. J.* **2020**, *392*, 123773. [CrossRef]
14. Assaad, N.; Sabeh, G.; Hmadeh, M. Defect Control in Zr-Based Metal–Organic Framework Nanoparticles for Arsenic Removal from Water. *ACS Appl. Nano Mater.* **2020**, *3*, 8997–9008. [CrossRef]
15. Weerasundara, L.; Ok, Y.-S.; Bundschuh, J. Selective Removal of Arsenic in Water: A Critical Review. *Environ. Pollut.* **2021**, *268*, 115668. [CrossRef] [PubMed]
16. Rahidul Hassan, H. A Review on Different Arsenic Removal Techniques Used for Decontamination of Drinking Water. *Env Pollut. Bioavail.* **2023**, *35*, 2165964. [CrossRef]
17. Fu, X.; Song, X.; Zheng, Q.; Liu, C.; Li, K.; Luo, Q.; Chen, J.; Wang, Z.; Luo, J. Frontier Materials for Adsorption of Antimony and Arsenic in Aqueous Environments: A Review. *Int. J. Environ. Res. Public Health* **2022**, *19*, 10824. [CrossRef]
18. Siddiqui, S.I.; Chaudhry, S.A. Iron Oxide and Its Modified Forms as an Adsorbent for Arsenic Removal: A Comprehensive Recent Advancement. *Process Saf. Environ. Prot.* **2017**, *111*, 592–626. [CrossRef]
19. Ishizaki, M.; Yanaoka, T.; Nakamura, M.; Hakuta, T.; Ueno, S.; Komuro, M.; Shibata, M.; Kitamura, T.; Honda, A.; Doy, M.; et al. Detection of Bis(Diphenylarsine)Oxide, Diphenylarsinic Acid and Phenylarsonic Acid, Compounds Probably Derived from Chemical Warfare Agents, in Drinking Well Water. *J. Health Sci.* **2005**, *51*, 130–137. [CrossRef]
20. Xie, X.; Cheng, H. Adsorption and Desorption of Phenylarsonic Acid Compounds on Metal Oxide and Hydroxide, and Clay Minerals. *Sci. Total Environ.* **2021**, *757*, 143765. [CrossRef]
21. Sparks, D.L. Toxic Metals in the Environment: The Role of Surfaces. *Elements* **2005**, *1*, 193–197. [CrossRef]
22. Ginder-Vogel, M.; Landrot, G.; Fischel, J.S.; Sparks, D.L. Quantification of Rapid Environmental Redox Processes with Quick-Scanning X-Ray Absorption Spectroscopy (Q-XAS). *Proc. Natl. Acad. Sci. USA* **2009**, *106*, 16124–16128. [CrossRef] [PubMed]
23. Siebecker, M.; Li, W.; Khalid, S.; Sparks, D. Real-Time QEXAFS Spectroscopy Measures Rapid Precipitate Formation at the Mineral–Water Interface. *Nat. Commun.* **2014**, *5*, 5003. [CrossRef]
24. Wang, S.; Mulligan, C.N. Speciation and Surface Structure of Inorganic Arsenic in Solid Phases: A Review. *Environ. Int.* **2008**, *34*, 867–879. [CrossRef] [PubMed]
25. Carabante, I.; Grahn, M.; Holmgren, A.; Kumpiene, J.; Hedlund, J. Adsorption of As (V) on Iron Oxide Nanoparticle Films Studied by in Situ ATR-FTIR Spectroscopy. *Colloids Surf. A Physicochem. Eng. Asp.* **2009**, *346*, 106–113. [CrossRef]
26. Li, Z.; Zhang, T.; Li, K. One-Step Synthesis of Mesoporous Two-Line Ferrihydrite for Effective Elimination of Arsenic Contaminants from Natural Water. *Dalton Trans.* **2011**, *40*, 2062–2066. [CrossRef]
27. Tanaka, M.; Togo, Y.S.; Yamaguchi, N.; Takahashi, Y. An EXAFS Study on the Adsorption Structure of Phenyl-Substituted Organoarsenic Compounds on Ferrihydrite. *J. Colloid Interface Sci.* **2014**, *415*, 13–17. [CrossRef]
28. Yan, L.; Chan, T.; Jing, C. Arsenic Adsorption on Hematite Facets: Spectroscopy and DFT Study. *Environ. Sci. Nano.* **2020**, *7*, 3927–3939. [CrossRef]
29. Chen, P.; Song, D.; Zhang, X.; Xie, Q.; Zhou, Y.; Liu, H.; Xu, L.; Chen, T.; Rosso, K.M. Understanding Competitive Phosphate and Silicate Adsorption on Goethite by Connecting Batch Experiments with Density Functional Theory Calculations. *Environ. Sci. Technol.* **2022**, *56*, 823–834. [CrossRef]
30. Cao, S.; Zhang, X.; Huang, X.; Wan, S.; An, X.; Jia, F.; Zhang, L. Insights into the Facet-Dependent Adsorption of Phenylarsonic Acid on Hematite Nanocrystals. *Environ. Sci. Nano.* **2019**, *6*, 3280–3291. [CrossRef]
31. Yang, H.; Zhang, Y.; Xia, S. Ferrihydrite/Ultrasound Activated Peroxymonosulfate for Humic Acid Removal. *Turk. J. Chem.* **2022**, *46*, 835–848. [CrossRef]
32. Das, S.; Hendry, M.J.; Essilfie-Dughan, J. Effects of Adsorbed Arsenate on the Rate of Transformation of 2-Line Ferrihydrite at PH 10. *Environ. Sci. Technol.* **2011**, *45*, 5557–5563. [CrossRef] [PubMed]
33. Chen, X.; Hossain, M.F.; Duan, C.; Lu, J.; Tsang, Y.F.; Islam, M.S.; Zhou, Y. Isotherm Models for Adsorption of Heavy Metals from Water—A Review. *Chemosphere* **2022**, *307*, 135545. [CrossRef] [PubMed]
34. Zhang, J.; Xie, L.; Ma, Q.; Liu, Y.; Li, J.; Li, Z.; Li, S.; Zhang, T. Ball Milling Enhanced Cr(VI) Removal of Zero-Valent Iron Biochar Composites: Functional Groups Response and Dominant Reduction Species. *Chemosphere* **2023**, *311*, 137174. [CrossRef] [PubMed]
35. Fan, H.; Ma, Y.; Wan, J.; Wang, Y. Removal of Gentian Violet and Rhodamine B Using Banyan Aerial Roots after Modification and Mechanism Studies of Differential Adsorption Behaviors. *Environ. Sci. Pollut. Res.* **2020**, *27*, 9152–9166. [CrossRef]
36. Wang, L.; Shi, C.; Wang, L.; Pan, L.; Zhang, X.; Zou, J.-J. Rational Design, Synthesis, Adsorption Principles and Applications of Metal Oxide Adsorbents: A Review. *Nanoscale* **2020**, *12*, 4790–4815. [CrossRef]

37. Sherlala, A.I.A.; Raman, A.A.A.; Bello, M.M.; Buthiyappan, A. Adsorption of Arsenic Using Chitosan Magnetic Graphene Oxide Nanocomposite. *J. Environ. Manag.* **2019**, *246*, 547–556. [CrossRef]
38. Wang, G.; Sun, T.; Sun, Z.; Hu, X. Preparation of Copper Based Metal Organic Framework Materials and Its Effective Adsorptive Removal of Ceftazidime from Aqueous Solutions. *Appl. Surf. Sci.* **2020**, *532*, 147411. [CrossRef]
39. Pan, T.; Liu, H.; Jiang, M.; Li, J.; Liu, W.; Jiao, Q.; Zhang, T. New Insights into the Adsorption Behavior of Thiacloprid at the Microfibers/Water Interface: Role of Humic Acid. *Chemosphere* **2023**, *311*, 136938. [CrossRef]
40. Ofomaja, A.E. Intraparticle Diffusion Process for Lead(II) Biosorption onto Mansonia Wood Sawdust. *Bioresour. Technol.* **2010**, *101*, 5868–5876. [CrossRef]
41. Ge, X.; Ma, Y.; Song, X.; Wang, G.; Zhang, H.; Zhang, Y.; Zhao, H. β-FeOOH Nanorods/Carbon Foam-Based Hierarchically Porous Monolith for Highly Effective Arsenic Removal. *ACS Appl. Mater. Interfaces* **2017**, *9*, 13480–13490. [CrossRef]
42. Yu, X.; Wei, Y.; Liu, C.; Ma, J.; Liu, H.; Wei, S.; Deng, W.; Xiang, J.; Luo, S. Ultrafast and Deep Removal of Arsenic in High-Concentration Wastewater: A Superior Bulk Adsorbent of Porous Fe_2O_3 Nanocubes-Impregnated Graphene Aerogel. *Chemosphere* **2019**, *222*, 258–266. [CrossRef] [PubMed]
43. Lin, Z.; Huan, Z.; Zhang, J.; Li, J.; Li, Z.; Guo, P.; Zhu, Y.; Zhang, T. CTAB-Functionalized δ-FeOOH for the Simultaneous Removal of Arsenate and Phenylarsonic Acid in Phenylarsenic Chemical Warfare. *Chemosphere* **2022**, *292*, 133373. [CrossRef] [PubMed]
44. Guan, X.; Dong, H.; Ma, J.; Jiang, L. Removal of Arsenic from Water: Effects of Competing Anions on As(III) Removal in $KMnO_4$–Fe(II) Process. *Water Res.* **2009**, *43*, 3891–3899. [CrossRef] [PubMed]
45. Wu, S.; Lu, J.; Ding, Z.; Li, N.; Fu, F.; Tang, B. Cr(VI) Removal by Mesoporous FeOOH Polymorphs: Performance and Mechanism. *RSC Adv.* **2016**, *6*, 82118–82130. [CrossRef]
46. Guo, T.; Wang, K.; Zhang, G.; Wu, X. A Novel α-Fe_2O_3@g-C_3N_4 Catalyst: Synthesis Derived from Fe-Based MOF and Its Superior Photo-Fenton Performance. *Appl. Surf. Sci.* **2019**, *469*, 331–339. [CrossRef]
47. Yang, K.; Peng, H.; Wen, Y.; Li, N. Re-Examination of Characteristic FTIR Spectrum of Secondary Layer in Bilayer Oleic Acid-Coated Fe_3O_4 Nanoparticles. *Appl. Surf. Sci.* **2010**, *256*, 3093–3097. [CrossRef]
48. Rahimi, S.; Moattari, R.M.; Rajabi, L.; Derakhshan, A.A.; Keyhani, M. Iron Oxide/Hydroxide (α,γ-FeOOH) Nanoparticles as High Potential Adsorbents for Lead Removal from Polluted Aquatic Media. *J. Ind. Eng. Chem.* **2015**, *23*, 33–43. [CrossRef]

molecules

MDPI

Article

Pb(II) Uptake from Polluted Irrigation Water Using Anatase TiO_2 Nanoadsorbent

Miguel A. Vasquez-Caballero [1], Yamerson Canchanya-Huaman [1], Angie F. Mayta-Armas [1], Jemina Pomalaya-Velasco [1], Noemi-Raquel Checca-Huaman [2], Yéssica Bendezú-Roca [1] and Juan A. Ramos-Guivar [3,*]

[1] Laboratorio de No Metálicos, Facultad de Ingeniería Química, Universidad Nacional del Centro del Perú (UNCP), Av. Mariscal Ramón Castilla N° 3909, El Tambo, Huancayo 12000, Peru; yessiben@yahoo.com (Y.B.-R.)

[2] Centro Brasileiro de Pesquisas Físicas, Rio de Janeiro 22290-180, RJ, Brazil

[3] Grupo de Investigación de Nanotecnología Aplicada para Biorremediación Ambiental, Energía, Biomedicina y Agricultura (NANOTECH), Facultad de Ciencias Físicas, Universidad Nacional Mayor de San Marcos, Av. Venezuela Cdra 34 S/N, Ciudad Universitaria, Lima 15081, Peru

* Correspondence: juan.ramos5@unmsm.edu.pe

Abstract: The adsorption characteristics of titanium dioxide nanoparticles (nano-TiO_2) for the removal of Pb(II) from irrigation water were investigated in this work. To accomplish this, several adsorption factors, such as contact time and pH, were tested to assess adsorption efficiencies and mechanisms. Before and after the adsorption experiments, commercial nano-TiO_2 was studied using X-ray diffraction (XRD), scanning and transmission electron microscopy (SEM and TEM), energy dispersive spectroscopy (EDS), and X-ray photoelectron spectroscopy (XPS). The outcomes showed that anatase nano-TiO_2 was remarkably efficient in cleaning Pb(II) from water, with a removal efficiency of more than 99% after only one hour of contact time at a pH of 6.5. Adsorption isotherms and kinetic adsorption data matched the Langmuir and Sips models quite well, showing that the adsorption process occurred at homogenous sites on the surface of nano-TiO_2 by forming a Pb(II) adsorbate monolayer. The XRD and TEM analysis of nano-TiO_2 following the adsorption procedure revealed a non-affected single phase (anatase) with crystallite sizes of 9.9 nm and particle sizes of 22.46 nm, respectively. According to the XPS data and analyzed adsorption data, Pb ions accumulated on the surface of nano-TiO_2 through a three-step mechanism involving ion exchange and hydrogen bonding mechanisms. Overall, the findings indicate that nano-TiO_2 has the potential to be used as an effective and long-lasting mesoporous adsorbent in the treatment and cleaning of Pb(II) from water bodies.

Keywords: water bodies recuperation; adsorbents post-exposure properties; anatase–lead complexation; water cleaning strategies

1. Introduction

Pollution of the environment with heavy metals has become a big issue across the world, affecting both industrialized and developing nations [1]. Water pollution and scarcity are the two most serious environmental issues threatening the existence of plants, animals, and people in the twenty-first century. Human activities have disrupted the natural recycling of water and its purifying mechanisms, resulting in a shortage of water for human use [2]. According to the 2018 United Nations Global Water Development Report, industrial and household water consumption will grow far faster than agricultural need, despite agriculture remaining the largest overall consumer [3]. Water is now utilized for agriculture 70% of the time, with irrigation being the most common application. Electroplating, metallurgy, steel, fertilizers, pesticides, textiles, paints, glass, synthetic materials, and mining are among the sectors that discharge heavy-metal-containing wastewater into the environment [4,5]. Even at low concentrations, these pollutants are very hazardous to human health; thus, they must be treated before disposal [6].

Lead, Pb(II), is one of the most damaging heavy metals; its influence on water contamination has sparked international concern [7]. Both natural weathering processes and human-made operations, such as mining, making batteries, and electroplating, can release Pb(II) into the environment [8]. Because of its bioaccumulation, toxicity, and persistence, Pb(II) is an extremely harmful chemical [9,10]. Infants, children under the age of six, the fetus, and pregnant women are the most vulnerable to poor health impacts [11]. The effects on the central nervous system can be especially severe, including cancer, anemia, and damage to the brain and other organ systems [12,13]. The World Health Organization (WHO) suggests keeping the amount of lead in drinking water to a maximum of 0.01 mg L^{-1} [11].

Several methods have been developed to reduce the levels of Pb(II) and other heavy metals in polluted water, including adsorption, biosorption, chemical precipitation, ion exchange, membrane separation, ion treatment, electrochemical treatment, recovery by evaporation, redox precipitation, and filtration [12,14–16]. Adsorption has numerous benefits over the other approaches [17,18]. It is easy to use and there are a wide range of adsorbents that can be reused after the desorption process in various adsorption cycles, allowing for the reasonably priced separation of pollutants [19]. Nowadays, the utilization of nanoparticles (NPs) as an absorbent for metal pollutants is a highly promising option [20]. Nanomaterials offer excellent characteristics for reducing pollutants in water. Due to their small size [21,22] and large surface area, NPs easily disperse in liquids where they can interact with a variety of chemical species [19].

Nanostructures have emerged as a very attractive alternative for water cleanup due to their huge surface-to-volume ratios, which are excellent for surface reactions [23]. Numerous studies have demonstrated that heavy metals, such as Pb(II) and Cd(II), can be effectively removed from aqueous solutions using nano-TiO_2. As a result, investigations on the use of nano-TiO_2 in wastewater treatment are extremely beneficial to the aquatic environment [24]. Moreover, numerous TiO_2-based nanocomposites have been investigated to increase these materials' adsorption capacity [25,26]. TiO_2 polymorphs include anatase, rutile, and brookite, which are the most prevalent. Each of these polymorphs has unique structural, physical, and chemical features that influence their ability for heavy metal adsorption. Anatase, for example, has a bigger surface area and porosity than rutile and brookite, allowing it to adsorb more metals. Yet, under specific pH and temperature circumstances, rutile and brookite are more stable than anatase [27,28]. Despite their remarkable Pb(II) removal properties in simulated aliquots, little literature is available concerning nano-TiO_2 adsorption performance in real waters. Certainly, diverse nano-TiO_2 polymorphs will exhibit different removal properties in river, lake, and sea samples containing heavy metals, and, more importantly, the conservation of the physicochemical properties of nano-TiO_2 will guarantee their environmental sustainability and reuse.

In that sense, the pollution of water bodies by Pb(II) is a current concern—taking in and getting rid of this metal is important and an interesting challenge. This work used nano-TiO_2 as an alternative to remove Pb(II) from irrigation water with a corrected starting concentration of 47.54 mg L^{-1}, and its removal effectiveness was evaluated. The water samples were collected from the CIMIRM irrigation canal in Peru's province of Jauja. Analytic techniques were used to evaluate nano-TiO_2 before and after Pb adsorption—this last to ensure their physicochemical and reusability properties. Multiparametric adsorption analysis, including adsorption kinetics, the effect of pH on adsorption, adsorbent dose on adsorption capacity, and the behavior of experimental data in relation to adsorption isotherms, were also investigated. The results indicate that anatase nano-TiO_2 is an alternative solution to the Pb(II) water cleaning problem, including modified surface water samples that exceed the permissible levels needed for agricultural irrigation, e.g., [29].

2. Results and Discussions

2.1. Characterization of Nano-TiO_2

2.1.1. XRD and Rietveld Analysis

Figure 1a,b shows the refined X-ray diffractograms of nano-TiO_2 with the corresponding refinement model after the Pb adsorption process with initial concentrations of 47.54 mg L^{-1} (TiO_2-PbC1) and 1.16 mg L^{-1} (TiO_2-PbC2). Only the pure anatase phase was observed; the refined parameters are summarized in Table 1. Considering anisotropic size broadening, the mean apparent crystallite sizes were calculated using Scherrer's formula, which may be expressed as a linear combination of spherical harmonics and is provided in the FullProf Suite program [30]. The data corroborated the presence of the tetragonal crystal structure of the anatase phase even after adsorption of Pb(II) for both initial concentrations. The average apparent size of anatase crystallites was 9.9 nm and 8.7 nm, evidencing that there was not much variation in the microstructural parameters (Table 1), even with increasing concentration of Pb. Anatase crystallite size values of 10.1 nm have been reported by Canchanya et al. and of 12.22 nm by Patidar and Jain, with no application in the adsorption process. In comparison with these values, the crystallite size of anatase showed a slight decrease after the Pb(II) adsorption process [31,32].

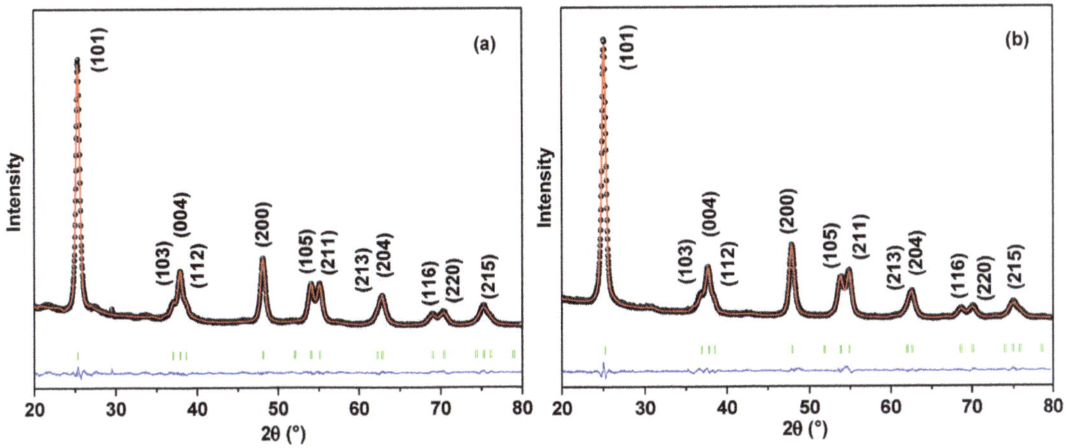

Figure 1. Rietveld refinement of X-ray diffractograms of TiO_2 anatase after adsorption of Pb(II) for C_o = 47.54 mg L^{-1} (**a**) and Pb C_o = 1.16 mg L^{-1} (**b**). The black dots (I_{obs}) represent the experimental data, the red lines (I_{cal}) are the calculated XRD diffractogram, the green vertical lines are the Bragg positions, and the blue lines represent the residual lines.

Table 1. Rietveld refinement parameters of the TiO_2-PbC1 and TiO_2-PbC2 samples: lattice parameters, cell volume, and Caglioti values. Agreement R-factors R_{exp} (%), R_p (%), R_{wp} (%), and goodness of fit, chi-square (χ^2).

Refined Parameters	TiO_2-PbC1 C_o = 47.54 mg L^{-1}	TiO_2-PbC2 C_o = 1.16 mg L^{-1}
$a\left(\text{Å}\right)$	3.782 (9)	3.798 (9)
$b\left(\text{Å}\right)$	3.782 (9)	3.798 (9)
$c\left(\text{Å}\right)$	9.488 (3)	9.527 (3)
$\alpha(°)$	90	90
$\beta(°)$	90	90
$\gamma(°)$	90	90
$V\left(\text{Å}^3\right)$	135.750 (6)	137.441 (6)

Table 1. *Cont.*

Refined Parameters	TiO_2-PbC1 C_o = 47.54 mg L^{-1}	TiO_2-PbC2 C_o = 1.16 mg L^{-1}
Y_{00}	−0.077 (7)	0.022 (4)
Y_{20}	0.675 (8)	0.972 (8)
Y_{40}	0.650 (9)	0.180 (9)
Y_{44+}	0.045 (6)	−0.127 (5)
Y_{44-}	−0.958 (2)	−0.585 (2)
Y_{60}	−0.120 (9)	0.554 (9)
Y_{64+}	−0.368 (4)	−0.184 (3)
Y_{64-}	−2.311 (2)	0.952 (2)
FWHM parameters		
U	4.464	4.342
V	−2.569	−2.549
W	0.907	0.919
Average max strain	108.48 (4)	98.63 (3)
Average size (nm)	9.9 (7)	8.7 (5)
R_{exp}(%)	4.52	4.54
R_p(%)	7.21	6.90
R_{wp}(%)	7.70	7.38
$\chi 2$	2.90	2.63

2.1.2. SEM and EDS Mapping Analysis

Figures 2 and S1 depict the SEM images for two magnifications (a and f), where characteristic granular morphologies of nano-TiO_2 powders exposed to two different Pb(II) concentrations are observed. EDS mapping images are observed in Figures 2 and S1b,g and the elemental images for O (c,h), Ti (d,i), and Pb (e,j) are shown. The presence of Pb(II) was corroborated at low and high initial concentrations. Table 2 summarizes the weight percentage for the identified elements after analyzing the EDS spectrum (see Figure S2).

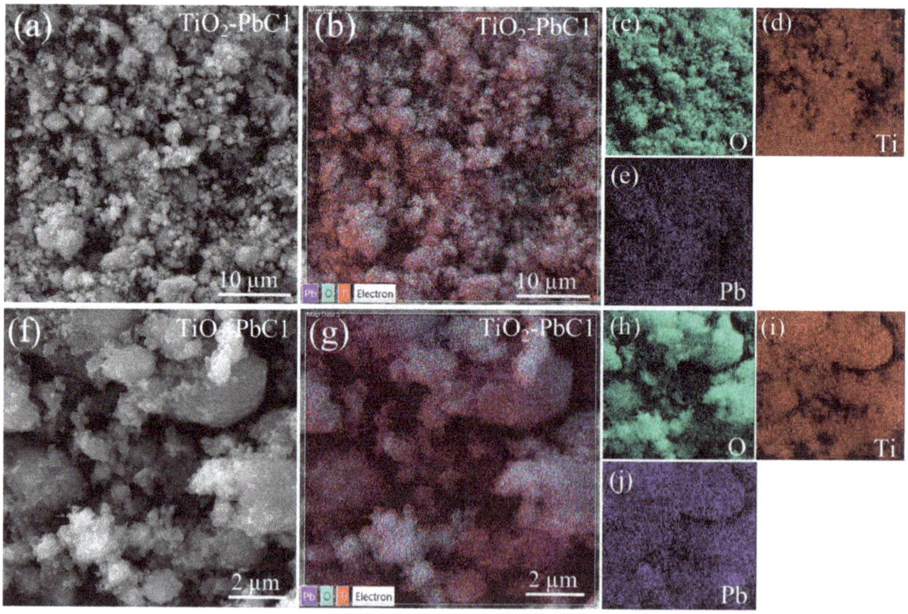

Figure 2. (**a**,**f**) SEM images, (**b**,**g**) EDS mapping images, and (**c**–**e**,**h**–**j**) elemental EDS images for nano-TiO_2 with C_0 (Pb(II)) = 47.54 mg L^{-1}.

Table 2. Elemental composition for the studied samples obtained from EDS spectrum analysis.

Sample	Ti (% wt)	O (% wt)	Pb (% wt)
TiO_2-PbC1	47.7 ± 0.2	47.5 ± 0.2	4.8 ± 0.1
TiO_2-PbC2	57.8 ± 0.2	37.3 ± 0.2	4.8 ± 0.1

2.1.3. TEM Analysis

Figures 3 and S3 show the effect of adsorption of Pb(II) on nano-TiO_2 morphology at different initial adsorbate concentrations of 47.54 mg L^{-1} and 1.16 mg L^{-1}, with the aim to study the morphology and particle size distribution (PSD). Figures 3a and S3a show that the nano-TiO_2 particles presented mostly spherical morphology [33]. In addition, the PSD was plotted from the count of 700 NPs, from which average particle sizes of 22.46 nm (Figure 3b) and 17.1 nm (Figure S3b) were estimated. The adsorption of Pb(II) with a higher initial concentration showed a slight increase in particle size compared to the TiO_2 NP size, without application in the adsorption process, of 17.9 nm reported by Canchanya et al. [31]. With a low initial concentration there was no variation in nano-TiO_2 particle size. The PSD histogram makes it obvious that the NPs had a wide PSD and that their distribution closely followed a normal curve. The selected area electron diffraction (SAED) patterns are presented in Figures 3d and S3d. The crystalline planes of nano-TiO_2 are specified by the Miller indices, which were identified from the pattern rings, which agree with the XRD results (Figure 1). Looking at the images, it can be suggested that the Pb adsorbed on nano-TiO_2 formed a polynanocrystalline structure [33,34]. Figures 3c and S3c show the crystal lattice planes of nano-TiO_2, for both samples; a crystal lattice plane (110) with a d-spacing of 0.35 nm was estimated. On the other hand, the crystallite domain size distribution gave values of 9.9 nm (Figure 3e) and 8.7 nm (Figure S3e).

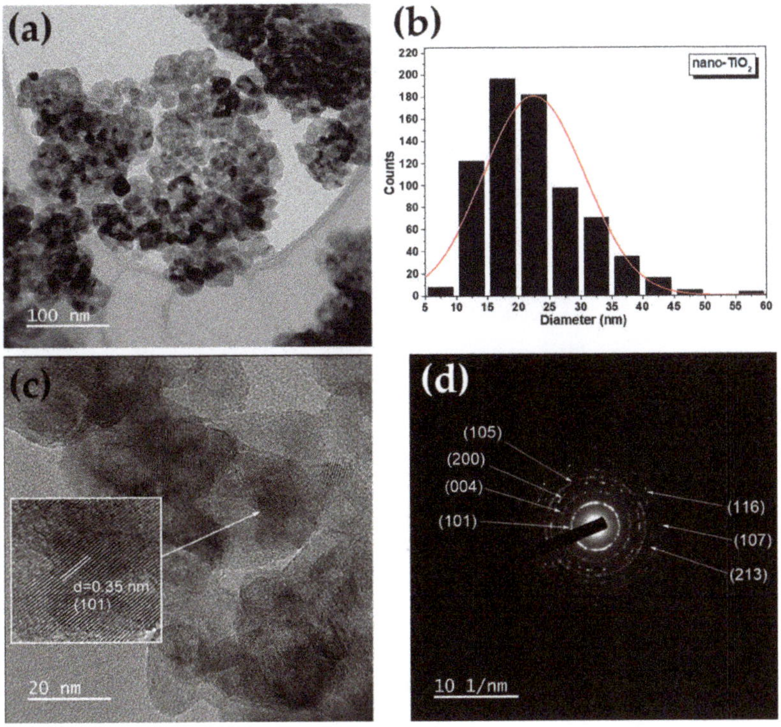

Figure 3. *Cont.*

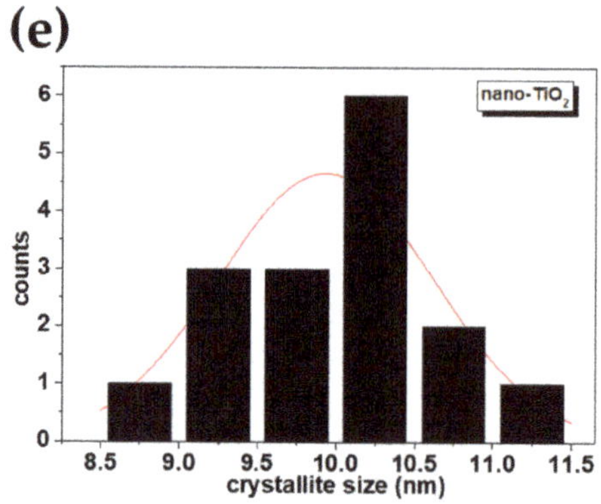

Figure 3. (**a**) TEM image of TiO_2-PbC1 NPs (bar length of 100 nm), (**b**) PSD histogram for nano-TiO_2-PbC1, (**c**) Zoomed TEM image of nano-TiO_2-PbC1, (**d**) SAED pattern of nano-TiO_2-PbC1, and (**e**) crystallite size distribution histogram for nano-TiO_2-PbC1 obtained from Rietveld refinement. C_0 (Pb(II)) = 47.54 mg L^{-1}.

2.1.4. XPS Analysis

Figures 4 and 5 show the XPS spectra for low- and high-concentrated Pb adsorbed onto the nano-TiO_2 sample—in principle, there was no alteration in the BE of Ti $2p_{3/2}$ with respect to both Pb concentrations. Furthermore, by increasing the concentration of the metal (Pb) of the aqueous solution (from 1.16 mg L^{-1} to 47.54 mg L^{-1}) in the nano-TiO_2, no increases in at%. Pb adsorbed were observed (Table 3). This indicates a saturation limit of adsorption of Pb atoms by the TiO_2 NPs. Regarding the energy analysis of Ti $2p_{3/2}$, this indicated that TiO_2 with a rutile or anatase structure (indistinguishable by XPS) [35] went through an oxidation process with respect to its initial structure. This was due to the adsorption process and was confirmed by the BE of O 1s at 531.7 eV, as suggested by Lu et al. [36].

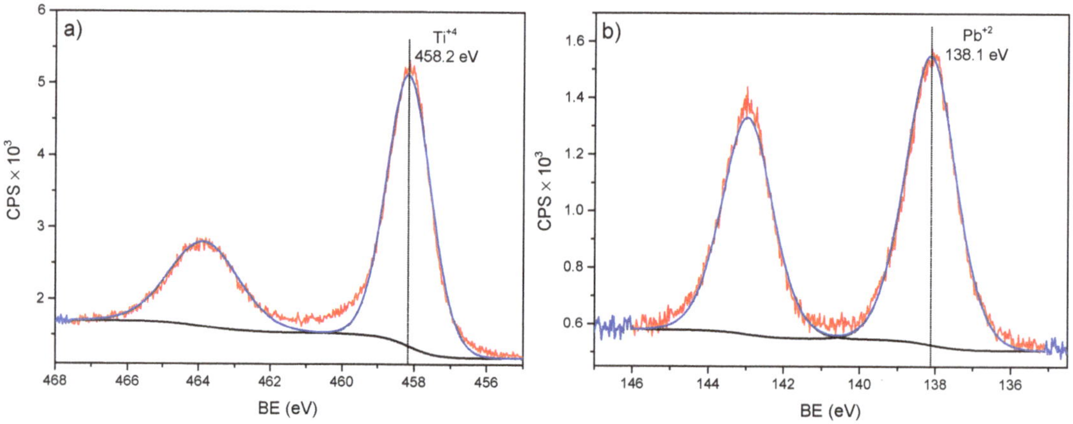

Figure 4. *Cont.*

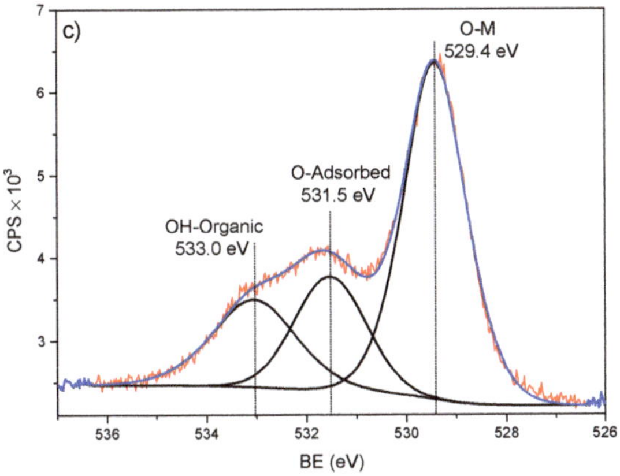

Figure 4. XPS spectra of the TiO_2-PbC1 sample, showing the deconvoluted peaks of: (**a**) Ti 2p, (**b**) Pb 4f, and (**c**) O 1s.

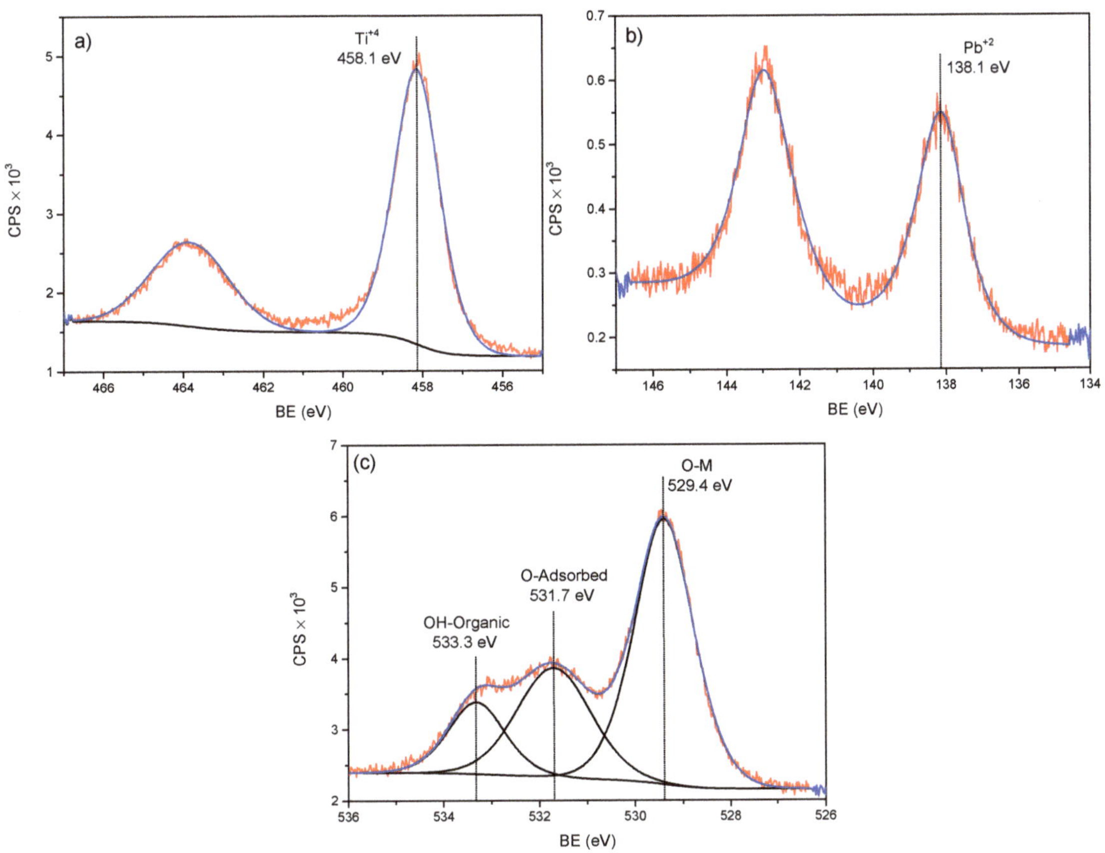

Figure 5. XPS spectra of the TiO_2Pb-C2 sample, showing the deconvoluted peaks of: (**a**) Ti 2p, (**b**) Pb 4f, and (**c**) O 1s.

Table 3. Result of the fit for the XPS peaks of the Pb(II) adsorbed nano-TiO_2 samples.

Sample	Level	BE (eV)	at.%	Bond Type	Ref.
TiO_2-Pb C2 1.16 mg L^{-1}	O 1s	529.4	45.2	O-Metal	[37]
	O 1s	531.7	22.7	O-adsorbed	[36]
	O 1s	533.3	11.6	OH-Organic	[38]
	Ti $2p_{3/2}$	458.1	18.4	Ti^{+4} (TiOx)	[35,36]
	Pb $4f_{7/2}$	138.1	2.1	Pb^{+2}	[36]
TiO_2-Pb C1 47.54 mg L^{-1}	O 1s	529.4	46.6	O-Metal	[37]
	O 1s	531.5	17.4	O-adsorbed	[36]
	O 1s	533.0	16.2	OH-Organic	[38]
	Ti $2p_{3/2}$	458.2	17.7	Ti^{+4} (TiOx)	[35,36]
	Pb $4f_{7/2}$	138.1	2.1	Pb^{+2}	[36]

2.2. Adsorption Kinetics

In order to find the time in which maximum removal is obtained, the adsorption kinetic performance was evaluated at pH 5.5, 298 K and a dose of 1 g L^{-1}. This removal percentage was calculated by applying the following Equation (1):

$$R(\%) = \frac{C_0 - C_f}{C_0} \times 100\% \quad (1)$$

Figure 6 shows that the percentage removal of Pb(II) on nano-TiO_2 was higher than 92% in all cases; with longer time, there were only slight variations. Therefore, the equilibrium time was set at 1 h with 99.18% Pb removal. This time frame was shorter than the value found by Poursani et al., who determined the equilibrium time to be 4 h at an adsorbent dose of 3 g L^{-1} [39]. Moreover, the study of Recillas et al. reported an equilibrium time of 0.62 h at 1 g L^{-1} [40]. Jyothi et al. achieved 70% Pb removal with a dose of 0.4 g L^{-1} [6]. Finally, Inquil and Valverde achieved an efficiency of 79.58% in river water with a concentration of Pb 2.36 mg L^{-1}, 0.6 g L^{-1} of nano-TiO_2, and 1 h of interaction [41]. These results indicate that the equilibrium time estimated in this work is smaller than that reported by other authors (Table 4).

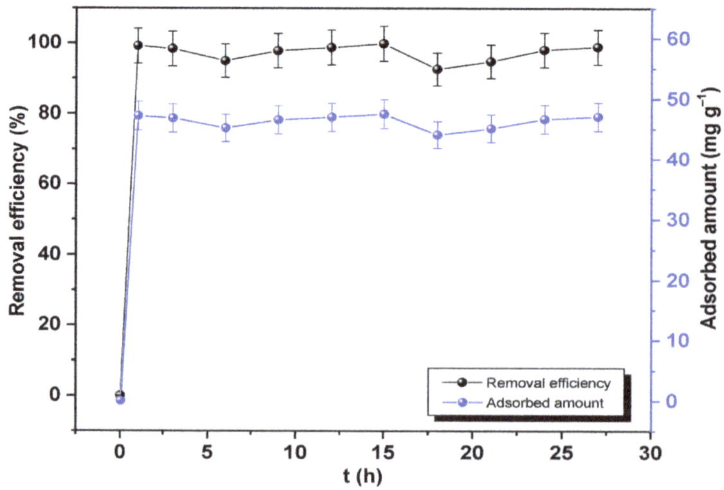

Figure 6. Evaluation of time on the efficiency and adsorbed amount of Pb(II) onto nano-TiO_2.

Therefore, it was shown that nano-TiO_2 has high potential for Pb(II) adsorption by reducing the concentration from 47.54 mg L^{-1} to 0.39 mg L^{-1}. Regarding the adsorbed amount, it was also constant from 1 h of interaction time, with the maximum adsorbed amount of 47.47 mg g^{-1} at 15 h (see Figure 6).

The maximum adsorption capacity determined in this work was 65.99 mg g^{-1}. This result is lower than the 194.28 mg g^{-1} obtained by Kanna et al., where an initial concentration of Pb(II) of 400 mg L^{-1}, pH 7, and an adsorbent dose of 0.50 g L^{-1}, were used [42].

Table 4. Maximum capacities and adsorption percentages of different adsorbents towards Pb(II) in aqueous solutions.

Adsorbent	Adsorption Capacity q_m (mg g^{-1})	pH	Time (h)	Dose (g L^{-1})	C_i (mg L^{-1})	Reference
nano-TiO_2	32.03	8	2	0.01	0.1	[43]
nano-TiO_2	7.41	6	4	4	25	[39]
nano-TiO_2	158.73	7	24	0.32	100	[40]
Anatase	194.28	7	0.25	2×10^{-3}	400	[42]
Titania nanofiber	2.56	5	4	0.05	0.50	[44]
Composed of pectin and nano-TiO_2	66.28	5.5	1	3	95.83	[4]
Graphene oxide and TiO_2 nanocomposite	65.6	5.6	12	0.025	50	[45]
Graphene oxide nanocomposite	35.6	5.6	6	0.025	50	[45]
nano-ZnO	6.7	6.6	72	1	1010	[46]
nano-CeO_2	9.2	7	NR	2	NR	[47]
Modified nano-Al_2O_3	100	5	1.5	1	50	[48]
Kaolinite	7.75	4.5	48	0.1	2000	[49]
Montmorillonite	31.10	5.7	3	2	50	[50]
Bentonite	51.19	NR	3	20	200	[51]
Biomass of *A. bisporus*	33.78	5.0	4	3	100	[52]
Biomass of *Aspergillus niger*	32.60	4.0	2	1	100	[53]
Anaerobic granular biomass	255.00	4.0–5.5	0.5	10	100	[54]
nano-TiO_2	65.99	6.5	1	1	47.54	This work.

NR: Indicates not reported.

In Table 4, a simple comparison of some adsorbents and nanoadsorbents for metal removal has been made. The adsorption capacity obtained in this work was higher than that of biomass-based adsorbents, except for anaerobic granular biomass [54]. Likewise, nanoadsorbents that have adsorption capacity values above nano-TiO_2 are modified nano-Al_2O_3 and pectin-TiO_2 nanocomposite. On the other hand, regarding the works carried out with nano-TiO_2, Recillas et al. obtained an adsorption capacity of 158.73 mg g^{-1}, which may be higher due to the contact time and initial concentration conditions, which were 24 h and 100 mg L^{-1}, respectively [40]. Subsequently, Kanna et al. found an adsorption capacity of 194.28 mg g^{-1} using anatase nanoparticles (neutral pH), a concentration of 400 mg L^{-1}, and a minimal dose of 2×10^{-3} g L^{-1}. Remarkably, our work employed a similar minimum dose of adsorbent and achieved comparable results, demonstrating the effectiveness of this low-dose approach [42]. It is worth noting that the advantages of TiO_2 NPs lie mainly in their high removal efficiency reported in previous studies, high specific surface area, low cost, availability, low environmental impact, and chemical activity, which means that modification of their adsorption capacity has been widely studied [5,19,55]. On the other hand, they should be considered as an eco-sustainable material because they can be reused by means of desorption. Evidence provided by Hu and Shipley indicates optimal desorption at pH 2 and up to eight desorption cycles; furthermore, an EDTA solution could be used for this purpose [56].

The kinetic behavior of Pb adsorption was investigated by analyzing the adsorption rate using linear and nonlinear PFO, PSO, E, and IDM, which are based on the adsorption equilibrium capacity (see Tables S1 and 5). In this work, linear and nonlinear models were compared since it is known that linear models can generate errors as a consequence of transformations from nonlinear to linear forms, implying that they can be inaccurate. Therefore, we prefer to compare with the nonlinear forms to produce accurate parameter results [57].

Table 5. Fit parameters of nonlinear models of adsorption kinetics of PFO, PSO, E, and IDM.

PFO Model		PSO Model		E Model	
q_e exp (mg g^{-1})	47.15	q_e (mg g^{-1})	46.29 (4)	β (g mg^{-1})	0.32 (4)
q_e (mg g^{-1})	46.28 (4)	k_2 (g mg^{-1} h^{-1})	12.74 (5)	α (mg h^{-1})	43,223,954.34 (1)
k_1 (h^{-1})	16.29	h (mg g^{-1} h^{-1})	27,298.58		
R^2	0.99	0.99		0.92	
RSS	11.38	11.38		126.33	
BIC	5.90	6.02		29.96	
IDM					
k_p (mg g^{-1} $h^{-0.5}$)	5.04 (2)				
C_1 (mg g^{-1})	26.28 (8)				
R^2	0.29				
RSS	1241.78				
BIC	52.82				

Four different kinetic adsorption models were applied to examine the kinetic behavior of Pb(II) adsorption (Figure S4). These models are based on equilibrium adsorption capacity, which is used to analyze the adsorption rate. The PFO equation's constant, k_1, yielded a value of 0.001 h^{-1}; however, the intercept was q_e = 0.69 mg g^{-1}, which was not the same as the value for q_e found in the kinetic adsorption tests (q_e = 47.15 mg g^{-1}). Additionally, the R^2 was close to 0, which further supports the lack of linearity. Therefore, it is concluded that the PFO, IDM, and E kinetic models cannot reproduce the results of kinetic adsorption. Instead, by applying the PSO model, the values k_2 = 0.58 g mg^{-1} h^{-1} and q_e = 46.29 mg g^{-1} were obtained, which correspond to the equilibrium adsorption capacity determined experimentally. The obtained R^2 value of 0.99 affirms the nearly perfect linearity depicted in Figure S4b. This result suggests that the Pb(II) adsorption kinetics mechanism is governed by the PSO equation.

The PSO lineal fit model had a capacity of 46.29 mg g^{-1}, equal to the value obtained by the nonlinear fit model. However, the velocity constant (k_2) for the linear model was very low at 0.58 g mg^{-1} h^{-1} compared to the nonlinear fit of 12.74 g mg^{-1} h^{-1}. Consequently, the linear model did not fit the experimental data well. On the other hand, the nonlinear models of PFO, PSO, and E presented a good fit to the experimental data with R^2 greater than 0.92, except for the IDM model (see Figure 7). Therefore, in order to find the best fit, the BIC was calculated with Equation (S12) in the Supplementary Materials [58]. The PFO and PSO models yielded the lowest BIC values, with a difference between the two values of less than 2 (Table 5), so both models can represent the adsorption process of Pb on TiO_2. In this regard, numerous authors have suggested that the PSO model best describes the adsorption procedure [39,59,60]; hence, the chemisorption process is the predominant one in the adsorption process.

2.3. *Effect of pH*

The pH is one of the most important adsorption parameters since it alters the charge on the adsorbent surface, causing a significant change in its adsorption capacity, in addition to affecting the existing form of Pb in the solution [61]. Figure 8 shows a low percentage of adsorption at low pH values of 2 to 3. In this range, the H^+ ions are found with a higher concentration in the solution and mobility on the surface, generating competition with the Pb^{2+} ions for the adsorption sites, generating a decrease in the adsorption percentage [61]. The gradual increase in pH causes a decrease in H^+ ions and an increase in hydroxyl groups on the nano-TiO_2 surface due to the protonation-dissociation equilibrium [62]. This increases the active adsorption sites available for interaction with Pb^{2+} ions and promotes the electrostatic attraction and ion exchange that occurs between the positive charges of the metal ions and the negative charges on the adsorbent surface [63]; these lead ions can form species such as $Pb(OH)^{+}$, $Pb_2(OH)^{3+}$, $Pb_3(OH)^{4+}$, and others [64].

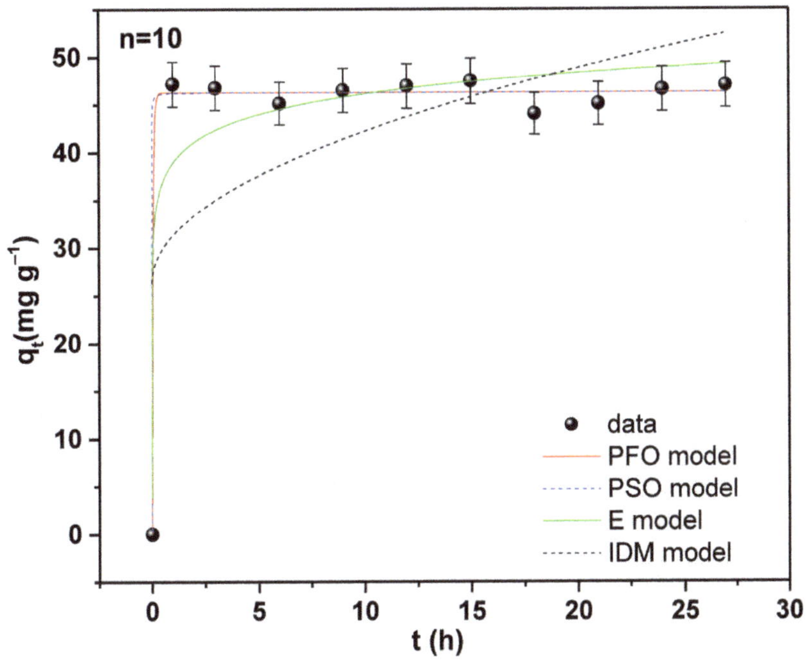

Figure 7. Nonlinear models of Pb(II) adsorption kinetics on nano-TiO_2. Adsorbent dose of 1 g L^{-1}, pH 5.5, C_o = 47.54 mg L^{-1}.

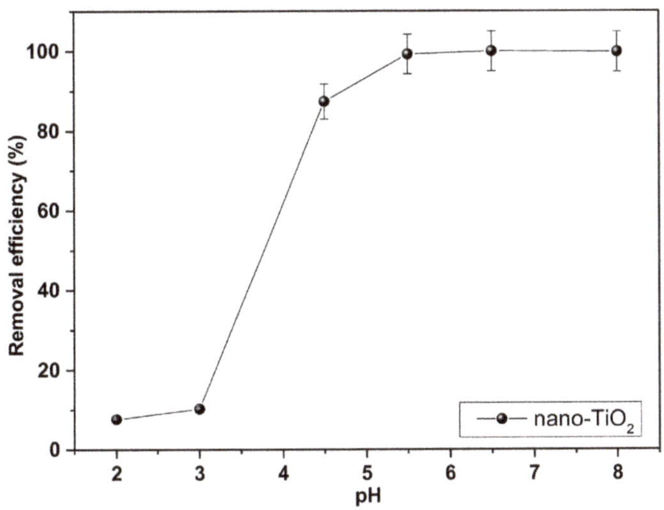

Figure 8. pH dependence of the Pb removal on nano-TiO_2.

Increasing the pH generated an increase in the adsorption efficiency, reaching a maximum percentage of adsorption at 6.5 with a value of 99.93%. In the same way, other studies obtained similar results for pH in the elimination of Pb with nano-TiO_2 [27,39,40,56]. At pH values above the point of zero charge (pH_{PZC}), for nano-TiO_2 located at 6.5 [65], there was no significant reduction in the adsorption percentage since, with increasing pH, the nano-TiO_2 surface increased its negative charge favoring adsorption [60]. However, the experiments were not conducted at pH levels greater than 8 due to the progressive precipitation of Pb and formation of complexes that prevented its adsorption [62].

2.4. Adsorbent Dose

Figure 9 illustrates the relationship between the Pb(II) removal efficiency and adsorbed amount against nano-TiO_2 adsorbent dose. It was observed that the percentage of adsorption increased as the absorbent dose increased, given that, with a greater amount of adsorbent, the number of adsorption active sites on the nano-TiO_2 surface increases and the competition between ions occupying the active sites reduces, as a result there is an increase in the adsorption of Pb ions [60,62]. On the other hand, the adsorbed amount decreases with increase in the dose since the amount of adsorbent is increased but the amount of solute in the solution remains constant [66]. This prevents reaching the maximum amount of adsorption in an active site. When the dose of the adsorbent coincides with the total amount of Pb^{2+} ions in the solution, the adsorbed amount reaches the highest possible value [59], in this case, 99.93% of Pb was removed from the water with a dose of 1 g L^{-1}; Giammar et al. reported similar results [27]. After that, by further increasing the adsorption dose, the removal percentage decreased slightly.

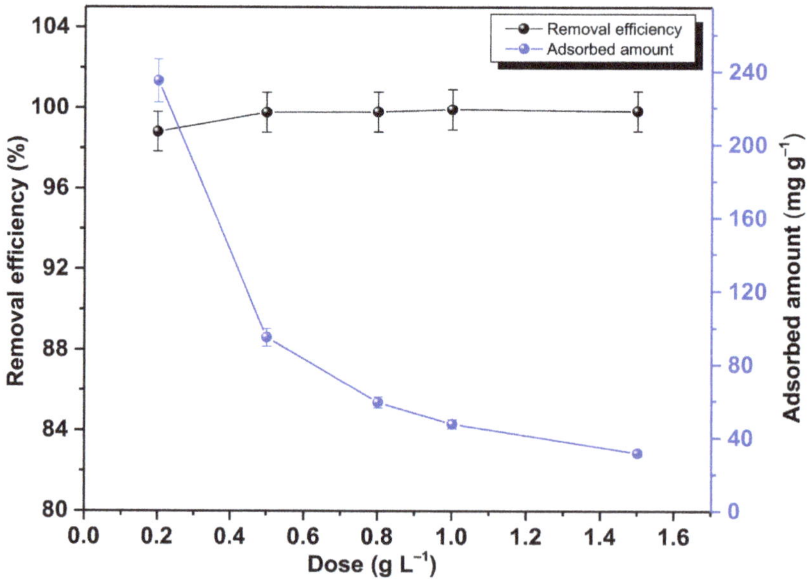

Figure 9. Adsorbent dose dependence of the Pb removal and adsorbed amount for nano-TiO_2, equilibrium time of 1 h.

2.5. Adsorption Isotherms

The adsorption isotherms obtained for the TiO_2 NPs are plotted in Figure 10. An increase in adsorption capacity can be seen with increasing initial adsorbate concentration. The optimal conditions for obtaining the adsorption experimental data were a pH of 6.5, a dose of 1 g L^{-1}, and 1 h of agitation at equilibrium. The adsorption parameters calculated from the fit, and the BIC and RSS values are presented in Table 6. Evidently, the Freundlich model yielded the lowest value of R^2, so it was ruled out, implying that the process is not governed by multilayer. The other models had high values close to each other of R^2. Therefore, another criterion by which to evaluate the models is the BIC value, where the lowest values were found for the Langmuir and Sips models; the difference between the BIC values of the two models was less than 2. In fact, both applied models could explain well the adsorption process of Pb(II) on nano-TiO_2 [67]. The Langmuir model implies that the process is characterized by adsorption on the homogeneous surface sites of the adsorbent by the formation of a monolayer of adsorbate, with all sites being identical and having equivalent energies [39]. Likewise, the Sips model supports the Langmuir model—this was inferred from the value of m_s = 0.85, close to unity, which confirms homogeneous adsorption [68]. It was also verified by the value of β = 0.97, close to 1, obtained by the Redlich–Peterson model, which indicates a monolayer reaction [69]. The maximum adsorption capacity predicted by the Langmuir model was 65.99 mg g^{-1}, close to the value computed by the Sips model of 69.22 mg g^{-1}. After comparing the adsorption capacity of TiO_2, it was confirmed that it was superior to other adsorbents reported in the literature (Table 4).

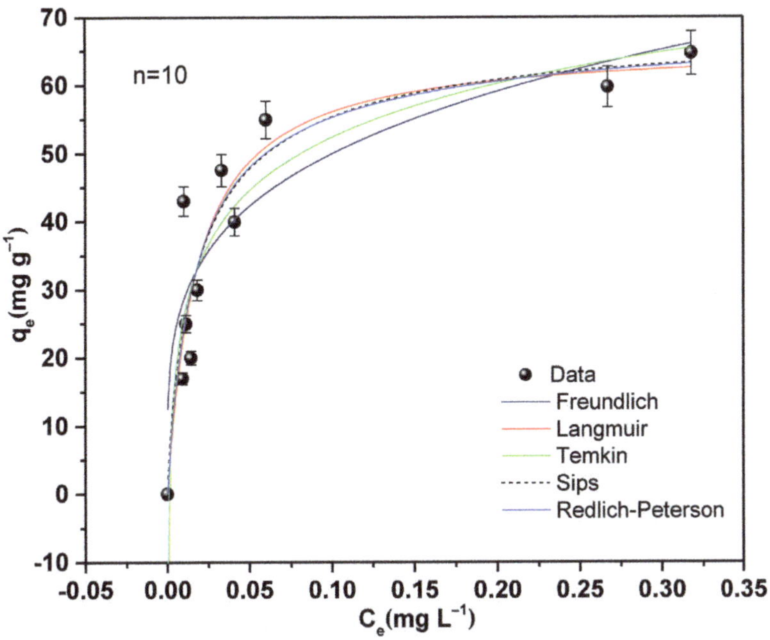

Figure 10. Pb(II) adsorption isotherms from irrigation water using nano-TiO_2.

Table 6. Parameters of the fitting of the adsorption isotherms of Langmuir, Freundlich, Temkin, Sips, and Redlich–Peterson models to the experimental data of the adsorption of Pb(II) on nano-TiO_2.

Langmuir		Freundlich		Temkin	
q_e exp (mg g^{-1})	47.5	k_F ((mg g^{-1})/(mg $L^{-1})^{1/n}$)	87.16 (1)	K_T (L g^{-1})	1049.45 (7)
q_m (mg g^{-1})	65.99 (6)	1/n	0.24 (5)	B_T (J mol^{-1})	219.85 (4)
k_L (L mg^{-1})	56.91 (2)	n	3.45 (7)		
R^2	0.84	0.69		0.83	
RSS	577.34	577.34		1260.61	
BIC	48.36	77.25		56.95	

Table 6. *Cont.*

Sips		Redlich-Peterson	
q_{m_s} (mg g^{-1})	69.22 (1)	A (L g^{-1})	4191.41 (3)
k_s $(L\ mg^{-1})^{m_s}$	50.71 (3)	B (L mg^{-1})	60.75 (3)
m_s	0.85 (5)	β	0.97 (2)
R^2	0.82	0.82	
RSS	570.48	575.87	
BIC	48.23	50.73	

2.6. Adsorption Mechanisms of Pb(II) on TiO_2

High Pb adsorption was seen at values of pH above pH_{PZC} = 6.5, because the adsorbent surface was negatively charged, which generated an increase in the Coulomb interaction between the ions and the adsorbent surface. In their study, Chen et al. showed how H_2O adsorbs to TiO_2 by bond breaking and formation, which is referred to as chemisorption. Their system demonstrated how the (O-H) bond in water dissociated and created new bonds that connected the H/O atom of H_2O and the O/Ti atom of TiO_2, which included forming the O-H bond on the surface of TiO_2 [70]. Therefore, the ion exchange mechanism described by Equations (2) and (3) can be used to explain the adsorption of Pb(II) on nano-TiO_2 as the binding of Pb^{2+}, $Pb(OH)^+$, and $Pb(OH)_2$ with the active sites available on the surface of the adsorbents.

$$2(-ROH) + Pb^{2+} \rightarrow 2(RO)Pb + 2H^+ \quad \text{Step 1} \tag{2}$$

$$-ROH + Pb(OH)^+ \rightarrow (-RO)Pb(OH) + H^+ \quad \text{Step 2} \tag{3}$$

or by hydrogen bonding given by Equation (4)

$$2(-ROH) + Pb(OH)_2 \rightarrow (-ROH)_2 + Pb(OH)_2 \quad \text{Step 3} \tag{4}$$

Since the oxygen of the (O-H) on the adsorbent's surface is a strong Lewis base by its characteristic vacant doublet electrons, the oxygen forms a coordination complex with the low electron chemical species, such as metal ions [60,71], as shown in Figure 11. Step 1 and step 2 reveal the ion exchange first stage (deprotonation) and the adsorption of Pb^{2+} and $Pb(OH)^+$ on the deprotonated surface active sites, respectively, expressed in Equations (2) and (3). Step 3 is the formation of hydrogen bonds, expressed in Equation (4).

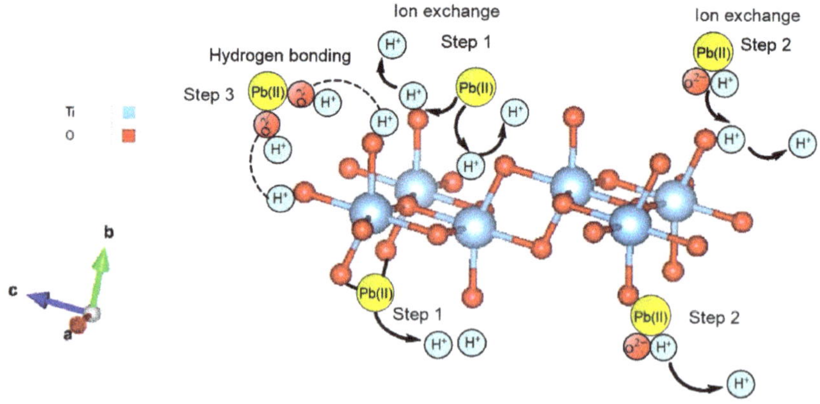

Figure 11. Pb(II) adsorption mechanisms on nano-TiO_2.

3. Perspectives

For further studies, we intend to continue with the application of nano-TiO_2 in water bodies with natural organic matter (NOM) since it has been found that nano-TiO_2 with ions such as Pb^{2+} and Ca^{2+} adsorbed provides more complexation sites, which may favor the adsorption of NOM through the formation of bridges between the nano-TiO_2 and NOM surface [72,73].

On the other hand, in the search to optimize the equilibrium time and the acceleration of the adsorption kinetics, it is intended to modify the surface properties of nano-TiO_2 by adding materials such as graphene oxide [45], pectin [4], chitosan [74], carbon nanotubes, montmorillonite, binary oxides, and ternary compounds with magnetite and silica [75], whose adsorbents have a higher adsorption capacity, reaching equilibrium times of less than 1 h.

Considering that, after the adsorption process, the nano-TiO_2 with Pb(II) ion species on its surface cannot be released into the environment due to its toxicity, the reusability of the nano-TiO_2 and the recovery of the adsorbed metal are important, making the adsorption process more cost-effective and environmentally friendly. Hu and Shipley [56] investigated the regeneration capacity of nano-TiO_2 for Pb(II) recovery using various salts, 95% of desorbed Pb(II) being obtained by use of $NaNO_3$. On the other hand, EDTA was used for optimal regeneration of nano-TiO_2, which can be reused even for removal in systems with various metals [56,76]. According to Engates and Shipley [43], nano-TiO_2 could be reused in eight cycles of Pb(II) adsorption, in which a constant and similar adsorption capacity was achieved in each cycle. With respect to the use of desorbed Pb(II) as $Pb(OH)_2$, this has applications in the field of chemical catalysis [77].

4. Materials and Methods

4.1. Materials and Chemicals

All the chemical reagents utilized were of analytical quality. TiO_2 NPs were acquired from Sigma Aldrich (Darmstadt, Germany) and used unpurified. SCP Science (Quebec, QC, Canada) provided a Pb(II) standard solution with a concentration of 999 ± 3 μg mL^{-1}. Merck (Darmstadt, Germany) brand hydrochloric acid (HCl) and sodium hydroxide (NaOH) reagents were used.

Water polluted with Pb was collected from the irrigation canal on the left bank of the Mantaro river (CIMIRM) at El Mantaro, Jauja, Peru, at 3320 m above sea level and 11°41′21.5″S 75°22′24.5″W. The original Pb(II) concentration of 0.0178 mg L^{-1} in the water was modified to 47.54 mg L^{-1}. In addition, other heavy metals were found in the water, whose concentrations were compared to the maximum concentration established (presented in parentheses) by the Food and Agriculture Organization (FAO) for irrigation water [78]: As: 0.0176 mg L^{-1} (0.1 mg L^{-1}), Cu: 0.021 mg L^{-1} (0.2 mg L^{-1}), Ni: 0.0032 mg L^{-1} (0.2 mg L^{-1}), Cd: 0.00047 mg L^{-1} (0.01 mg L^{-1}), Cr: 0.0026 mg L^{-1} (0.1 mg L^{-1}), Zn: 0.1305 mg L^{-1} (2 mg L^{-1}) and Fe: 2.8477 mg L^{-1} (5 mg L^{-1}). As can be seen, there were no concentrations above the limits established by FAO.

4.2. Characterization of Nano-TiO_2

4.2.1. X-ray Diffraction (XRD)

XRD was used to characterize the materials using an Empyrean diffractometer with CuK_α radiation at wavelength = 1.5406 Å. The X-ray diffractograms were taken following a step scanning arrangement with a range of 2θ = 20–80° and a resolution of 0.01° per step. Match V3 software was used to identify the crystallographic phase of anatase nano-TiO_2, with the tetragonal crystalline structure, space group I41/amd, and cell parameters a = 3.7892 Å and c = 9.5370 Å as an initial matching candidate, with the crystallographic information files (CIF) #5000223 for the anatase phase. The peak's profile function Thompson–Cox–Hastings pseudo-Voigt axial divergence asymmetry was used for the Rietveld refinement using the program FullProf Suite (version January 2021). Furthermore, the aluminum oxide Al_2O_3 standard was used to determine the diffractometer's instrumental resolution function (IRF), which yielded Caglioti coefficients including U = 0.0093, V = 0.0051, and W = 0.0013 [31].

4.2.2. Scanning Electron Microscopy (SEM), Energy Dispersive Spectroscopy (EDS), and Transmission Electron Microscopy (TEM)

SEM and TEM techniques were employed to estimate the mean size of particles and the TiO_2 morphologies. To assess the atomic composition, the elemental compositions of the materials were examined using EDS mapping.

4.2.3. X-ray Photoelectron Spectroscopy (XPS)

Chemical surface examinations on the studied materials were carried out utilizing the SPECS PHOIBOS 100/150 equipment with a hemispheric analyzer spectrometer, operating at 1486.6 eV of Al Kα. The XPS spectra were obtained using a high-resolution polychromatic X-ray source with a 0.02 eV energy step. The SPECS company's Casa-XPS software was used to change the peak locations of the Ti 2p, O 1s, and Pb 4f levels to compute the chemical binding energies (BE) of produced species, as well as the relative atomic amounts on the sample surfaces. The spectra were calibrated using adventitious carbon (B.E. reference) at C 1 s = 284.6 eV after being calibrated with an electron flood cannon at 12 μA and 1 eV.

4.3. *Pb(II) Removal Experiments with Nano-TiO_2*

The Pb(II) removal experiments were carried out in real water samples contaminated with Pb, taken from the CIMIRM-Junín-Perú irrigation canal, whose initial concentration was modified from 0.0178 mg L^{-1} to 47.54 mg L^{-1}. To modify the concentration of this metal, a dilution was carried out from a standard solution of Pb at an initial concentration of 999 ± 3 μg mL^{-1}. Pb quantification, in all cases, was performed using the inductively coupled plasma mass spectrometry (ICP-MS) analysis technique.

4.3.1. Adsorption Kinetics

Pb(II) adsorption kinetics experiments with nano-TiO_2 were performed in order to determine the equilibrium time and evaluate the kinetic models.

Ten experiments were carried out depending on the contact time, in a range from 1 to 27 h, at 47.54 mg L^{-1} of initial concentration of Pb(II), using a nano-TiO_2 dose of 1 g L^{-1}, at pH 5.5, under constant stirring at 300 rpm, at room temperature.

4.3.2. pH Effect

In order to know the effect of pH on the removal of Pb(II) using nano-TiO_2, experiments were carried out controlling pH in the range of 2 to 8, for a stirring time of 1 h (equilibrium time determined by the experiments of adsorption kinetics), with an adsorbent dose of 1 g L^{-1}, maintaining constant stirring at 300 rpm at room temperature. The control of the previously indicated pH range was performed by adding drops of HCl and NaOH solutions in concentrations of 0.05 M, 0.1 M, 1 M, and 4 M.

4.3.3. Adsorbent Dose

The adsorbent dose experiments were carried out on real water samples at an initial Pb(II) concentration of 47.54 mg L^{-1}. The nano-TiO_2 dose range considered in this work was from 0.2 g L^{-1} to 1.5 g L^{-1}. The other experimental parameters were: pH 6.5, constant stirring time of 1 h at 300 rpm at room temperature.

4.3.4. Adsorption Isotherms

Pb(II) adsorption isotherms with nano-TiO_2 were constructed using the previously determined experimental conditions: the equilibrium time, the optimum pH, and the adsorbent dose. Utilizing these values, ten tests were conducted on irrigation canal water samples with beginning concentrations ranging from 3 to 65 mg L^{-1}. To analyze the data obtained, the Langmuir, Freundlich, Redlich–Peterson, Sips and Temkin adsorption models were used.

The theoretical description of the kinetic and isotherm models is given in the Supplementary Materials.

5. Conclusions

The removal efficiency of Pb(II) using nano-TiO_2 in irrigation water was evaluated at a modified initial concentration of 47.54 mg L^{-1}, in the presence of other metals. This nanomaterial was characterized by XRD after adsorption, confirming the presence of the tetragonal crystalline structure of the anatase phase. The mean crystallite size was obtained by Rietveld refinement giving a value of 9.9 nm; the mean particle size was determined by TEM, being 22.46 nm. The XPS studies indicated that there was no alteration of the absorption bands with respect to increase in the concentration of Pb(II). The SEM images showed changes in the surface morphology of the nano-TiO_2, confirming the occurrence of metal adsorption. In the Pb(II) removal experiments, more than 99% Pb(II) elimination was achieved by determining the optimal contact time, pH, and dose of adsorbent, having values of 1 h, 6.5, and 1 g L^{-1}, suggesting that there was no significant influence of the prevalence of other metal ions. The experimental data were evaluated with linear and nonlinear models of adsorption kinetics. The best fit was found to correspond to the PSO model, indicating that the process was governed by chemisorption. This result was supported by evaluation of the adsorption isotherms, where the Langmuir and Sips models provided the best fit, evidencing that adsorption occurred in the homogeneous sites of the nano-TiO_2 surface through the formation of a adsorbate monolayer. In addition, a high adsorption capacity of nano-TiO_2 for Pb(II) of 65.99 mg g^{-1} was estimated, higher than the values reported in other studies. In general, the results of this study indicate that nano-TiO_2 is a potential adsorbent for application in the treatment of water contaminated with Pb(II).

Supplementary Materials: The following supporting information can be downloaded at: https://www.mdpi.com/article/10.3390/molecules28124596/s1, Supporting section: Theoretical background of kinetic and isotherm adsorption models. Figure S1: (a,f) SEM images, (b,g) EDS mapping images, and (c,d,e,h,i,j) elemental EDS images for nano-TiO_2 with C_0 (Pb(II)) = 1.16 mg L^{-1}. Figure S2: EDS spectrum for both nano-TiO_2 samples exposed to two different Pb(II) concentrations. Figure S3: (a) TEM image of TiO_2-PbC2 NPs (bar length of 100 nm), (b) PSD for nano-TiO_2-PbC2, (c) Zoomed TEM image of nano-TiO_2-PbC2, (d) SAED pattern of nano-TiO_2-PbC2, and (e) crystallite size distribution histogram for nano-TiO_2-PbC2 obtained from Rietveld refinement. C_0 (Pb(II)) = 1.16 mg L^{-1}. Figure S4: Linear models of Pb (II) adsorption kinetics on nano-TiO_2. Adsorbent dose of 1 g L^{-1}, pH 5.5. C_o = 47.54 mg L^{-1}. Table S1: Fit parameters of linear models of adsorption kinetics of PFO, PSO, E, and IDM. References [79–90] are cited in Supplementary Materials.

Author Contributions: Conceptualization, M.A.V.-C., Y.C.-H., A.F.M.-A., J.P.-V., N.-R.C.-H., Y.B.-R. and J.A.R.-G.; methodology, M.A.V.-C., Y.C.-H., A.F.M.-A., J.P.-V., N.-R.C.-H., Y.B.-R. and J.A.R.-G.; software, M.A.V.-C., Y.C.-H., A.F.M.-A., J.P.-V., N.-R.C.-H., Y.B.-R. and J.A.R.-G.; validation, M.A.V.-C., Y.C.-H., A.F.M.-A., J.P.-V., N.-R.C.-H., Y.B.-R. and J.A.R.-G.; formal analysis, M.A.V.-C., Y.C.-H., A.F.M.-A., J.P.-V., N.-R.C.-H., Y.B.-R. and J.A.R.-G.; investigation, M.A.V.-C., Y.C.-H., A.F.M.-A., J.P.-V., N.-R.C.-H., Y.B.-R. and J.A.R.-G.; resources, M.A.V.-C., Y.C.-H., A.F.M.-A., J.P.-V., N.-R.C.-H., Y.B.-R. and J.A.R.-G.; data curation, M.A.V.-C., Y.C.-H., A.F.M.-A., J.P.-V., N.-R.C.-H., Y.B.-R. and J.A.R.-G.; writing—original draft preparation, M.A.V.-C., Y.C.-H., A.F.M.-A., J.P.-V., N.-R.C.-H., Y.B.-R. and J.A.R.-G.; writing—review and editing, M.A.V.-C., Y.C.-H., A.F.M.-A., J.P.-V., N.-R.C.-H., Y.B.-R. and J.A.R.-G.; visualization, M.A.V.-C., Y.C.-H., A.F.M.-A., J.P.-V., N.-R.C.-H., Y.B.-R. and J.A.R.-G.; supervision, Y.B.-R. and J.A.R.-G.; project administration, Y.B.-R. and J.A.R.-G.; funding acquisition, J.A.R.-G. All authors have read and agreed to the published version of the manuscript.

Funding: This research was supported by Vicerrectorado de Investigación y Posgrado (VRIP) de la Universidad Nacional Mayor de San Marcos (UNMSM)—RR N° 09412-R-21 and project number B21130021-MINEDU-2021. The APC was funded by VRIP-UNMSM.

Data Availability Statement: The original data related to this research can be obtained at any time via the corresponding author's email: juan.ramos5@unmsm.edu.pe.

Acknowledgments: We thank the Investigación y Posgrado (VRIP) de la Universidad Nacional Mayor de San Marcos (UNMSM)—RR N° 09412-R-21 and project number B2113002i-MINEDU-2021 for financially supporting this work.

Conflicts of Interest: The authors declare no conflict of interest.

Sample Availability: Not applicable.

References

1. Lu, F.; Astruc, D. Nanomaterials for removal of toxic elements from water. *Coord. Chem. Rev.* **2018**, *356*, 147–164. [CrossRef]
2. Baruah, S.; Najam Khan, M.; Dutta, J. Perspectives and applications of nanotechnology in water treatment. *Environ. Chem. Lett.* **2016**, *14*, 1–14. [CrossRef]
3. World Water Development Report 2018 | UN-Water. Available online: https://www.unwater.org/publications/world-water-development-report-2018 (accessed on 19 March 2023).
4. Bok-Badura, J.; Jakóbik-Kolon, A.; Karoń, K.; Mitko, K. Sorption studies of heavy metal ions on pectin-nano-titanium dioxide composite adsorbent. *Sep. Sci. Technol.* **2018**, *53*, 1034–1044. [CrossRef]
5. Bhatia, M.; Satish Babu, R.; Sonawane, S.H.; Gogate, P.R.; Girdhar, A.; Reddy, E.R.; Pola, M. Application of nanoadsorbents for removal of lead from water. *Int. J. Environ. Sci. Technol.* **2017**, *14*, 1135–1154. [CrossRef]
6. Jyothi, B.; Singa, P.M.; Desai, N.N. Performance of TiO_2 Nano powder in Removal of Lead from Synthetic Solution. *Int. Res. J. Eng. Technol.* **2015**, *2*, 1396–1399.
7. Sepehri, S.; Kanani, E.; Abdoli, S.; Rajput, V.D.; Minkina, T.; Asgari Lajayer, B. Pb(II) Removal from Aqueous Solutions by Adsorption on Stabilized Zero-Valent Iron Nanoparticles—A Green Approach. *Water* **2023**, *15*, 222. [CrossRef]
8. Chanthapon, N.; Sarkar, S.; Kidkhunthod, P.; Padungthon, S. Lead removal by a reusable gel cation exchange resin containing nano-scale zero valent iron. *Chem. Eng. J.* **2018**, *331*, 545–555. [CrossRef]
9. Wang, C.; Wang, H.; Cao, Y. Pb(II) sorption by biochar derived from Cinnamomum camphora and its improvement with ultrasound-assisted alkali activation. *Colloids Surf. A Physicochem. Eng. Asp.* **2018**, *556*, 177–184. [CrossRef]
10. Moustafa, H.; Isawi, H.; Abd El Wahab, S.M. Utilization of PVA nano-membrane based synthesized magnetic GO-Ni-Fe_2O_4 nanoparticles for removal of heavy metals from water resources. *Environ. Nanotechnol. Monit. Manag.* **2022**, *18*, 100696. [CrossRef]
11. World Health Organization. *Guidelines for Drinking-Water Quality*; World Health Organization: Geneva, Switzerland, 2011; ISBN 9789241548151.
12. Long, Y.; Jiang, J.; Hu, J.; Hu, X.; Yang, Q.; Zhou, S. Removal of Pb(II) from aqueous solution by hydroxyapatite/carbon composite: Preparation and adsorption behavior. *Colloids Surf. A Physicochem. Eng. Asp.* **2019**, *577*, 471–479. [CrossRef]
13. Van Tran, C.; Quang, D.V.; Nguyen Thi, H.P.; Truong, T.N.; La, D.D. Effective Removal of Pb(II) from Aqueous Media by a New Design of Cu-Mg Binary Ferrite. *ACS Omega* **2020**, *5*, 7298–7306. [CrossRef] [PubMed]
14. Wang, Y.Y.; Liu, Y.X.; Lu, H.H.; Yang, R.Q.; Yang, S.M. Competitive adsorption of Pb(II), Cu(II), and Zn(II) ions onto hydroxyapatite-biochar nanocomposite in aqueous solutions. *J. Solid State Chem.* **2018**, *261*, 53–61. [CrossRef]
15. Carolin, C.F.; Kumar, P.S.; Saravanan, A.; Joshiba, G.J.; Naushad, M. Efficient techniques for the removal of toxic heavy metals from aquatic environment: A review. *J. Environ. Chem. Eng.* **2017**, *5*, 2782–2799. [CrossRef]
16. Azimi, A.; Azari, A.; Rezakazemi, M.; Ansarpour, M. Removal of Heavy Metals from Industrial Wastewaters: A Review. *Chem. Bio. Eng. Rev.* **2017**, *4*, 37–59. [CrossRef]
17. Bolisetty, S.; Peydayesh, M.; Mezzenga, R. Sustainable technologies for water purification from heavy metals: Review and analysis. *Chem. Soc. Rev.* **2019**, *48*, 463–487. [CrossRef] [PubMed]
18. Lata, S.; Singh, P.K.; Samadder, S.R. Regeneration of adsorbents and recovery of heavy metals: A review. *Int. J. Environ. Sci. Technol.* **2015**, *12*, 1461–1478. [CrossRef]
19. Santhosh, C.; Velmurugan, V.; Jacob, G.; Jeong, S.K.; Grace, A.N.; Bhatnagar, A. Role of nanomaterials in water treatment applications: A review. *Chem. Eng. J.* **2016**, *306*, 1116–1137. [CrossRef]
20. Hashem, M.; Al-Karagoly, H. Synthesis, characterization, and cytotoxicity of titanium dioxide nanoparticles and in vitro study of its impact on lead concentrations in bovine blood and milk. *J. Biotech Res.* **2021**, *12*, 93–105.
21. Motlochová, M.; Slovák, V.; Pližingrová, E.; Lidin, S.; Šubrt, J. Highly-efficient removal of Pb(II), Cu(II) and Cd(II) from water by novel lithium, sodium and potassium titanate reusable microrods. *RSC Adv.* **2020**, *10*, 3694–3704. [CrossRef]
22. Govan, J. Recent advances in magnetic nanoparticles and nanocomposites for the remediation of water resources. *Magnetochemistry* **2020**, *6*, 49. [CrossRef]
23. Hornyak, G.L. *Introduction to Nanoscience and Nanotechnology*; CRC Press: Boca Raton, FL, USA, 2008; ISBN 9781420048056.
24. Zhu, Y.; Liu, X.; Hu, Y.; Wang, R.; Chen, M.; Wu, J.; Wang, Y.; Kang, S.; Sun, Y.; Zhu, M. Behavior, remediation effect and toxicity of nanomaterials in water environments. *Environ. Res.* **2019**, *174*, 54–60. [CrossRef]
25. George, R.; Bahadur, N.; Singh, N.; Singh, R.; Verma, A.; Shukla, A.K. Environmentally Benign TiO_2 Nanomaterials for Removal of Heavy Metal Ions with Interfering Ions Present in Tap Water. *Mater. Today Proc.* **2016**, *3*, 162–166. [CrossRef]
26. Song, H.; Kuang, X.; Wei, X.; Luo, S.; Zeng, Q.; Peng, L. The effect of TiO_2 nanoparticles size on Cd (II) removal by the paddy crusts from waterbody. *J. Environ. Chem. Eng.* **2022**, *10*, 107883. [CrossRef]

27. Giammar, D.E.; Maus, C.J.; Xie, L. Effects of particle size and crystalline phase on lead adsorption to titanium dioxide nanoparticles. *Environ. Eng. Sci.* **2007**, *24*, 85–95. [CrossRef]
28. Prasai, B.; Cai, B.; Underwood, M.K.; Lewis, J.P.; Drabold, D.A. Properties of amorphous and crystalline titanium dioxide from first principles. *J. Mater. Sci.* **2012**, *47*, 7515–7521. [CrossRef]
29. Custodio, M.; Álvarez, D.; Cuadrado, W.; Montalvo, R.; Ochoa, S. Potentially toxic metals and metalloids in surface water intended for human consumption and other uses in the Mantaro River watershed, Peru. *Soil Water Res.* **2020**, *15*, 237–245. [CrossRef]
30. Popa, N.C. The (hkl) dependence of diffraction-line broadening caused by strain and size for all laue groups in rietveld refinement. *J. Appl. Crystallogr.* **1998**, *31*, 176–180. [CrossRef]
31. Canchanya-Huaman, Y.; Mayta-Armas, A.F.; Pomalaya-Velasco, J.; Bendezú-Roca, Y.; Guerra, J.A.; Ramos-Guivar, J.A. Strain and grain size determination of CeO_2 and TiO_2 nanoparticles: Comparing integral breadth methods versus Rietveld, μ-Raman, and TEM. *Nanomaterials* **2021**, *11*, 2311. [CrossRef]
32. Patidar, V.; Jain, P. Green Synthesis of TiO_2 Nanoparticle Using Moringa Oleifera Leaf Extract. *Int. Res. J. Eng. Technol.* **2017**, *4*, 470–473.
33. Thamaphat, K.; Limsuwan, P.; Ngotawornchai, B. Phase Characterization of TiO_2 Powder by XRD and TEM. *Nat. Sci.* **2008**, *42*, 357–361.
34. Zhang, S.; Shi, Q.; Chou, T.M.; Christodoulatos, C.; Korfiatis, G.P.; Meng, X. Mechanistic study of Pb(II) removal by TiO_2 and effect of PO_4. *Langmuir* **2020**, *36*, 13918–13927. [CrossRef]
35. Dementjev, A.P.; Ivanova, O.P.; Vasilyev, L.A.; Naumkin, A.V.; Nemirovsky, D.M.; Shalaev, D.Y. Altered layer as sensitive initial chemical state indicator. *J. Vac. Sci. Technol. A Vac. Surf. Film.* **1994**, *12*, 423–427. [CrossRef]
36. Lu, C.J.; Kuang, A.X.; Huang, G.Y. X-ray photoelectron spectroscopy study on composition and structure of sol-gel derived $PbTiO_3$ thin films. *J. Appl. Phys.* **1996**, *80*, 202–206. [CrossRef]
37. Moulder, J.F.; Stickle, W.F.; Sobol, P.E.; Bomben, K.D. *Handbook of X-ray Photoelectron Spectroscopy: A Reference Book of Standard Spectra for Identification and Interpretation of XPS Data*; Perkin-Elmer Corporation: Waltham, MA, USA, 1992.
38. Beamson, G.; Briggs, D. High Resolution XPS of Organic Polymers: The Scienta ESCA300 Database. *J. Chem. Educ.* **1993**, *70*, 778. [CrossRef]
39. Poursani, A.S.; Nilchi, A.; Hassani, A. The synthesis of nano TiO_2 and its use for removal of lead ions from aqueous solution. *J. Water Resour. Prot.* **2016**, *08*, 438–448. [CrossRef]
40. Recillas, S.; García, A.; González, E.; Casals, E.; Puntes, V.; Sánchez, A.; Font, X. Use of CeO_2, TiO_2 and Fe_3O_4 nanoparticles for the removal of lead from water. Toxicity of nanoparticles and derived compounds. *Desalination* **2011**, *277*, 213–220. [CrossRef]
41. Inquil Ayquipa, L.; Valverde Flores, J. Adsorption of lead and iron present in the waters of the Santa River using titanium dioxide nanoparticles (TiO_2). *J. Nanotechnol.* **2020**, *4*, 8–10. [CrossRef]
42. Kanna, M.; Wongnawa, S.; Sherdshoopongse, P. Adsorption behavior of some metal ions on hydrated amorphous titanium dioxide surface. *Songklanakarin J. Sci. Technol.* **2005**, *27*, 1017–1026.
43. Engates, K.E.; Shipley, H.J. Adsorption of Pb, Cd, Cu, Zn, and Ni to titanium dioxide nanoparticles: Effect of particle size, solid concentration, and exhaustion. *Environ. Sci. Pollut. Res.* **2011**, *18*, 386–395. [CrossRef]
44. Dixit, A.; Mishra, P.K.; Alam, M.S. Titania Nanofibers: A Potential Adsorbent for Mercury and Lead Uptake. *Int. J. Chem. Eng. Appl.* **2017**, *8*, 75–81. [CrossRef]
45. Lee, Y.C.; Yang, J.W. Self-assembled flower-like TiO_2 on exfoliated graphite oxide for heavy metal removal. *J. Ind. Eng. Chem.* **2012**, *18*, 1178–1185. [CrossRef]
46. Kikuchi, Y.; Qian, Q.; Machida, M.; Tatsumoto, H. Effect of ZnO loading to activated carbon on Pb(II) adsorption from aqueous solution. *Carbon* **2006**, *44*, 195–202. [CrossRef]
47. Cao, C.Y.; Cui, Z.M.; Chen, C.Q.; Song, W.G.; Cai, W. Ceria hollow nanospheres produced by a template-free microwave-assisted hydrothermal method for heavy metal ion removal and catalysis. *J. Phys. Chem. C* **2010**, *114*, 9865–9870. [CrossRef]
48. Afkhami, A.; Saber-Tehrani, M.; Bagheri, H. Simultaneous removal of heavy-metal ions in wastewater samples using nano-alumina modified with 2,4-dinitrophenylhydrazine. *J. Hazard Mater.* **2010**, *181*, 836–844. [CrossRef]
49. Shahmohammadi-Kalalagh, S.; Babazadeh, H.; Manshouri, M. Isotherm and kinetic studies on adsorption of Pb, Zn and Cu by Kaolinite. *Casp. J. Environ. Sci.* **2010**, *9*, 243–255.
50. Sen Gupta, S.; Bhattacharyya, K.G. Immobilization of Pb(II), Cd(II) and Ni(II) ions on kaolinite and montmorillonite surfaces from aqueous medium. *J. Environ. Manag.* **2008**, *87*, 46–58. [CrossRef] [PubMed]
51. Al-Jlil, S.A. Kinetic study of adsorption of chromium and lead ions on bentonite clay using novel internal series model. *Trends Appl. Sci. Res.* **2015**, *10*, 38–53. [CrossRef]
52. Vimala, R.; Das, N. Biosorption of cadmium (II) and lead (II) from aqueous solutions using mushrooms: A comparative study. *J. Hazard Mater.* **2009**, *168*, 376–382. [CrossRef]
53. Dursun, A.Y. A comparative study on determination of the equilibrium, kinetic and thermodynamic parameters of biosorption of copper(II) and lead(II) ions onto pretreated Aspergillus niger. *Biochem. Eng. J.* **2006**, *28*, 187–195. [CrossRef]
54. Hawari, A.H.; Mulligan, C.N. Biosorption of lead(II), cadmium(II), copper(II) and nickel(II) by anaerobic granular biomass. *Bioresour. Technol.* **2006**, *97*, 692–700. [CrossRef]

55. Singh, S.; Barick, K.C.; Bahadur, D. Functional oxide nanomaterials and nanocomposites for the removal of heavy metals and dyes. *Nanomater. Nanotechnol.* **2013**, *3*, 3–20. [CrossRef]
56. Hu, J.; Shipley, H.J. Regeneration of spent TiO_2 nanoparticles for Pb (II), Cu(II), and Zn(II) removal. *Environ. Sci. Pollut. Res.* **2013**, *20*, 5125–5137. [CrossRef]
57. López-Luna, J.; Ramírez-Montes, L.E.; Martinez-Vargas, S.; Martínez, A.I.; Mijangos-Ricardez, O.F.; González-Chávez, M.d.C.A.; Carrillo-González, R.; Solís-Domínguez, F.A.; Cuevas-Díaz, M.d.C.; Vázquez-Hipólito, V. Linear and nonlinear kinetic and isotherm adsorption models for arsenic removal by manganese ferrite nanoparticles. *SN Appl. Sci.* **2019**, *1*, 950. [CrossRef]
58. Cao, Y.; Dong, S.; Dai, Z.; Zhu, L.; Xiao, T.; Zhang, X.; Yin, S.; Soltanian, M.R. Adsorption model identification for chromium(VI) transport in unconsolidated sediments. *J. Hydrol.* **2021**, *598*, 126228. [CrossRef]
59. Zhao, X.; Jia, Q.; Song, N.; Zhou, W.; Li, Y. Adsorption of Pb(II) from an aqueous solution by titanium dioxide/carbon nanotube nanocomposites: Kinetics, thermodynamics, and isotherms. *J. Chem. Eng. Data* **2010**, *55*, 4428–4433. [CrossRef]
60. Mostafa, N.G.; Yunnus, A.F.; Elawwad, A. Adsorption of Pb(II) from water onto ZnO, TiO_2, and Al_2O_3: Process study, adsorption behaviour, and thermodynamics. *Adsorpt. Sci. Technol.* **2022**, *2022*, 7582756. [CrossRef]
61. Shu, Y.; Huang, R.; Wei, X.; Liu, L.; Jia, Z. Pb(II) Removal using TiO_2-Embedded Monolith Composite Cryogel as an alternative wastewater treatment method. *Water Air Soil Pollut.* **2017**, *228*, 375. [CrossRef]
62. Karapinar, H.S.; Kilicel, F.; Ozel, F.; Sarilmaz, A. Fast and effective removal of Pb(II), Cu(II) and Ni(II) ions from aqueous solutions with TiO_2 nanofibers: Synthesis, adsorption-desorption process and kinetic studies. *Int. J. Environ. Anal. Chem.* **2021**, *103*, 1–22. [CrossRef]
63. Maleki, A.; Hayati, B.; Najafi, F.; Gharibi, F.; Joo, S.W. Heavy metal adsorption from industrial wastewater by PAMAM/TiO_2 nanohybrid: Preparation, characterization and adsorption studies. *J. Mol. Liq.* **2016**, *224*, 95–104. [CrossRef]
64. Tao, Y.; Ye, L.; Pan, J.; Wang, Y.; Tang, B. Removal of Pb(II) from aqueous solution on chitosan/TiO_2 hybrid film. *J. Hazard Mater.* **2009**, *161*, 718–722. [CrossRef]
65. Mendoza-Villa, F.; Checca-Huaman, N.R.; Ramos-Guivar, J.A. Ecotoxicological properties of titanium dioxide nanomorphologies in daphnia magna. *Nanomaterials* **2023**, *13*, 927. [CrossRef] [PubMed]
66. Bhargavi, R.J.; Maheshwari, U.; Gupta, S. Synthesis and use of alumina nanoparticles as an adsorbent for the removal of Zn(II) and CBG dye from wastewater. *Int. J. Ind. Chem.* **2014**, *6*, 31–41. [CrossRef]
67. Lima, E.C.; Sher, F.; Guleria, A.; Reza, M.; Anastopoulos, I.; Nguyen, H.; Hosseini-bandegharaei, A. Is One Performing the Treatment Data of Adsorption Kinetics Correctly? *J. Environ. Chem. Eng.* **2021**, *9*, 104813. [CrossRef]
68. Ramos Guivar, J.A.; Bustamante, D.A.; Gonzalez, J.C.; Sanches, E.A.; Morales, M.A.; Raez, J.M.; López-Muñoz, M.J.; Arencibia, A. Adsorption of arsenite and arsenate on binary and ternary magnetic nanocomposites with high iron oxide content. *Appl. Surf. Sci.* **2018**, *454*, 87–100. [CrossRef]
69. Razzaz, A.; Ghorban, S.; Hosayni, L.; Irani, M.; Aliabadi, M. Chitosan nanofibers functionalized by TiO_2 nanoparticles for the removal of heavy metal ions. *J. Taiwan Inst. Chem. Eng.* **2016**, *58*, 333–343. [CrossRef]
70. Chen, B.; Li, L.; Liu, L.; Cao, J. Molecular simulation of adsorption properties of thiol-functionalized titanium dioxide (TiO_2) nanostructure for heavy metal ions removal from aqueous solution. *J. Mol. Liq.* **2022**, *346*, 118281. [CrossRef]
71. Taty-Costodes, V.C.; Fauduet, H.; Porte, C.; Delacroix, A. Removal of Cd(II) and Pb(II) ions, from aqueous solutions, by adsorption onto sawdust of Pinus sylvestris. *J. Hazard Mater.* **2003**, *105*, 121–142. [CrossRef]
72. Wang, D.; Wang, P.; Wang, C.; Ao, Y. Effects of interactions between humic acid and heavy metal ions on the aggregation of TiO_2 nanoparticles in water environment. *Environ. Pollut.* **2019**, *248*, 834–844. [CrossRef]
73. Ramirez, L.; Gentile, R.S.; Zimmerman, S.; Stoll, S. Behavior of TiO_2 and CeO_2 nanoparticles and polystyrene nanoplastics in bottled mineral, drinking and Lake Geneva Waters. Impact of water hardness and natural organic matter on nanoparticle surface properties and aggregation. *Water* **2019**, *11*, 721. [CrossRef]
74. Spoială, A.; Ilie, C.-I.; Dolete, G.; Croitoru, A.-M.; Surdu, V.-A.; Trușcă, R.-D.; Motelica, L.; Oprea, O.-C.; Ficai, D.; Ficai, A.; et al. Preparation and characterization of chitosan/TiO_2 composite membranes as adsorbent materials for water purification. *Membranes* **2022**, *12*, 804. [CrossRef]
75. Ashraf, S.; Siddiqa, A.; Shahida, S.; Qaisar, S. Titanium-based nanocomposite materials for arsenic removal from water: A review. *Heliyon* **2019**, *5*, e01577. [CrossRef] [PubMed]
76. Hu, J.; Shipley, H.J. Evaluation of desorption of Pb(II), Cu(II) and Zn(II) from titanium dioxide nanoparticles. *Sci. Total Environ.* **2012**, *431*, 209–220. [CrossRef] [PubMed]
77. Perera, W.N.; Hefter, G.; Sipos, P.M. An investigation of the lead (II)-Hydroxide System. *Inorg. Chem.* **2001**, *40*, 3974–3978. [CrossRef] [PubMed]
78. Ayers, R.S.; Westcot, D.W. *Water Quality for Agriculture*; Food and Agriculture Organization of the United Nations: Rome, Italy, 1985; ISBN 92-5-102263-1.
79. Ho, Y.S. Citation review of Lagergren kinetic rate equation on adsorption reactions. *Scientometrics* **2004**, *59*, 171–177. [CrossRef]
80. Tseng, R.L.; Wu, F.C.; Juang, R.S. Characteristics and applications of the Lagergren's first-order equation for adsorption kinetics. *J. Taiwan Inst. Chem. Eng.* **2010**, *41*, 661–669. [CrossRef]
81. Ho, Y.S.; McKay, G. Pseudo-second order model for sorption processes. *Process Biochem.* **1999**, *34*, 451–465. [CrossRef]
82. Hamidpour, M.; Hosseini, N.; Mozafari, V.; Heshmati, M. Removal of Cd(II) and Pb(II) from aqueous solutions by pistachio hull waste. *Rev. Int. de Contam.* **2018**, *34*, 307–316. [CrossRef]

83. Wu, F.C.; Tseng, R.L.; Juang, R.S. Characteristics of Elovich equation used for the analysis of adsorption kinetics in dye-chitosan systems. *Chem. Eng. J.* **2009**, *150*, 366–373. [CrossRef]
84. Wang, J.; Guo, X. Rethinking of the Intraparticle Diffusion Adsorption Kinetics Model: Interpretation, Solving Methods and Applications. *Chemosphere* **2022**, *309*, 1–14. [CrossRef] [PubMed]
85. Langmuir, I. The adsorption of gases on plane surfaces of glass, mica and platinum. *J. Am. Chem. Soc.* **1918**, *40*, 1361–1403. [CrossRef]
86. Skopp, J. Derivation of the Freundlich Adsorption Isotherm from Kinetics. *J. Chem. Educ.* **2009**, *86*, 1341–1343. [CrossRef]
87. Chu, K.H. Revisiting the Temkin Isotherm: Dimensional Inconsistency and Approximate Forms. *Ind. Eng. Chem. Res.* **2021**, *60*, 13140–13147. [CrossRef]
88. Sips, R. On the structure of a catalyst surface. *J. Chem. Phys.* **1948**, *16*, 490–495. [CrossRef]
89. Redlich, O.; Peterson, D.L. A useful adsorption isotherm. *J. Phys. Chem.* **1959**, *63*, 1024. [CrossRef]
90. James, G.; Witten, D.; Hastie, T.; Tibshirani, R. *An Introduction to Statistical Learning*; Springer: New York, NY, USA, 2017. [CrossRef]

molecules

MDPI

Article

β-CD-Induced Precipitation of Eriochrome Black T Recovered via CTAB-Assisted Foam Fractionation for Adsorption of Trace Cu(II)

Yunkang Chang [1,2], Chengsong Cao [2], Yuhuan Li [2], Yitong Yin [2], Yangjing Liu [2], Rui Li [2,*] and Yimin Zhu [1,*]

1 Institute of Environmental Remediation, Dalian Maritime University, Dalian 116026, China; cyk@dlmu.edu.cn
2 School of Biological Science, Jining Medical University, Rizhao 276826, China
* Correspondence: ruili061289@163.com (R.L.); ntp@dlmu.edu.cn (Y.Z.)

Abstract: In order to remove and reuse the ecotoxic dye Eriochrome black T (EBT) from dyeing wastewater, we used a process called cetyltrimethylammonium bromide (CTAB)-assisted foam fractionation. By optimizing this process with response surface methodology, we achieved an enrichment ratio of 110.3 ± 3.8 and a recovery rate of 99.1 ± 0.3%. Next, we prepared composite particles by adding β-cyclodextrin (β-CD) to the foamate obtained through foam fractionation. These particles had an average diameter of 80.9 μm, an irregular shape, and a specific surface area of 0.15 m^2/g. Using these β-CD-CTAB-EBT particles, we were able to effectively remove trace amounts of Cu^{2+} ions (4 mg/L) from the wastewater. The adsorption of these ions followed pseudo-second-order kinetics and Langmuir isotherm models, and the maximal adsorption capacities at different temperatures were 141.4 mg/g at 298.15 K, 143.1 mg/g at 308.15 K, and 144.5 mg/g at 318.15 K. Thermodynamic analysis showed that the mechanism of Cu^{2+} removal via β-CD-CTAB-EBT was spontaneous and endothermic physisorption. Under the optimized conditions, we achieved a removal ratio of 95.3 ± 3.0% for Cu^{2+} ions, and the adsorption capacity remained at 78.3% after four reuse cycles. Overall, these results demonstrate the potential of β-CD-CTAB-EBT particles for the recovery and reuse of EBT in dyeing wastewater.

Keywords: Eriochrome black T; separation; β-CD-CTAB-EBT particles; reuse; Cu(II) ions

1. Introduction

As society has developed rapidly, the consumption of dyes in various industries such as textiles, printing, dyeing, and medicine has increased [1]. However, many dyes have low fixation ability, meaning that they have to be discharged with a large volume of wastewater at the end of the production process [2,3]. This dye wastewater can cause serious environmental pollution, making its treatment a significant issue in the field of environmental protection [4]. One commonly found dye in wastewater is Eriochrome black T (EBT), an acidic and reactive dye with a conjugated azo (-N=N-) group as a chromophore [5,6]. EBT can harm the environment and aquatic organisms, as well as pose a risk to human health [7,8]. It is also resistant to biological treatment [9,10], making it important to find effective methods for removing EBT from water bodies.

Currently, there are two main strategies for removing EBT from dye wastewater: degradation and separation [3]. Fenton oxidation and photocatalytic oxidation are commonly used for degrading EBT [11–13], with the latter being preferred due to its minimal environmental impact [14]. However, the low light transmittance of dyeing wastewater can reduce the efficiency of photocatalytic degradation [15], and recovering nano-photocatalysts from wastewater can also be challenging [16]. As an alternative, separating EBT from dye wastewater for reuse may be a better option. Adsorption separation has been widely applied to remove EBT from simulated wastewater in the laboratory [17–19], but real wastewater often contains surfactants [20] that can interact with dyes, especially those with opposite charges, and decrease their efficiency in adsorption separation [21]. Foam fractionation may be a cost-effective separation method for removing EBT in the presence of surfactants, as these agents help to adsorb dye molecules at the gas–liquid interface [22]. However, there have been few reports on the further treatment of recovered surfactant–dye complexes, possibly because they have more complex structures than dyes alone [23,24]. This lack of research means that the application of foam fractionation for removing EBT from dye wastewater containing surfactants is still largely limited.

In a previous study, we discovered that cetyltrimethylammonium bromide (CTAB)-EBT complexes can be precipitated via β-cyclodextrin (β-CD) to form insoluble particles [25]. Additionally, EBT and β-CD have been shown to interact with Cu(II) ions [26,27]. Based on this information, we aimed to use CTAB-assisted foam fractionation to separate EBT from dye wastewater and then use β-CD to precipitate the recovered CTAB-EBT complexes and finally utilize the insoluble particles to remove Cu(II) ions from wastewater through adsorption. In this way, we hoped to overcome the limitations of using foam fractionation to treat EBT-containing wastewater. More importantly, the costs of the wastewater treatment will be reduced effectively by such a cost-effective separation method and the reuse of EBT. This study is divided into three main sections: (1) optimizing the foam fractionation of EBT in the presence of CTAB using response surface methodology (RSM) with a Box–Behnken design (BBD) approach, a statistical experimental design technique used in the field of optimization and process improvement [28]; (2) preparing and characterizing β-CD-CTAB-EBT particles; and (3) using β-CD-CTAB-EBT particles to remove Cu(II) ions through adsorption.

2. Results and Discussion

2.1. Optimization of CTAB-Assisted Foam Fractionation of EBT Using RSM

To optimize the CTAB-assisted foam fractionation (FF) of EBT for effective recovery from simulated wastewater, we used a Box–Behnken design for the RSM with four independent variables: volumetric air flow rate (*A*), liquid loading volume (*B*), CTAB-EBT molar ratio (*C*), and pH (*D*). The response items were the enrichment ratio (E_f, Y_1) and recovery percentage (R_f, Y_2) of EBT. Each variable had three levels, resulting in a total of 27 experimental runs, the results of which are shown in Table 1.

Table 1. Experimental scheme obtained from Box–Behnken design and corresponding result data.

Run	Coded (Uncoded) Variable Level				Y_1, Enrichment Ratio (E_f)	Y_2, Recovery Percentage (R_f/%)
	A, Volumetric Air Flow Rate (mL/min)	*B*, Liquid Loading Volume (mL)	*C*, CTAB-EBT Molar Ratio	*D*, pH		
1	1 (400)	−1 (200)	0 (1)	0 (7)	45.7	95.8
2	−1 (100)	−1 (200)	0 (1)	0 (7)	107.5	94.5
3	0 (250)	−1 (200)	−1 (0.5)	0 (7)	0	0
4	0 (250)	1 (600)	1 (1.5)	0 (7)	53.9	99
5	1 (400)	0 (400)	1 (1.5)	0 (7)	35.2	98.7
6	1 (400)	1 (600)	0 (1)	0 (7)	42.9	93.2
7	−1 (100)	0 (400)	1 (1.5)	0 (7)	87.6	98.1
8	−1 (100)	0 (400)	0 (1)	1 (9)	129.5	93.7
9	0 (250)	1 (600)	0 (1)	1 (9)	58.6	93.5
10	0 (250)	0 (400)	1 (1.5)	−1 (5)	68.1	98.8
11	0 (250)	0 (400)	1 (1.5)	1 (9)	53.2	98.8

Table 1. *Cont.*

Run	Coded (Uncoded) Variable Level				Y_1, Enrichment Ratio (E_f)	Y_2, Recovery Percentage (R_f/%)
	A, Volumetric Air Flow Rate (mL/min)	*B*, Liquid Loading Volume (mL)	*C*, CTAB-EBT Molar Ratio	*D*, pH		
12	1 (400)	0 (400)	−1 (0.5)	0 (7)	0	0
13	1 (400)	0 (400)	0 (1)	−1 (5)	49.4	94.1
14	0 (250)	−1 (200)	0 (1)	−1 (5)	99.5	95.6
15	0 (250)	0 (400)	0 (1)	0 (7)	78.6	95.5
16	−1 (100)	0 (400)	−1 (0.5)	0 (7)	0	0
17	0 (250)	1 (600)	0 (1)	−1 (5)	74.1	98.7
18	0 (250)	1 (600)	−1 (0.5)	0 (7)	0	0
19	0 (250)	−1 (200)	1 (1.5)	0 (7)	67.1	98.7
20	0 (250)	0 (400)	0 (1)	0 (7)	69.2	96.8
21	0 (250)	0 (400)	−1 (0.5)	1 (9)	0	0
22	0 (250)	0 (400)	−1 (0.5)	−1 (5)	0	0
23	0 (250)	−1 (200)	0 (1)	1 (9)	63.5	95.6
24	−1 (100)	0 (400)	0 (1)	−1 (5)	109.7	98.1
25	−1 (100)	1 (600)	0 (1)	0 (7)	89.5	96.4
26	1 (400)	0 (400)	0 (1)	1 (9)	55.6	98.1
27	0 (250)	0 (400)	0 (1)	0 (7)	78.5	97.6

The data in Table 1 were used to fit the dependency of both response items on the four independent variables into a quadratic regression model. The resulting polynomial equations, in terms of the coded factors shown in Table 2, are given in Equations (1) and (2).

$$Y_1 = 75.43 - 24.58A + 30.43C - 45.85C^2 \quad (1)$$

$$Y_2 = 96.63 + 49.34C - 1.13AB + 2.1AD - 1.3BD - 46.78C^2 \quad (2)$$

Table 2. Variables and their levels used for Box–Behnken design.

Variables	Label	Levels		
		−1	0	1
A	Volumetric air flow rate (mL/min)	100	250	400
B	Liquid loading volume (mL)	200	400	600
C	CTAB-EBT molar ratio	0.5	1	1.5
D	pH	5	7	9

Table 3 presents a summary of the ANOVA for the two regression models. The *F* values for both models (15.06 and 4148.97) are significant at $p < 0.0001$, indicating their high statistical significance. For Y_1, variables *A*, *C*, and C^2 are significant model terms ($p < 0.0001$). For Y_2, variables *C*, *AB*, *AD*, *BD*, and C^2 are significant ($p < 0.05$). The *p* values for "lack of fit" are 0.1401 and 0.7613, which are higher than 0.05, indicating that they are non-significant relative to pure error and thus confirm the goodness of fit [29]. The determination coefficients (R2) for Y_1 and Y_2 (0.9461 and 0.9998, respectively) indicate that there is only a 5.49% and 0.02% chance that the models with four factors in the selected ranges cannot explain the variations in Y_1 and Y_2, respectively [30]. The adjusted R2 values for Y_1 and Y_2 (0.8833 and 0.9996, respec-tively) are close to the corresponding R2 values, suggesting that both regression models can effectively explain the variations in the response items. The model for Y_2 has a par-ticularly good fit. The statistical validity of the developed models is confirmed by cal-culating the coefficients of variation (CV), with lower values indicating more signifi-cant models [31]. As shown in Table 3, the CV for both models is low (22.75% and 1.15%, respectively), indicating the accuracy of the experimental results. In addition, the precision of both models is higher than 4 (12.368 and 154.634, respectively), indi-cating an appropriate signal-to-noise ratio [32]. Based on these results, we conclude that both models can be used to design experiments.

Table 3. ANOVA for the response surface quadratic regression models.

Source	Y_1			Y_2		
	Mean Square	*F*-Value	*p*-Value	Mean Square	*F*-Value	*p*-Value
Model	2458.9	15.06	<0.0001	3117.78	4148.97	<0.0001
A	7252.08	44.41	<0.0001	0.0675	0.0898	0.7695
B	344.54	2.11	0.172	0.03	0.0399	0.845
C	11,108.17	68.03	<0.0001	29,215.2	38,878	<0.0001
D	136.01	0.8329	0.3794	2.61	3.48	0.0868
AB	57.76	0.3537	0.5631	5.06	6.74	0.0234
AC	686.44	4.2	0.0628	0.09	0.1198	0.7353
AD	46.24	0.2832	0.6043	17.64	23.47	0.0004
BC	43.56	0.2668	0.6149	0.0225	0.0299	0.8655
BD	105.06	0.6434	0.4381	6.76	9	0.0111
CD	55.5	0.3399	0.5707	0	0	1
A^2	35.59	0.218	0.649	2.64	3.52	0.0852
B^2	73.18	0.4481	0.5159	2.37	3.15	0.1011
C^2	11,213.89	68.67	<0.0001	11,670.88	15,530.98	<0.0001
D^2	71.38	0.4371	0.521	0.1481	0.1971	0.6649
Residual	163.29			0.7515		
Lack of Fit	190.12	6.52	0.1401	0.6771	0.6027	0.7613
Pure Error	29.14			1.12		
R^2	0.9461			0.9998		
Adj. R^2	0.8833			0.9996		
Variation coefficient	22.75%			1.15%		
Adequate precision	12.368			154.634		

The perturbation plots in Figure 1 demonstrate the influence of the independent variables on the E_f and R_f of EBT at a specific point in the design space.

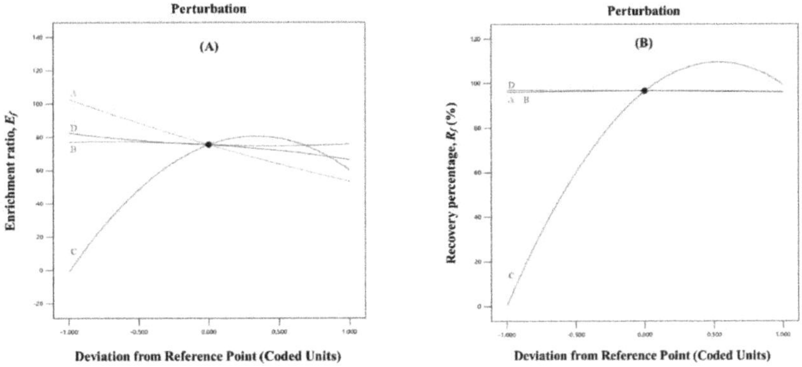

Figure 1. Perturbation plots that show the trends in E_f (**A**) and R_f (**B**) of EBT on volumetric air flow rate—*A*, liquid loading volume—*B*, CTAB-EBT molar ratio—*C*, and pH—*D* obtained at volumetric air flow rate of 250 mL/min, liquid loading volume of 400 mL, CTAB-EBT molar ratio of 1:1, and pH of 7.0.

The sensitivity of the response variables to a particular factor is typically proportional to the slope (curvature) of the plot [33]. In general, the CTAB-EBT molar ratio had the greatest impact on both response variables. As the CTAB-EBT molar ratio increased, the enrichment ratio and recovery of EBT also increased, reached a maximum, and then decreased. These results suggest that high values of both response variables could be achieved simultaneously at the same CTAB-EBT molar ratio. In addition, the volumetric air flow rate (*A*) and liquid loading volume (*B*) slowly decreased the E_f and had a negligible effect on the R_f. Furthermore, the pH of the solution (*D*) did not significantly affect either of the response variables. Using the enrichment ratio, it was possible to determine optimal values for the volumetric air flow rate, liquid loading volume, and pH.

To optimize the experimental conditions for CTAB-assisted foam fractionation, we considered both the E_f and R_f of EBT using the Design Expert software. The best performance was predicted using a volumetric air flow rate of 101 mL/min, a liquid loading volume of 447 mL, a CTAB-EBT molar ratio of 1.1:1, and a pH of 5.0, resulting in an enrichment ratio of 108.5 and an EBT recovery of 106.3%. To confirm these predictions, we conducted experiments under the proposed conditions in triplicate and found that the enrichment ratio was 110.3 $\pm$ 3.8 and the EBT recovery was 99.1 $\pm$ 0.3%, which are close to the predicted values.

2.2. Characterization of β-CD-CTAB-EBT Particles

2.2.1. Morphology of β-CD-CTAB-EBT Particles

The CTAB-EBT mixture was obtained via foam fractionation using the experimental conditions described in Section 2.1. β-CD was then combined with this foamate to prepare β-CD-CTAB-EBT particles using the procedure described in Section 3.4. SEM images in Figure 2A,B show that the β-CD-CTAB-EBT particles had an irregular shape due to the accumulation of irregular cones. Additionally, when dispersed in water, the particles appeared flocculent (see Figure 2C,D).

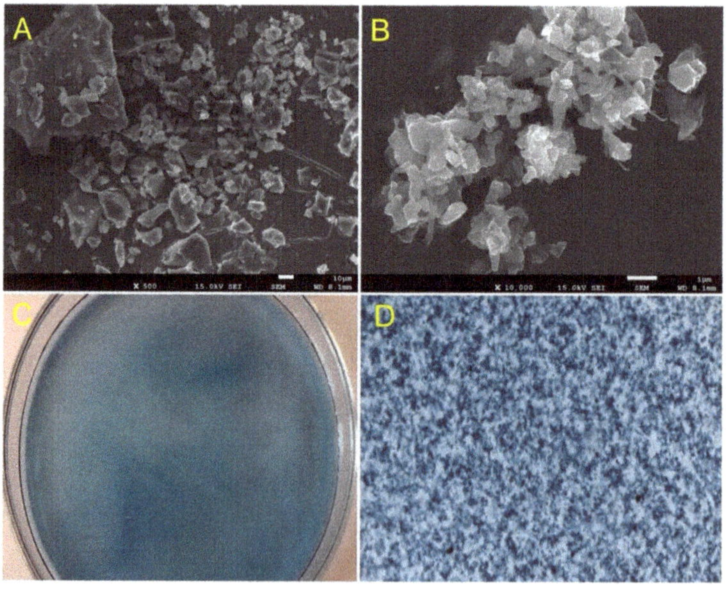

Figure 2. SEM images of β-CD-CTAB-EBT particles ((**A**) amplification × 500 and (**B**) amplification × 10,000) and optical images of β-CD-CTAB-EBT particles dispersed in water ((**C**) without amplification and (**D**) amplification × 40).

2.2.2. Size and Zeta Potential of β-CD-CTAB-EBT Particles

To gain further insights into the properties of the novel β-CD-CTAB-EBT particles, the size distribution and zeta potential in water were analyzed. As shown in Figure 3A, the particle size ranged from 2.59 μm to 336 μm with a volume density between 0.01 and 7.45%. The most abundant particles had a diameter of 66.4 μm and a volume density of 7.45%. Using these data, the mean volume-related equivalent diameter (D[4,3]) and the mean surface-area-related equivalent diameter (D[3,2]) were calculated via Mastersizer 3000 and found to be 80.9 μm and 39.5 μm, respectively [33]. The mean specific surface area of the particles was also determined to be 0.15 m^2/g. The large differences in particle size and the wide particle size distribution are consistent with their irregular shape [34].

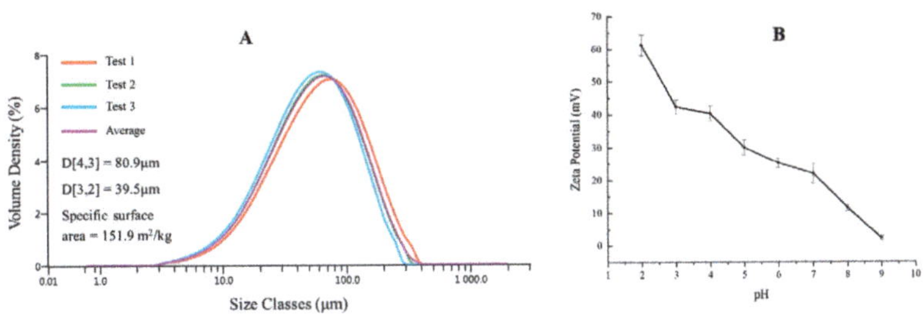

Figure 3. The results of three consecutive measurements of β-CD-CTAB-EBT particles size distribution with calculated particle size parameter (**A**), and variation in the zeta potential of particles with pH (**B**).

The charge of an adsorbent plays a crucial role in the adsorption of metal cations from a solution [35]. Therefore, we measured the zeta potential of the β-CD-CTAB-EBT particles in a pH range from 2.0 to 9.0. The results in Figure 3B show that the zeta potential decreases from 61.3 ± 3.1 mV to 2.2 ± 0.6 mV as the pH value increases. The positive zeta potential at all pH values suggests that the β-CD-CTAB-EBT particles are positively charged and may have electrostatic repulsion with Cu(II), which could potentially limit their adsorption efficiency.

2.2.3. SEM-EDS Analysis of Cu(II) Ions@β-CD-CTAB-EBT Particles

We performed SEM-EDS analyses of three random sites on the Cu@β-CD-CTAB-EBT surface to confirm the adsorption of Cu^{2+} ions. The results, shown in Figure 4, indicate the presence of copper at all examined sites. This suggests that β-CD-CTAB-EBT can be used for the adsorption of Cu^{2+}, even though it has a positive zeta potential at the pH of the experiment (pH = 4.0).

2.3. Adsorption of Cu(II) Ions Using β-CD-CTAB-EBT Particles

2.3.1. Adsorption Kinetics

To study the kinetics of Cu^{2+} adsorption at the surface of β-CD-CTAB-EBT particles, we monitored the time-dependent changes in the adsorption capacity at three temperatures (298.15, 308.15, and 318.15 K). As shown in Figure 5A, the adsorption capacity increased quickly from 0 to 20 min, and then slowly approached equilibrium from 20 to 100 min.

Additionally, higher temperatures increased the adsorption of Cu^{2+} ions. We then fitted these data to PFO and PSO kinetics models to determine which model better describes the adsorption kinetics. The results in Figure 5B,C show that both models are statistically significant ($p < 0.05$), but the correlation coefficients were higher for the PSO model. Furthermore, the ANOVA results of linear fitting showed that the F values for the PSO model fitting (523.8 at 298.15 K, 771.2 at 308.15 K, and 1257.8 at 318.15 K) were much higher than those for the PFO model fitting (66.1 at 298.15 K, 126.7 at 308.15 K and 82.7 at 318.15 K). It is indicated that t/q_t vs. t was more likely to be linear than $\ln(q_e - q_t)$ vs. t. Therefore, the PSO model better described the variation in the adsorption capacity of Cu(II) at β-CD-CTAB-EBT particles with time. According to this model, the equilibrium capacities of β-CD-CTAB-EBT particles in a 4 mg/L Cu(II) solution are 74.1 mg/g at 298.15 K, 85.0 mg/g at 308.15 K, and 99.9 mg/g at 318.15 K.

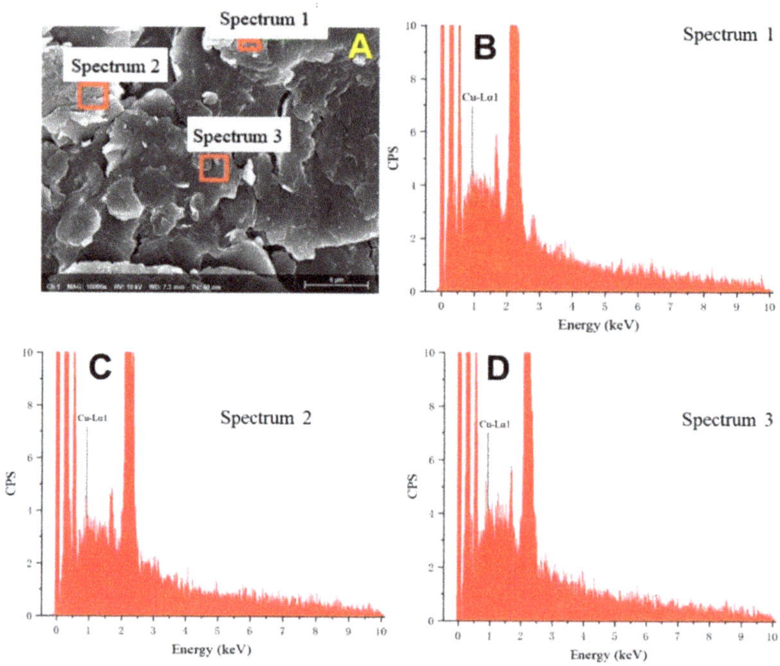

Figure 4. SEM image of scanning sites for EDS (**A**) and corresponding spectra of Cu(II) ions@β-CD-CTAB-EBT particles (**B–D**).

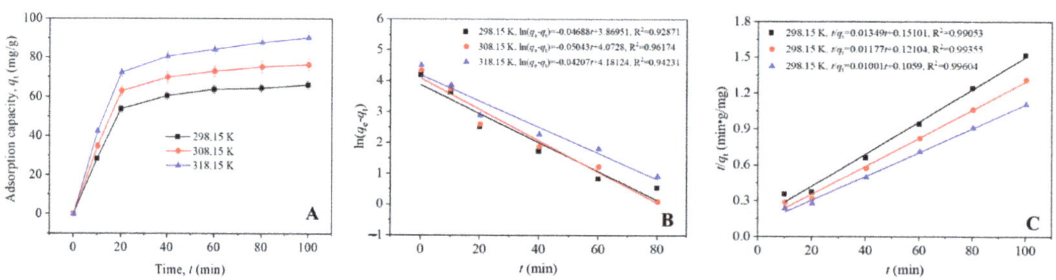

Figure 5. The changes in the adsorption capacity of β-CD-CTAB-EBT particles for Cu(II) ions over time at 298.15, 308.15, and 318.15 K (**A**). The data are also plotted with linear fits according to the PFO kinetics model (**B**) and the PSO kinetics model (**C**).

2.3.2. Adsorption Isotherms

We examined the effect of temperature and starting concentration of Cu^{2+} ions on the equilibrium adsorption capacity (q_e) of β-CD-CTAB-EBT. As shown in Figure 6A, the q_e increased with the concentration of Cu^{2+} ions in the solution, and this increase was more pronounced at lower concentrations. In addition, temperature increased q_e at each equilibrium Cu(II) concentration. To gain a better understanding of the adsorption process, the data from Figure 6A were fitted to Langmuir and Freundlich models.

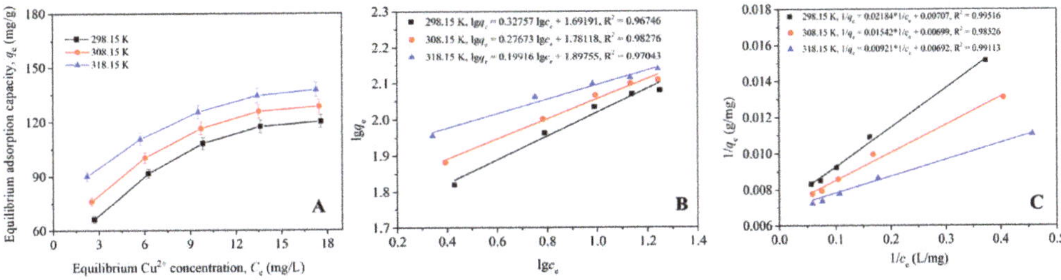

Figure 6. The dependence of the equilibrium adsorption capacity of β-CD-CTAB-EBT on the equilibrium concentration of Cu^{2+} in aqueous solution is shown at three temperatures: 298.15, 308.15, and 318.15 K (**A**). The data are also plotted with fits to the Freundlich adsorption isotherm (**B**) and the Langmuir adsorption isotherm (**C**).

The results in Figure 6B,C show that both models are statistically significant ($p < 0.05$). Thus, both Langmuir and Freundlich models could be used to explain the results in Figure 6A. The ANOVA results of linear fitting presented that the F values for the Langmuir model fitting were 823.1, 268.4, and 448.0 at 298.15, 308.15, and 318.15 K, higher than those for the Freundlich model fitting (119.9, 229.4, and 132.3, respectively). Thus, $1/q_e$ vs. C_e had higher probabilities in linear relationships than $\lg q_e$ vs. $\lg C_e$. Furthermore, the Langmuir model had a higher correlation coefficient at each temperature. This suggests that Cu(II) likely adsorbs at the surface of β-CD-CTAB-EBT particles by forming a single layer. Based on the Freundlich model, the maximal adsorption capacities at 298.15, 308.15, and 318.15 K were calculated as 141.4, 143.1, and 144.5 mg/g, respectively.

2.3.3. Adsorption Thermodynamics

The thermodynamic parameters of the adsorption of Cu^{2+} ions onto β-CD-CTAB-EBT particles were calculated based on the temperature changes in the q_e vs. C_e profiles (Figure 6A).

As shown in Table 4, ΔG had negative values at all temperatures, indicating that the process is spontaneous [36]. Additionally, the positive values of ΔH indicate that the adsorption is endothermic [37]. The positive ΔS values of 0.37 kJ/(mol·K) suggest that the randomness of the system slightly increases upon the adsorption of Cu(II) ions [36]. These observations suggest that the interactions between Cu^{2+} and the β-CD-CTAB-EBT surface obey the laws of physisorption [38] and β-CD and EBT were responsible for the adsorption of Cu^{2+}. Their interactions with Cu^{2+} may be driven by Van der Waals forces, covalent binding, or surface precipitation because of unfavorable electrostatic interactions between cations and the positively charged adsorbent surface [26,27,39].

Table 4. Thermodynamic parameters for the adsorption of Cu (II) ions at β-CD-CTAB-EBT particles.

T/K	ΔG (KJ/mol)	ΔH (KJ/mol)	ΔS (KJ/(mol T))
298.15	−31.84		
308.15	−35.54	78.48	0.37
318.15	−39.24		

2.3.4. Effect of pH on the Removal Ratio of Cu^{2+} Ions and Recyclability of β-CD-CTAB-EBT Particles

The positive charge of β-CD-CTAB-EBT particles at low pH (Figure 3A) allows for the easy desorption of physically adsorbed copper cations from the surface, enabling the reuse of the adsorbent. The effect of pH on the adsorption efficiency of Cu(II) ions was investigated by changing the pH value from 2.0 to 6.5 while keeping the other parameters constant. As shown in Figure 7A, the Cu^{2+} removal ratio increased from 5.4 ± 0.5% at pH 2.0 to 95.3 ± 3.0% at pH 6.0, with little additional change at pH 6.5. This trend may be attributed to the decreased amount of positive charge on the surface of β-CD-CTAB-EBT particles at higher pH values (Figure 3B), which allows for stronger interactions with copper ions and therefore a higher removal efficiency.

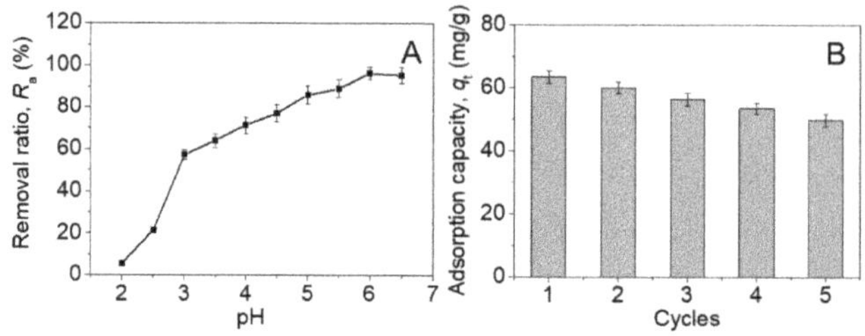

Figure 7. The effect of pH on removal ratio of Cu^{2+} (**A**) and recyclability of β-CD-CTAB-EBT particles (**B**).

To recycle the β-CD-CTAB-EBT adsorbent, we used a 10 mM solution of sulfuric acid to desorb Cu(II) ions. The initial adsorption capacity (q_e) of β-CD-CTAB-EBT for Cu^{2+} ions was 63.5 ± 2.0 mg/g (Figure 7B). After five consecutive cycles of reuse, the q_e decreased by 21.7% to 49.7 ± 1.9 mg/g. Despite this decrease, the β-CD-CTAB-EBT particles demonstrated a relatively high and reversible adsorption capacity, making them a promising candidate for the removal of Cu^{2+} from wastewater. In a word, EBT was effectively separated and reused by the CTAB-assisted foam fractionation and subsequent β-CD-induced precipitation. However, there was still a common drawback for this technology that the treatment of by-products after the adsorption of Cu^{2+} ions would need lots of effort.

3. Materials and Methods

3.1. Chemicals

The *analytical-grade* EBT, CTAB, and β-CD were ordered from Sinopharm (Beijing, China). EBT was dissolved in a 10 mM Na_2HPO_4-H_3PO_4 solution at pH 7.0 to a concentration of 0.22 mM. A 6 M H_3PO_4 or NaOH solution was used to adjust the pH of each solution, which was measured using a PHS-3C pH meter (Shanghai Yifen Scientific Instrument Co., Ltd., Shanghai, China). Ultrapure water (18.2 MΩ) was used in all experiments.

3.2. Equipment for EBT Batch Foam Fractionation

A simulated dye wastewater containing surfactant was prepared by mixing the EBT solution with a certain amount of CTAB. The CTAB-EBT mixture was then separated from this wastewater using batch foam fractionation (FF) equipment, as shown in Figure 8.

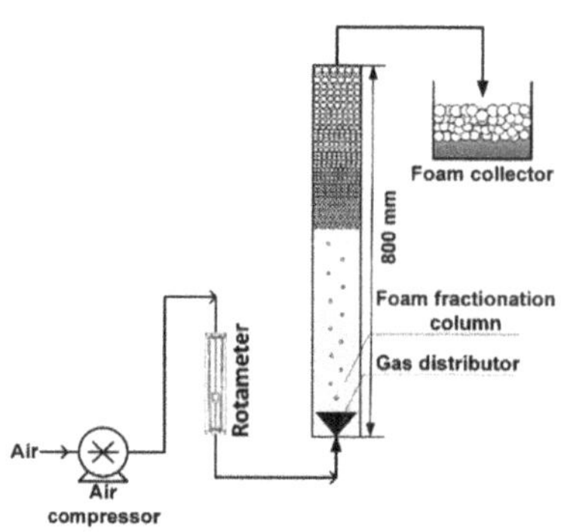

Figure 8. Diagram that illustrates batch foam fractionation of EBT.

The FF column was made of transparent plexiglass and had dimensions of 800 mm × 50 mm (height × inner diameter). A gas distributor with a mean pore diameter of 180 μm was placed at the exit of the column. An air compressor (ACO-018A from Guangdong Haili Group Co., Ltd., in Rudong, China) was used to pump air through the column at a flow rate between 60 and 600 mL/min that was controlled using an air rotameter (LZB-W from Shanghai Automation Instrument Co., Ltd., in Shanghai, China). The resulting foam was stored for further use. The performance of the FF process was studied by calculating the enrichment ratio (E_f) and recovery percentage (R_f) using Equations (3) and (4), respectively.

$$E_f = \frac{A_f}{Ao} \tag{3}$$

$$R_f = \frac{V_f A_f}{V_o Ao} \times 100\% \tag{4}$$

where A_f and A_o are absorbances of the foamate and the feed solution, respectively, measured using a colorimeter (CN60M/SD9011, Westmid, Birmingham, UK); V_f and V_o are the volumes of the foamate and the feed solution (in mL).

3.3. *Optimization of EBT Batch Foam Fractionation via RSM*

Volumetric air flow rate, liquid loading volume, surfactant–dye molar ratio, and pH have been shown to significantly affect the surfactant-assisted foam fractionation of dyes [40,41]. Therefore, we used a Box–Behnken design of the response surface methodology (BBD-RSM) to optimize the CTAB-assisted foam fractionation of the EBT mixture, with these four factors as independent variables at three levels, and the E_f and R_f as response items. The starting points for the optimization were the results of single-factor experiments (not presented here). The specific levels of the four factors used in the optimization are listed in Table 1, and Table 2 shows the base run of 27 experiments designed using the software Design Expert (Version 13.0.1.0. Stat-Ease Inc., Minneapolis, MN, USA) and the corresponding results.

3.4. *Preparation of β-CD-CTAB-EBT Particles*

The preparation of β-CD-CTAB-EBT particles is depicted in Figure 9. The process begins by adding β-CD to the foamate in a molar ratio of EBT to β-CD of 1:1.2, resulting in the formation of β-CD-CTAB-EBT particles. These particles are then separated from the turbid solution through centrifugation at 5000 rpm for 10 min. The particles are subsequently washed 3× with ultrapure water and dried at 60 degrees Celsius to complete the process.

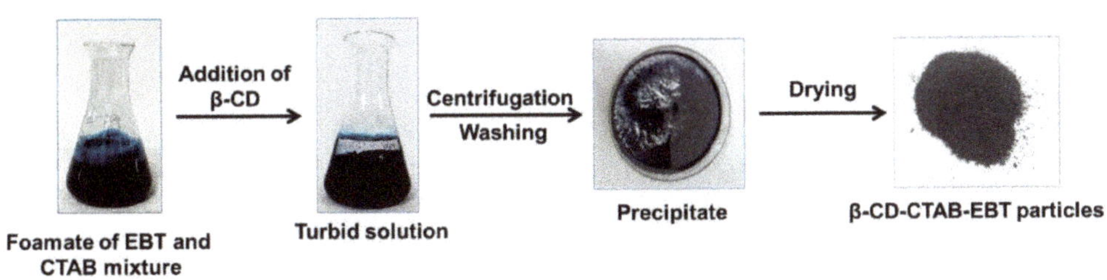

Figure 9. Process flow diagram for the preparation of β-CD-CTAB-EBT particles from the foamate of EBT and CTAB mixture.

3.5. *Adsorption Removal of Cu(II) Ions Using β-CD-CTAB-EBT Particles*

3.5.1. Evaluation of Adsorption Performance

In the adsorption experiments of Cu^{2+} ions using β-CD-CTAB-EBT particles, a batch mode was used with a water area oscillator (SHA-B, Shanghai Enyi Medical Technology

Development Co., Ltd., Shanghai, China). The effectiveness of the adsorption process was assessed through the calculation of the adsorption capacity (q_t, mg/g) and removal ratio (R_a) using Equations (5) and (6), respectively.

$$q_t = \frac{(c_0 - c_t) \times V}{m} \tag{5}$$

$$R_a = \frac{c_0 - c_t}{c_0} \times 100\% \tag{6}$$

The Cu^{2+} ion concentration in the feed and residual solution after adsorption (in mg/L) is represented by c_0 and c_t, respectively. The volume of the feed solution (in L) is represented by V, and the mass of the added β-CD-CTAB-EBT particles (in g) is represented by m.

3.5.2. Adsorption Kinetics

In the experimental setup, 20 mg of β-CD-CTAB-EBT particles was added to a 1000 mL copper sulfate solution with a Cu^{2+} ion concentration of 4 mg/L and a pH of 6.0. This mixture was placed in a 1000 mL conical flask and placed in a water area oscillator, which was set to shake at 200 rpm at three temperatures (298.15, 308.15, and 318.15 K). Aliquots of the mixture (2 mL) were taken at intervals of 10 min, and the β-CD-CTAB-EBT particles were removed by filtering through a 0.22 μm cellulose acetate filter mounted on a glass syringe. The concentration of Cu^{2+} ions in the filtrate was measured, and the corresponding adsorption capacity (q_t) was calculated. The kinetics of the adsorption of Cu^{2+} ions at the β-CD-CTAB-EBT particles were evaluated based on the variation in q_t over time using pseudo-first-order (PFO) and pseudo-second-order (PSO) models, as defined in Equations (7) and (8), respectively [42].

$$\ln(q_e - q_t) = \ln q_e - k_1 t \tag{7}$$

$$\frac{t}{q_t} = \frac{1}{k_2 q_e^2} + \frac{t}{q_e} \tag{8}$$

In these equations, q_e (mg/g) represents the equilibrium absorption capacity of Cu(II) ions, t (min) represents time, and k_1 (1/min) and k_2 (L/(mol × min)) represent the PFO and PSO rate constants, respectively.

3.5.3. Adsorption Isotherms

The experimental procedures in this section were similar to those described in Section 3.5.2, with the exception that the starting concentrations of Cu^{2+} ions were set at 4, 8, 12, 16, and 20 mg/L. In each experiment, q_e and the equilibrium concentration of Cu^{2+} ions (C_e, mg/L) in the residual solution were calculated by fitting the data to the PSO model. The Langmuir and Freundlich isotherm adsorption models were then used to evaluate the adsorption of Cu(II) ions in the β-CD-CTAB-EBT particles using the data of q_e vs. C_e. These models are represented by Equations (9) and (10), respectively [43].

$$q_e = \frac{q_{\max} k_L C_e}{1 + k_L C_e} \Rightarrow \frac{1}{q_e} = \frac{1}{q_{\max} k_L} * \frac{1}{c_e} + \frac{1}{q_{\max}} \tag{9}$$

$$q_e = k_F C_e^{1/n} \Rightarrow \lg q_e = \lg k_F + \frac{1}{n} \lg c_e \tag{10}$$

In these equations, $q_{\max}$ (mg/g) denotes the theoretical maximal absorption capacity of Cu(II) ions, k_L (L/mg) represents the Langmuir adsorption constant, k_F represents the Freundlich adsorption constant, and n is a dimensionless coefficient.

3.5.4. Adsorption Thermodynamics

The thermodynamics of the adsorption of Cu^{2+} ions at the β-CD-CTAB-EBT particles was analyzed by calculating the ΔG, ΔH, and ΔS values. The data obtained in Section 3.5.3 were fitted into the Van't Hoff and the Gibbs free energy equation, as defined in Equations (11) and (12), respectively [44].

$$\ln k_d = \frac{-\Delta H}{RT} + \frac{\Delta S}{R} \tag{11}$$

$$\Delta G = \Delta H - T\Delta S \tag{12}$$

In these equations, k_d represents the distribution coefficient for the Cu(II) ions adsorption, as defined in Equation (13) [45], T (K) is temperature, and R (8.314 J/(mol·K)) represents the molar gas constant.

$$k_d = \frac{q_e}{C_e} \times 1000 \tag{13}$$

3.6. Characterization of β-CD-CTAB-EBT Particles

The size distribution and zeta potential of β-CD-CTAB-EBT particles were determined using the Mastersizer 3000 and Zetasizer Pro instruments from Malvern Panalytical (Malvern, UK). The microstructure of the particles was examined using an SEM (JSM-7610F, JEOL, Tokyo, Japan) and a high-definition electronic microscope (BC1201E, Shenzhen Bosheng Electronic Technology, Shenzhen, China). The adsorption of Cu^{2+} ions on the surface of the β-CD-CTAB-EBT particles was analyzed using a ZEISS EVO 18 SEM (Carl Zeiss Microscopy, Aalen, Germany) equipped with EDS.

3.7. Measurement of Cu(II) Ion Concentration

The concentration of Cu^{2+} ions in an aqueous solution was determined using a spectrophotometric detection kit for copper provided by Hangzhou Luhang Seng Science and Technology (Hangzhou, China) and a U-3900H UV/VIS spectrophotometer from Hitachi (Tokyo, Japan). A linear relationship between the absorbance at 560 nm and the concentration of copper ions was established in the range from 0 to 5 mg/L, with the calibration curve equation A_{560} = 0.098 C_{cu} + 0.0157 (R^2 = 0.98702).

3.8. Statistical Analysis

The experiments were carried out in triplicates, and the final result was presented as the mean ± standard deviation. An analysis of variance (ANOVA) at $p \leq 0.05$ was carried out in Microsoft Excel Professional Plus 2013 (v. 15).

4. Conclusions

In this work, β-CD-CTAB-EBT particles were prepared by adding β-CD to the foamate of the EBT and CTAB mixture recovered from a process of the CTAB-assisted foam fractionation of EBT and then used for the adsorption of Cu^{2+} from wastewater. These particles had an irregular shape with a specific surface area of 0.15 m^2/g and interacted with Cu^{2+} ions via nonionic forces. The kinetics of the adsorption process followed the PSO rate law, and the data fit well with the Langmuir isotherm, suggesting that the main interaction between the sorbent and sorbate was physisorption. The q_e increased from 141.4 mg/g to 144.5 mg/g with an increase in temperature from 298.15 K to 318.15 K, and the process was spontaneous and endothermic. Under the optimized conditions, the removal ratio of copper ions reached 95.3%, and 78.3% of the initial q_e was retained after four cycles of reuse. Thus, the β-CD-CTAB-EBT particles are a recyclable, ecofriendly material for the removal of Cu^{2+} ions from wastewater. Overall, the CTAB-assisted foam fractionation and subsequent β-CD-induced precipitation were effective in the separation and reuse of EBT. These results will have practical importance to the treatment of EBT-containing wastewater.

Author Contributions: Conceptualization, Y.C., R.L. and Y.Z.; Methodology, Y.C.; Software, C.C. and Y.L. (Yangjing Liu); Validation, Y.L. (Yangjing Liu); Formal analysis, C.C.; Investigation, Y.L. (Yuhuan Li) and Y.Y.; Resources, R.L. and Y.Z.; Data curation, C.C.; Writing—original draft, Y.C.; Writing—review & editing, R.L. and Y.Z.; Visualization, Y.L. (Yuhuan Li) and Y.Y.; Supervision, R.L. All authors have read and agreed to the published version of the manuscript.

Funding: This work was supported by the Chinese Ministry of Education's Cooperative Education Project (220505421031716), the Chinese Medicine Science and Technology Development Project of Shandong Province (Q-2022148), the Natural Science Foundation of Rizhao (RZ2021ZR21), the Education Science Planning Project for the Subject of Innovation Capability of Shandong Province (2022CYB210), and the Innovation and Entrepreneurship Training Projects for College Students of Shandong Province (S202110443016).

Institutional Review Board Statement: Not applicable.

Informed Consent Statement: Not applicable.

Data Availability Statement: The data are contained within the article.

Conflicts of Interest: The authors declare no conflict of interest.

References

1. Liu, Z.; Alam Khan, T.; Islam, A.; Tabrez, U. A review on the treatment of dyes in printing and dyeing wastewater by plant biomass carbon. *Bioresour. Technol.* **2022**, *354*, 127168. [CrossRef] [PubMed]
2. Azam, K.; Shezad, N.; Shafiq, I.; Akhter, P.; Akhtar, F.; Jamil, F.; Shafique, S.; Park, Y.-K.; Hussain, M. A review on activated carbon modifications for the treatment of wastewater containing anionic dyes. *Chemosphere* **2022**, *306*, 135566. [CrossRef] [PubMed]
3. Chen, H.; Yu, X.; Wang, X.; He, Y.; Zhang, C.; Xue, G.; Liu, Z.; Lao, H.; Song, H.; Chen, W.; et al. Dyeing and finishing wastewater treatment in China: State of the art and perspective. *J. Clean. Prod.* **2021**, *326*, 129353. [CrossRef]
4. Zhang, C.; Chen, H.; Xue, G.; Liu, Y.; Chen, S.; Jia, C. A critical review of the aniline transformation fate in azo dye wastewater treatment. *J. Clean. Prod.* **2021**, *321*, 128971. [CrossRef]
5. Gomes, A.S.; Leal, M.V.G.; Tolosa, G.R.; Cabrera, F.C.; Dognani, G.; Job, A.E. Cationic dialdehyde cellulose microfibers for efficient removal of eriochrome black T from aqueous solution. *Bioresour. Technol.* **2023**, *380*, 129096. [CrossRef] [PubMed]
6. Kotnala, S.; Bhushan, B.; Nayak, A. Fabrication of nano-biocomposite for the removal of Eriochrome Black T and malachite green from aqueous solution: Isotherm and kinetic studies. *Environ. Sci. Pollut. Res.* **2023**, *30*, 27846–27862. [CrossRef]
7. Honarmand, M.; Golmohammadi, M.; Hafezi-Bakhtiari, J. Synthesis and characterization of SnO2 NPs for photodegradation of eriochrome black-T using response surface methodology. *Environ. Sci. Pollut. Res.* **2021**, *28*, 7123–7133. [CrossRef]
8. De, I.; Pahuja, M.; Wani, H.M.U.D.; Dey, A.; Dube, T.; Ghosh, R.; Kankan, N.; Mishra, J.; Panda, J.J.; Maruyama, T.; et al. In-vitro toxicity assessment of a textile dye Eriochrome Black T and its nano-photocatalytic degradation through an innovative approach using Mf-NGr-CNTs-SnO2 heterostructures. *Ecotoxicol. Environ. Saf.* **2022**, *243*, 113985. [CrossRef]
9. Solís, M.; Solís, A.; Pérez, H.I.; Manjarrez, N.; Flores, M. Microbial decolouration of azo dyes: A review. *Process. Biochem.* **2012**, *47*, 1723–1748. [CrossRef]
10. Mishra, S.; Maiti, A. The efficacy of bacterial species to decolourise reactive azo, anthroquinone and triphenylmethane dyes from wastewater: A review. *Environ. Sci. Pollut. Res.* **2018**, *25*, 8286–8314. [CrossRef]
11. Lanjwani, M.F.; Khuhawar, M.Y.; Khuhawar, T.M.J.; Lanjwani, A.H.; Memon, S.Q.; Soomro, W.A.; Rind, I.K. Photocatalytic Degradation of Eriochrome Black T Dye by ZnO Nanoparticles Using Multivariate Factorial, Kinetics and Isotherm Models. *J. Clust. Sci.* **2022**, *34*, 1121–1132. [CrossRef]
12. Waghchaure, R.H.; Adole, V.A.; Jagdale, B.S. Photocatalytic degradation of methylene blue, rhodamine B, methyl orange and Eriochrome black T dyes by modified ZnO nanocatalysts: A concise review. *Inorg. Chem. Commun.* **2022**, *143*, 109764. [CrossRef]
13. Dos Santos, A.J.; de Lima, M.D.; da Silva, D.R.; Garcia-Segura, S.; Martínez-Huitle, C.A. Influence of the water hardness on the performance of electro-Fenton approach: Decolorization and mineralization of Eriochrome Black T. *Electrochim. Acta* **2016**, *208*, 156–163. [CrossRef]
14. Koe, W.S.; Lee, J.W.; Chong, W.C.; Pang, Y.L.; Sim, L.C. An overview of photocatalytic degradation: Photocatalysts, mechanisms, and development of photocatalytic membrane. *Environ. Sci. Pollut. Res.* **2020**, *27*, 2522–2565. [CrossRef]
15. Ani, I.; Akpan, U.; Olutoye, M.; Hameed, B. Photocatalytic degradation of pollutants in petroleum refinery wastewater by TiO_2- and ZnO-based photocatalysts: Recent development. *J. Clean. Prod.* **2018**, *205*, 930–954. [CrossRef]
16. Ramalingam, G.; Perumal, N.; Priya, A.; Rajendran, S. A review of graphene-based semiconductors for photocatalytic degradation of pollutants in wastewater. *Chemosphere* **2022**, *300*, 134391. [CrossRef]
17. Kaur, Y.; Jasrotia, T.; Kumar, R.; Chaudhary, G.R.; Chaudhary, S. Adsorptive removal of eriochrome black T (EBT) dye by using surface active low cost zinc oxide nanoparticles: A comparative overview. *Chemosphere* **2021**, *278*, 130366. [CrossRef]
18. Bansal, M.; Patnala, P.K.; Dugmore, T. Adsorption of Eriochrome Black-T(EBT) using tea waste as a low cost adsorbent by batch studies: A green approach for dye effluent treatments. *Curr. Res. Green Sustain. Chem.* **2020**, *3*, 100036. [CrossRef]

19. Balouchi, H.; Baziar, M.; Dehghan, A.; Alidadi, H.; Shams, M. Combination of electrocoagulation and MOF adsorption systems for EBT removal from water. *Int. J. Environ. Anal. Chem.* **2020**, *102*, 1307–1317. [CrossRef]
20. Mahbub, S.; Shahriar, I.; Iqfath, M.; Hoque, A.; Halim, M.A.; Khan, M.A.; Rub, M.A.; Asiri, A.M. Influence of alcohols/electrolytes on the interaction of reactive red dye with surfactant and removal of dye from solutions. *J. Environ. Chem. Eng.* **2019**, *7*, 103364. [CrossRef]
21. Janoš, P.; Šmídová, V. Effects of surfactants on the adsorptive removal of basic dyes from water using an organomineral sorbent—Iron humate. *J. Colloid Interface Sci.* **2005**, *291*, 19–27. [CrossRef] [PubMed]
22. Buckley, T.; Xu, X.; Rudolph, V.; Firouzi, M.; Shukla, P. Review of foam fractionation as a water treatment technology. *Sep. Sci. Technol.* **2022**, *57*, 929–958. [CrossRef]
23. Sachin, K.; Karpe, S.A.; Singh, M.; Bhattarai, A. Study on surface properties of sodiumdodecyl sulfate and dodecyltrimethylammonium bromide mixed surfactants and their interaction with dyes. *Heliyon* **2019**, *5*, e01510. [CrossRef] [PubMed]
24. Noor, S.; Taj, M.B.; Naz, I. Comparative solubilization of reactive dyes in single and mixed surfactants. *J. Dispers. Sci. Technol.* **2022**, *43*, 2058–2068. [CrossRef]
25. Chang, Y.; Dou, N.; Liu, M.; Jiang, M.; Men, J.; Cui, Y.; Li, R.; Zhu, Y. Efficient removal of anionic dyes from aqueous solution using CTAB and β-cyclodextrin-induced dye aggregation. *J. Mol. Liq.* **2020**, *309*, 113021. [CrossRef]
26. He, J.; Li, Y.; Wang, C.; Zhang, K.; Lin, D.; Kong, L.; Liu, J. Rapid adsorption of Pb, Cu and Cd from aqueous solutions by β-cyclodextrin polymers. *Appl. Surf. Sci.* **2017**, *426*, 29–39. [CrossRef]
27. Masoud, M.S.; Hammud, H.H.; Beidas, H. Dissociation constants of eriochrome black T and eriochrome blue black RC indicators and the formation constants of their complexes with Fe(III), Co(II), Ni(II), Cu(II), Zn(II), Cd(II), Hg(II), and Pb(II), under different temperatures and in presence of different solvents. *Thermochim. Acta* **2002**, *381*, 119–131. [CrossRef]
28. Amiri, M.J. Synthesis and optimization of spherical nZVI (20–60 nm) immobilized in bio-apatite-based material for efficient removal of phosphate: Box-Behnken design in a fixed-bed column. *Environ. Sci. Pollut. Res.* **2022**, *29*, 67751–67764. [CrossRef]
29. Paswan, M.; Prajapati, V.; Dholakiya, B.Z. Optimization of biodegradable cross-linked guar-gum-PLA superabsorbent hydrogel formation employing response surface methodology. *Int. J. Biol. Macromol.* **2022**, *223*, 652–662. [CrossRef]
30. Behera, S.K.; Meena, H.; Chakraborty, S.; Meikap, B. Application of response surface methodology (RSM) for optimization of leaching parameters for ash reduction from low-grade coal. *Int. J. Min. Sci. Technol.* **2018**, *28*, 621–629. [CrossRef]
31. Chaudhary, A.; Sharma, S.; Verma, A. Optimization of WEDM process parameters for machining of heat treated ASSAB '88 tool steel using Response surface methodology (RSM). *Mater. Today Proc.* **2022**, *50*, 917–922. [CrossRef]
32. Liu, W.; Liu, D.; Yin, H.; Yang, C.; Lu, K. Foam fractionation for the separation of SDBS from its aqueous solution: Process optimization and property test. *Sep. Purif. Technol.* **2021**, *262*, 118305. [CrossRef]
33. Xing, Z.; Zhang, M.; Wang, Y.; Yang, G.; Han, L.; Loor, J. A decrease in diameter of milk fat globules accompanies milk fat depression induced by conjugated linoleic acid supplementation in lactating dairy cows. *J. Dairy Sci.* **2020**, *103*, 5143–5147. [CrossRef] [PubMed]
34. Kwak, B.-M.; Lee, J.E.; Ahn, J.-H.; Jeon, T.-H. Laser diffraction particle sizing by wet dispersion method for spray-dried infant formula. *J. Food Eng.* **2009**, *92*, 324–330. [CrossRef]
35. Samadi, A.; Xie, M.; Li, J.; Shon, H.; Zheng, C.; Zhao, S. Polyaniline-based adsorbents for aqueous pollutants removal: A review. *Chem. Eng. J.* **2021**, *418*, 129425. [CrossRef]
36. Fang, Z.; Suhua, H.; Xu, L.; Jian, F.; Qi, L.; Zhiwei, W.; Chuanchang, L.; Yuanlai, X. Adsorption kinetics and thermodynamics of rare earth on Montmorillonite modified by sulfuric acid. *Colloids Surf. A Physicochem. Eng. Asp.* **2021**, *627*, 127063. [CrossRef]
37. Ahamad, T.; Naushad, M.; Al-Shahrani, T.; Al-hokbany, N.; Alshehri, S.M. Preparation of chitosan based magnetic nanocomposite for tetracycline adsorption: Kinetic and thermodynamic studie. *Int. J. Biol. Macromol.* **2020**, *147*, 258–267. [CrossRef]
38. Cimirro, N.F.; Lima, E.C.; Cunha, M.R.; Thue, P.S.; Grimm, A.; dos Reis, G.S.; Rabiee, N.; Saeb, M.R.; Keivanimehr, F.; Habibzadeh, S. Removal of diphenols using pine biochar. Kinetics, equilibrium, thermodynamics, and mechanism of uptake. *J. Mol. Liq.* **2022**, *364*, 119979. [CrossRef]
39. Fenti, A.; Chianese, S.; Iovino, P.; Musmarra, D.; Salvestrini, S. Cr(VI) Sorption from Aqueous Solution: A Review. *Appl. Sci.* **2020**, *10*, 6477. [CrossRef]
40. Jia, L.; Liu, W.; Cao, J.; Wu, Z.; Yang, C. Modified multi-walled carbon nanotubes assisted foam fractionation for effective removal of acid orange 7 from the dyestuff wastewater. *J. Environ. Manag.* **2020**, *262*, 110260. [CrossRef]
41. Soylu, M.; Gökkuş, Ö.; Özyonar, F. Foam separation for effective removal of disperse and reactive dyes from aqueous solutions. *Sep. Purif. Technol.* **2020**, *247*, 116985. [CrossRef]
42. Simonin, J.P. On the comparison of pseudo-first order and pseudo-second order rate laws in the modeling of adsorption kinetics. *Chem. Eng. J.* **2016**, *300*, 254–263. [CrossRef]
43. Zhuang, S.; Chen, R.; Liu, Y.; Wang, J. Magnetic COFs for the adsorptive removal of diclofenac and sulfamethazine from aqueous solution: Adsorption kinetics, isotherms study and DFT calculation. *J. Hazard. Mater.* **2020**, *385*, 121596. [CrossRef] [PubMed]

44. Khandaker, S.; Toyohara, Y.; Saha, G.C.; Awual, R.; Kuba, T. Development of synthetic zeolites from bio-slag for cesium adsorption: Kinetic, isotherm and thermodynamic studies. *J. Water Process. Eng.* **2020**, *33*, 101055. [CrossRef]
45. Lima, E.C.; Hosseini-Bandegharaei, A.; Moreno-Piraján, J.C.; Anastopoulos, I. A critical review of the estimation of the thermodynamic parameters on adsorption equilibria. Wrong use of equilibrium constant in the Van't Hoof equation for calculation of thermodynamic parameters of adsorption. *J. Mol. Liq.* **2019**, *273*, 425–434. [CrossRef]

molecules

MDPI

Article

Electron-Rich Triazine-Conjugated Microporous Polymers for the Removal of Dyes from Wastewater

Bao-Ning Li [1,*,†], Xing-Long Zhang [1,†], Xiao-Hui Bai [1,†], Zhen-Jie Liang [2], Jian Li [1] and Xiao-Yong Fan [1]

[1] School of Chemistry and Chemical Engineering, Yulin University, Yulin 719000, China; bxh0727@163.com (X.-H.B.); lijian5220@163.com (J.L.); fanxy@yulinu.edu.cn (X.-Y.F.)

[2] School of Chemistry, Sun Yat-Sen University, Guangzhou 510006, China

* Correspondence: lbning@yulinu.edu.cn

† These authors contributed equally to this work.

Abstract: Conjugated microporous polymers (CMP) as porous functional materials have received considerable attention due to their unique structures and fascinating properties for the adsorption and degradation of dyes. Herein, a triazine-conjugated microporous polymer material with rich N-donors at the skeleton itself was successfully synthesized via the Sonogashira–Hagihara coupling by a one–pot reaction. These two polymers had Brunauer–Emmett–Teller (BET) surface areas of 322 and 435 m^2g^{-1} for triazine-conjugated microporous polymers (T-CMP) and T-CMP-Me, respectively. Due to the porous effects and the rich N-donor at the framework, it displayed a higher removal efficiency and adsorption performance compared to cationic-type dyes and selectivity properties for (methylene blue) MB^+ from a mixture solution of cationic-type dyes. Furthermore, the T-CMP-Me could quickly and drastically separate MB^+ and (methyl orange) MO^- from the mixed solution within a short time. Their intriguing absorption behaviors are supported by ^{13}C NMR, UV−vis absorption spectroscopy, scanning electron microscopy, and X-ray powder diffraction studies. This work will not only improve the development of porous material varieties, but also demonstrate the adsorption or selectivity of porous materials for dyes from wastewater.

Keywords: conjugated microporous polymers; adsorption; dyes; wastewater

1. Introduction

Recently, porous organic materials have become a hot topic due to their numerous advantages, such as diversified structure, huge surface, tunable porosity, et al. [1–5]. Over the past few years, metal–organic frameworks (MOFs) [6], covalent organic frameworks (COFs) [7], and porous organic polymers (POPs) [8] have developed rapidly. MOFs and COFs have robust and ordered structures but low physicochemical stability, while conjugated microporous polymers (CMP) have high stability and an easy preparation method [9]. Thus, designing a functional porous material that can combine the advantages of different porous materials, called triazine-conjugated microporous polymers (T-CMP), may be more meaningful for the development of porous organic materials [10–13].

Nowadays, water pollution has directly affected human health [14,15], owing to numerous dyes released by factories, and the different types of dyes have become different sources of pollution; therefore, effective methods are needed to deal with the toxic chemical dyes from the wastewater [2]. Different methods, such as precipitation, ultra-filtration, ion exchange, coagulation, and electro-catalytic degradation, have been adopted for the adsorption and degradation of these organic pollutants [16–20]. Apart from these methods, adsorption, photocatalytic degradation, and chemical degradation are considered the most economical and efficient ones to control water pollution from dyes [21].

Conjugated microporous polymers are considered the best candidates employed in various environmental and energy-related systems for their stable chemical/physical structure, pore walls, pore volume, pore surface area, pore skeleton, and pore technology [12,13]. In past years, different terminologies were used [20], but since an important class of porous materials, CMP–1, was obtained by the Sonogashira reaction between 1,3,5-triethynylbenzene and 1,4-diiodobenzene in 2007 [22], the word "CMP" has become the most popular and well known. CMPs are synthesized under kinetic and/or thermodynamic control with rigid *p*-conjugate structures and three-dimensional networks [23,24]. Additionally, ionic functionalized CMP porous materials also exhibit unique properties, such as high porosity, broad and interconnected pores, and the ability to quickly absorb and desorb ionic dyes, making them more ideal [12,25]. Therefore, the preparation and application of ion-functionalized CMPs have attracted more and more research interest [26,27]. Along with these effects, a series of porous polyionic liquids have been successfully synthesized by template or no template methods [25], revealing the high ion density, high polarization, and strong electrostatic field, which have been widely used in catalysis, adsorption, separation, sensors, actuators, and other fields [14,28,29]. In the past few decades, people have developed various preparation strategies to construct new ionic types of CMPs [30,31]. One of the traditional methods is the cross-coupling reaction between the conjugated building unit and ionic functional monomers [32]. Even if it is an effective method to obtain ionic CMP, it is difficult to make organic salts participate in cross-coupling reactions. Compared with previous methods, the post-synthetic modification (PSM) method has the advantages of high yield, simple monomers, simple operation, and good selectivity, which plays an important role in the construction of the ionic CMP material [33]. The ionic porous nanostructured polymer adsorbents have quick adsorption and ion exchange processes, resulting in high capacity and analyte selectivity for ionic dyes [17,34]. In the process of material preparation, designing specific building blocks with similar functions to construct CMPs has received increasing attention [16,21,35].

Herein, we report two functional triazine-conjugated microporous polymers, which were synthesized by the Sonogashira–Hagihara coupling reaction. The functional polymer materials show excellent selectivity and adsorption for various types of dyes.

2. Results and Discussion

2.1. Synthesis of T-CMP and T-CMP-Me

As shown in Scheme 1, the T-CMP network is synthesized by polycondensation via a Sonogashira–Hagihara coupling reaction [36]. In a typical procedure, the molecules for 2,4,6-tri(p-bromophenyl)-1,3,5-triazine (TBPT) and 2,4,6-tris(4-ethynylphenyl)-1,3,5-triazine (TEPT) are synthesized and characterized (Figures S1 and S2). The general procedure for the synthesis of T-CMP followed the methods in the literature [36]. An ice-cooled mixture solution of 20 mL THF and 20 mL diisopropylamine was added to (0.05 mmol, 27.0 mg) TBPT, (0.05 mmol, 19.0 mg) TEPT, (15.0 mg, 0.08 mmol) CuI, and (84.0 mg, 0.12 mmol) $Pd(PPh_3)_2Cl_2$, and stirred at room temperature for 20 min. After this procedure, the mixture was stirred at 65 °C for another 48 h. Then, the hot dark-yellow solution was filtered and washed with $CHCl_3$, CH_3OH, and H_2O to remove the unreacted reactants as well as catalysts, and then the polymer was further purified by Soxhlet extraction with acetone for three days. There was a yield of 98% for T-CMP. In addition, the T-CMP-Me was synthesized by refluxing a mixture of T-CMP and excess MeI in CH_3CN (40 mL) under an N_2 atmosphere, according to the literature. Finally, the brown-yellow solid powder was obtained and then dried in vacuo. Additionally, T-CMP-Me was prepared through protonation modification by iodomethane under conventional refluxing conditions [37].

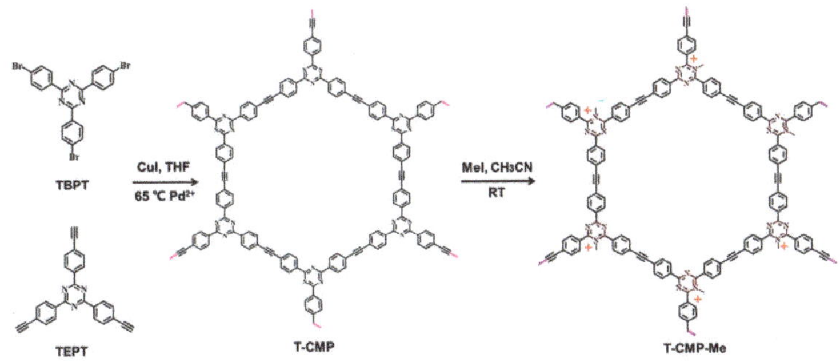

Scheme 1. Synthetic strategy for T-CMP and T-CMP-Me.

2.2. Characterization of T-CMP and T-CMP-Me

The as-prepared polymers T-CMP and T-CMP-Me are further well characterized by FT-IR, ^{13}C-NMR, UV/vis, SEM, and TEM, respectively, as shown in Figure 1, Figures S3–S5, and Table S1. The successful connection between TBPT and TEPT is confirmed by the absence of the C-H stretching band of terminal alkyne at ca. 3288 cm^{-1} and the C-Br vibration band at ca. 1060 cm^{-1} in the IR spectrum. In addition, the peaks at 670–751, 1069–1421, and 1583–1596 cm^{-1} are assigned to the wagging and torsion vibrations of benzene triazine stretching vibrations and the −C=N group stretching vibrations of benzene triazine [38], while benzene ring typical absorptions are also observed at 1341 and 1597 cm^{-1}, as shown in Figure 1a. The solid UV-visible reflectance spectra show a broader and stronger absorption with a larger reflection edge of 1000 nm compared to the molecules TEPT and TBPT, as shown in Figure 1b, as well as displays the highly conjugated structure of the triazine polymers. From the solid ^{13}C-NMR spectra (Figure 1d,e), the alkyne groups, as the important connecting group of the networks, have two obvious peaks at δ = 70.78 ppm; meanwhile, the response at δ = 128.6 ppm could be assigned to the aromatic carbon atoms of benzene. In comparison, the additional peaks at δ = 170.5 ppm are the characteristic signal of benzene triazine rings. Importantly, the resonances at δ = 30.0 ppm belong to the functional methyl group; all these results indicate the successful formation of the network polymers.

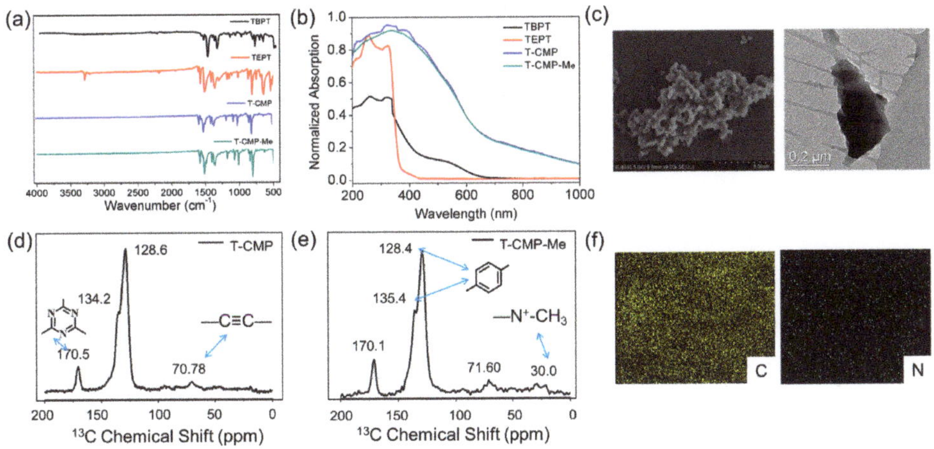

Figure 1. The (**a**) FT-IR and (**b**) solid-state UV/vis diffuse reflectance spectra of T-CMP, T-CMP-Me, and the corresponding monomers. (**c**,**f**) STEM-HAADF and elemental-mapping images of T-CMP. (**d**,**e**) The ^{13}C CP/MAS NMR spectra of T-CMP and T-CMP-Me.

The morphologies of the polymer materials T-CMP and T-CMP-Me are further characterized by scanning electron microscopy (SEM) and transmission electron microscope (TEM) (Figure 1c,f and Figure S5). It reveals that T-CMP and T-CMP-Me showed an amorphous ellipsoidal-like in morphology features, respectively. In addition, inductively coupled plasma atomic emission spectroscopy analysis of T-CMP revealed that the C and N contents are 14.29 wt% and 4.21 wt%, respectively, which are close to the theoretical values, as shown in Figure S6 and Table S1. The powder X-ray diffraction (PXRD) patterns give no definite evident peaks (Figure S7), revealing their amorphous nature [28]. Among them, the X-ray powder diffraction pattern of T-CMP shows obvious diffraction peaks at around 7.75° and 20.86°, and the diffraction peaks appearing around 20.86° are attributed to the tendency of interlayer accumulation, according to the previously reported conjugated microporous polymers [1,7,12]. It estimates that the layer spacing of T-CMP is about 4.3 Å through the Bragg equation; at the same time, the X-ray powder diffraction pattern of T-CMP-Me only shows a wide diffraction peak of nearly 22°, which indicates that the interlayer accumulation is affected and the degree of crystallization decreased after methylation. Therefore, we speculate that after methylation, the charge effect is enhanced, which leads to strong repulsion between each of the layers.

3. Porosity Measurements and Gas Adsorption Studies

The porous structures of T-CMP and T-CMP-Me are further confirmed by nitrogen sorption isotherm measurements at 77 K, which are shown in Figure 2a,b, respectively. The solid materials exhibited typical type I isotherm curves, confirming its microporous framework. Moreover, the TG analysis of the polymers revealed good thermal stability at about 400 °C (Figure S8). The Brunauer–Emmett–Teller (BET) surface areas were 322 and 435 m^2g^{-1} for T-CMP and T-CMP-Me, with average pore widths of 1.55 and 1.50 nm (Figures S9 and S10), respectively. T-CMP-Me has a larger surface area than T-CMP, which is attributed to the steric hindrance of methyl groups or lamellar charge repulsion [39]. Although the surface area of T-CMP is not large enough, the triazine-conjugated microporous polymers contain a large number of bare nitrogen atoms, which may enhance the adsorption of CO_2 and some other gases due to the excess electrons. First, the H_2 absorption at 77 K (Figure S11) was measured, which shows an uptake of 46.88 cm^3g^{-1}. Furthermore, the CO_2 uptake absorptions at different temperatures were measured, which show uptakes of 27.12 cm^3g^{-1} at 273 K and 0.39 cm^3g^{-1} at 298 K (Figure 2c). This revealed that the conjugated polymer would be a candidate for the capture of CO_2, accounting for the triazine pyrimidine rings, which have good bond energy to CO_2. Finally, as shown in Figure 2d, the T-CMP porous materials could also be a sorbent for C_2H_4; however, it displays a lower absorption for the olifant gases. All in all, the favorable capacity of the various gases was due to the abundant basic pyrimidine rings or the conjugated microporosity of the polymer skeleton.

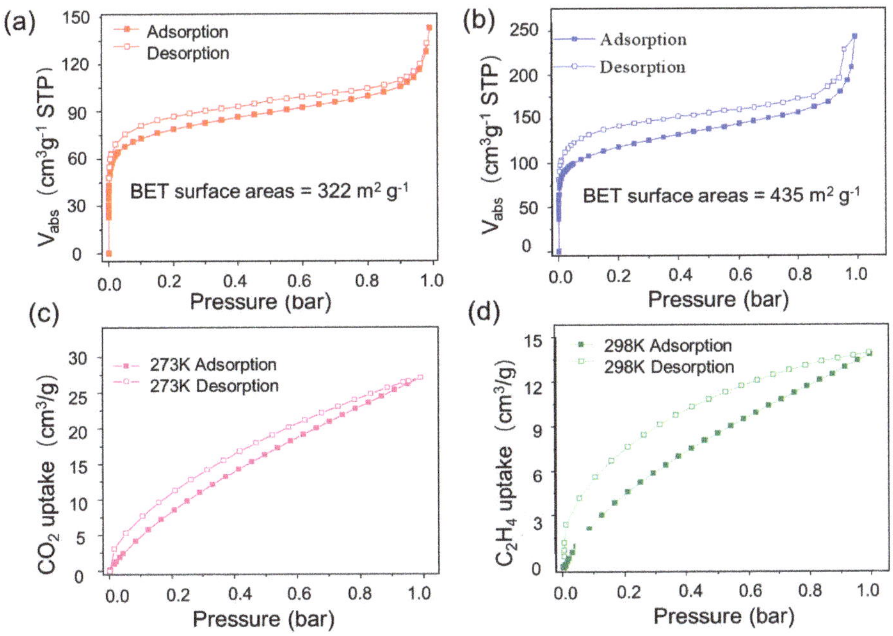

Figure 2. (**a**,**b**) N_2 adsorption–desorption isotherms (77 K) of T-CMP and T-CMP-Me, respectively. (**c**) CO_2 adsorption–desorption isotherms (273 K) of T-CMP. (**d**) C_2H_4 adsorption–desorption isotherms (298 K) of T-CMP.

4. Dye Adsorption Performance

The T-CMP and T-CMP-Me have a large number of N-donor sites and pore space inside, which lead to the ability to adsorb dye pollutants from water well. Herein, the material was considered a candidate for the capture of cation dyes from the colorful water solution (Figure 3 and Figures S12–S19). Thus, seven kinds of dyes, namely, cationic-type crystal violet (CV^+), methylene blue (MB^+), rhodamine B (RB^+) and basic red (BR^+), anionic-type methyl orange (MO^-) and acid orange (AO^-), and neutral-type dimethyl yellow (DY) were chosen as the target adsorbent dyes (Table S2). In order to further verify the hydrophilic and hydrophobic properties (Figure S19), the contact angle tests were measured. It can be observed that T-CMP has a better hydrophobic ability, which is consistent with the absorption of dyes. To check the adsorption abilities of T-CMP and T-CMP-Me, T-CMP and T-CMP (3 mg) were added to 20 mL of dye (CV^+, MB^+, RB^+, and BR^+) solution in water (20 mg/L^{-1}). Obviously, the color and the concentration of CV^+, MB^+, RB^+, and BR^+ changed at different rates (Figure 3). As shown in Figure S17, the near removal efficiency of the cation dyes are in the order of $BR^+ < RB^+ < MB^+ < CV^+$, which may be caused by the molecular size or the interaction between the molecules and the T-CMP [40]; the above phenomenon demonstrated that T-CMP would only adsorb cationic-type dyes. Furthermore, for the monoanionic dye MO^- or AO^-, almost no dye was adsorbed by the T-CMP material for a long time (Figures S18 and S19). However, the cationic T-CMP-Me shows obvious adsorbed properties (Figures S15 and S16) due to the electric charge and physisorption effect. For neutral dyes, DY (Figure S19) takes two times as long as the cationic types, owing to their porous effect.

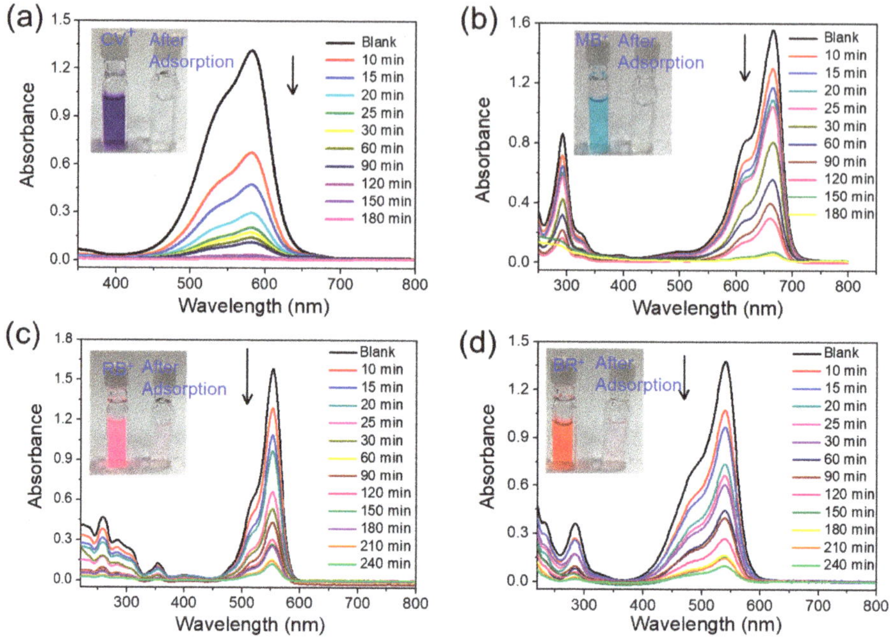

Figure 3. Time-dependent adsorption study of 20 mg/L^{-1} for each kind of cationic dye in 20 mL aqueous solution: (**a**) CV^+, (**b**) MB^+, (**c**) RB^+, and (**d**) BR^+ for T-CMP (3 mg). Inset photographs show the color change before and after dye adsorption.

The above results of UV/vis absorption spectra for different dyes show that the T-CMP and T-CMP-Me materials like to adsorb various types of dyes because the skeleton of the material itself contains N-donors. This is due to the fact that the polymers are rich in electrons, which are conducive to adsorbing the molecular dyes with electron deficiency. Meanwhile, the triazine-conjugated microporous polymers also have natural adsorption properties, which is consistent with the conclusion of gas adsorption.

For further study of the phenomenon, we also studied the aqueous solutions containing mixed dyes of MB^+/CV^+, MB^+/BR^+, MB^+/RB^+, MB^+/AO^-, and MB^+/DY, as shown in Figure 4. Owing to the rich N-donor porous framework of T-CMP, obviously, the homologous cationic dye colors of the MB^+/CV^+, MB^+/BR^+, and MB^+/CV^+ mixtures changed from deep blue to light blue, from modena to slightly pink, and from bluish violet to lavender, respectively, in Figure 4a–c. Checking the adsorption selectivity, it was found that only the mixture of MB^+/BR^+ changed significantly; the highest absorption peaks at 655 nm of MB^+ drops quickly compared to the BR^+; however, the removal efficiency for the mixture of MB^+/CV^+ and MB^+/CV^+ is changed homogenously. From the above conclusions, the selective cationic dye adsorption suggests the T-CMP could adsorb homologous dyes in a mixture of water pollution.

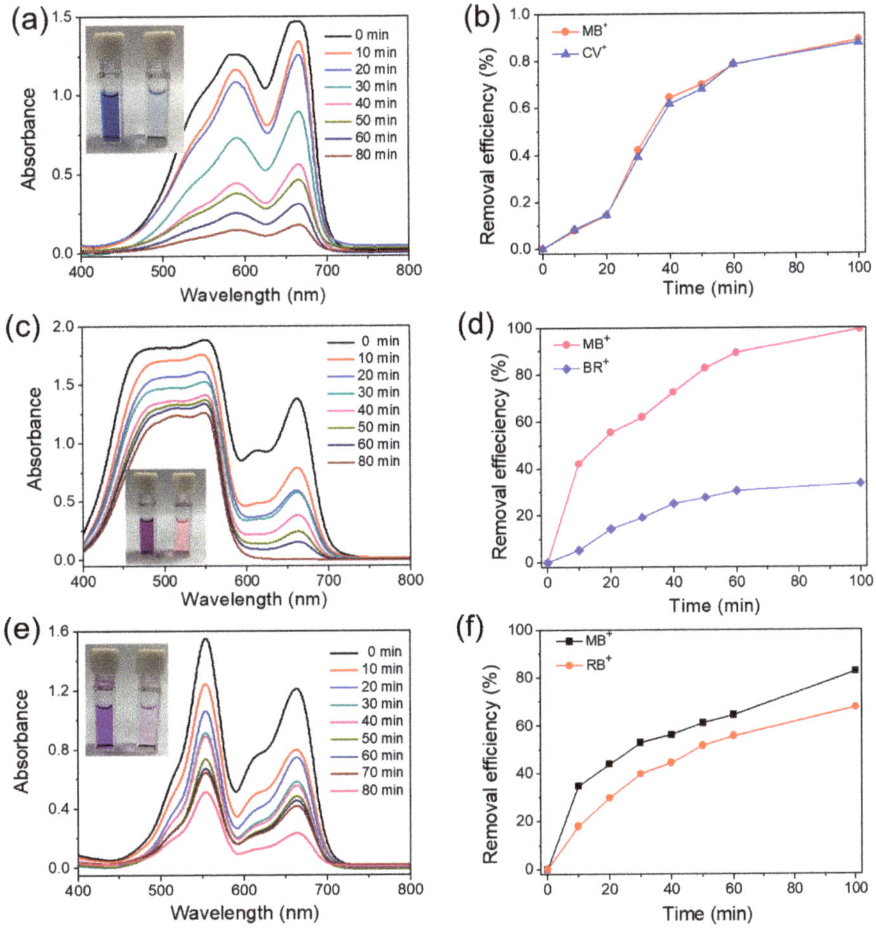

Figure 4. Time-dependent adsorption study of 20 mg/L^{-1} for each kind of cationic dye in 20 mL aqueous solution: (**a**) MB^+/CV^+, (**b**) MB^+/BR^+, and (**c**) MB^+/RB^+ for T-CMP (5 mg). The removal efficiency of each aqueous mixture: (**d**) MB^+/CV^+, (**e**) MB^+/BR^+, and (**f**) MB^+/RB^+. Inset photographs show the color change before and after dye adsorption (removal efficiency = $(C_0 - C_t)/C_t$; C_0: the original concentration; C_t: the concentration at the moment t).

Meanwhile, it was found that the T-CMP could selectively adsorb mixtures of MB^+/AO^- and MB^+/DY by the meaningful structure; see Figure 5a–d. It was found that T-CMP only absorbs cationic-type dyes from the MB^+/AO^- mixture solution; see Figure 5a. The solution color turned from dark green to light yellow, and the absorption peak at 655 nm dropped more violently; however, the peak at 460 nm remained unchanged, so the selective adsorption of anion and cationic dyes might have been achieved, as shown in Figure 5c. The mixture color of MB^+/DY always changed from blue-green to colorless; therefore, the T-CMP could adsorb dyes from the solution for a long time, but has no selectivity for cationic or neutral dyes; see Figure 5b,d.

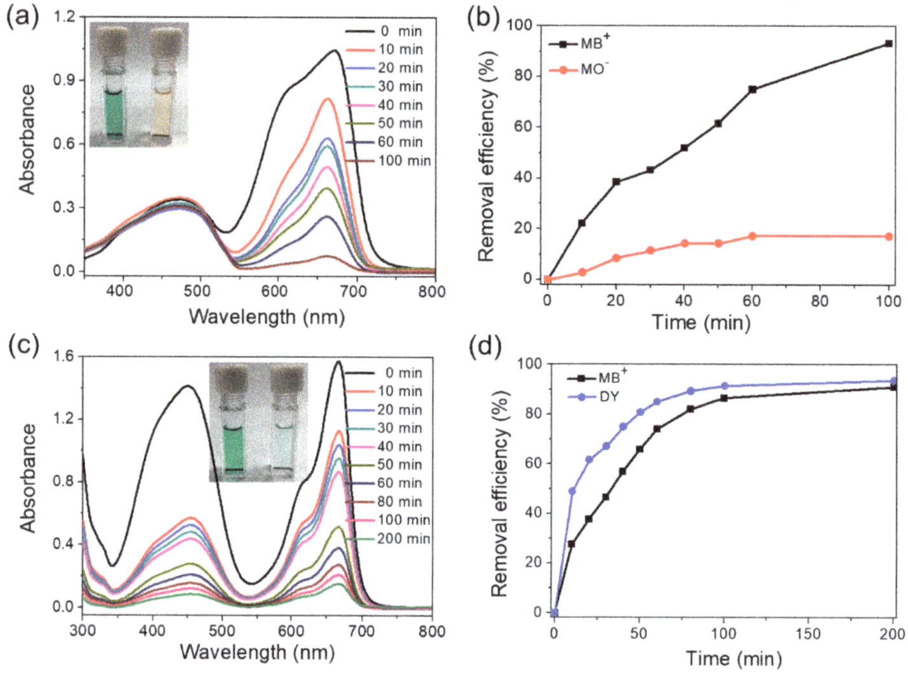

Figure 5. Time-dependent adsorption study of 20 mg/L^{-1} for each kind of cationic dye in 20 mL aqueous solution: (**a**) MB^+/MO^- and (**b**) MB^+/DY. The removal efficiency of each aqueous mixture: (**c**) MB^+/MO^- and (**d**) MB^+/DY. Inset photographs show the color change before and after dye adsorption (removal efficiency = $(C_0 - C_e)/C_e$; C_0: the original concentration; C_t: the concentration at the moment t).

For further study and in order to establish the amount of the dye uptake capacities of T-CMP, the dye adsorption amounts are calculated using the equation R (mg g^{-1}) = 20 mL × $(C_0 - C_t)$/3 mg, in which R is the dye uptake, C_0 is the initial concentration (mg L^{-1}), and C_e is the equilibrium concentration (mg L^{-1}). As a result, the maximum uptake amounts of T-CMP were determined to be 132.47 mg g^{-1} for CV^+, 128.32 mg g^{-1} for MB^+, 123.29 mg g^{-1} for RB^+, 123.60 mg g^{-1} for BR^+, and 116.03 mg g^{-1} for DY, as shown in Figure S21. It was found that the T-CMP materials have large uptake capacities for cationic-type dyes by the rich N-donor and the microporous effect, or molecular interactions. This excellent performance might develop the T-CMP itself to be more advanced in the treatment of water pollution for target dyes.

According to the results of the pseudo-first-order kinetic simulation and the pseudo-second-order kinetic simulation [41–43], shown in the Supplementary Information (Figures 6 and S22 and Table S3), it is found that the adsorption of MB by T-CMP and T-CMP-Me was more consistent with the quasi-second-order kinetic simulation. Considering that the adsorption rate is proportional to the number of unoccupied adsorption sites in the quasi-second-order kinetic model, the adsorption mechanism is assumed to be controlled by chemisorption. This result reveals that the porous materials T-CMP and T-CMP-Me adsorbed MB mainly by chemical adsorption. Considering that the T-CMP, T-CMP-Me, and organic dye molecules all have conjugate structures, we believe that the T-CMP and T-CMP-Me mainly adsorbed dye molecules by molecular interactions. By comparing the simulation results of the adsorption of MB by the T-CMP and T-CMP-Me with the pseudo-second-order kinetic constant k_2 and the equilibrium adsorption capacity Q_e, it is found that T-CMP-Me has slightly faster adsorption but a smaller equilibrium adsorption capacity than T-CMP (Figure S23). The reason is that T-CMP-Me with positive charges is less stacked than T-CMP layers (Figure S4). The former increases the specific surface area and improves the adsorption rate, while the latter repulses dye molecules and reduces the adsorption sites.

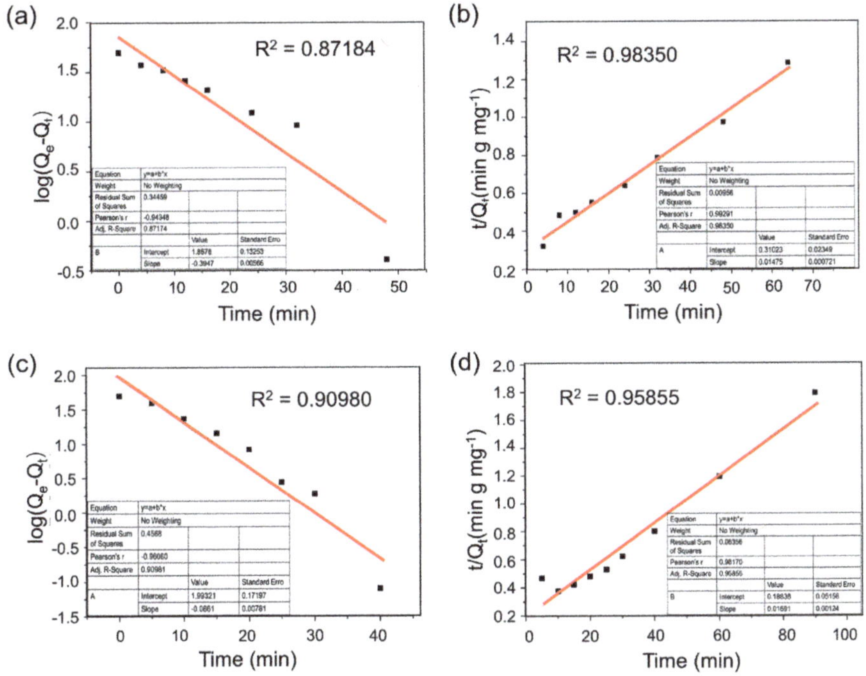

Figure 6. Pseudo-first-order kinetics (**a**) and pseudo-second-order kinetics (**b**) for adsorption of MB in the presence of T-CMP. Pseudo-first-order kinetics (**c**) and pseudo-second-order kinetics (**d**) for adsorption of MB in the presence of T-CMP-Me.

Furthermore, we carried out the adsorbent recycling experiment and found that both the polymer materials T-CMP and T-CMP-Me have good regeneration abilities (Figure 7). Subsequently, T-CMP-Me was selected as an adsorbent for adsorption in the mixed aqueous solution of MO and MB (Figure 7a), and it was found that it could selectively adsorb MB (Figure 7a,b). Finally, in order to simulate industrial sewage adsorption treatment, we built a set of simple flow adsorption devices (Figure 7c,d) to realize the selective adsorption of MB in a mixed aqueous solution of MO and MB, revealing that the material has good selective cyclic adsorption capacity.

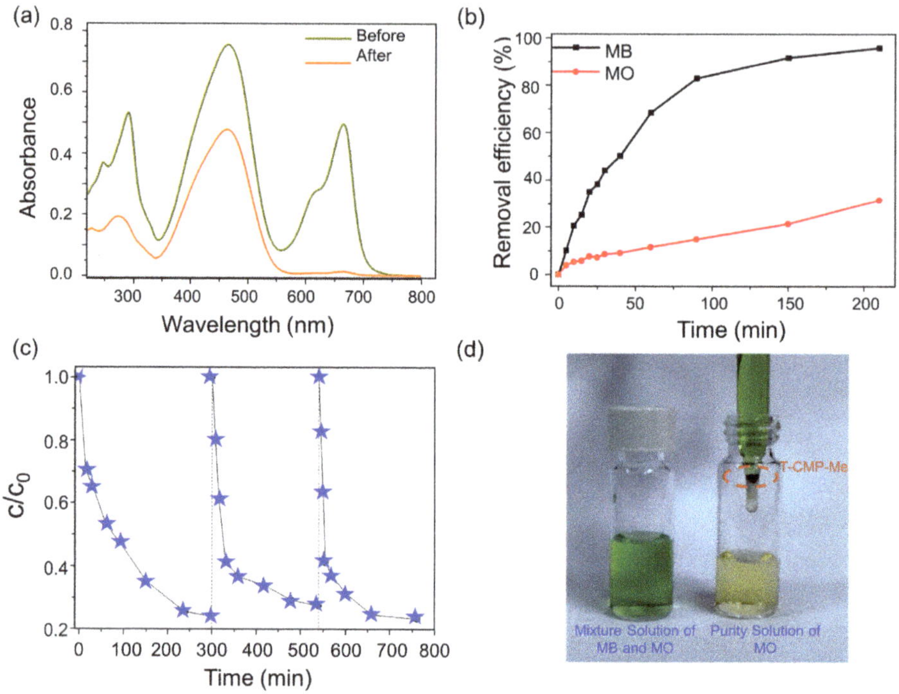

Figure 7. UV-vis absorption spectra (**a**) and adsorption rates (**b**) of the mixed aqueous solutions of MO and MB in the presence of T-CMP-Me. (**c**) The recycling performance of T-CMP-Me in the mixture. (**d**) The flow adsorption equipment and the photograph of flow adsorption equipment.

5. Conclusions

In summary, triazine-conjugated microporous polymers with good dye adsorption properties were synthesized and characterized by the Sonogashira–Hagihara coupling reaction. The network polymers are found to have potential value for gas adsorption. All the materials underwent dye adsorption via the quasi-second-order kinetic simulation to give selective adsorption performance; more importantly, they displayed efficient removal abilities to adsorb cationic-type dyes and selectivity for the mixture solution. Finally, a set of simple flow adsorption devices was constructed, which will improve the triazine-conjugated microporous polymers (T-CMP and T-CMP-Me) as a new method for water pollution treatment.

Supplementary Materials: The following are available online at https://www.mdpi.com/article/10.3390/molecules28124785/s1. Figure S1. ^{1}H-NMR spectrum of TBPT. Figure S2. ^{1}H-NMR spectrum of TEPT. Figure S3. Partial SEM images of T-CMP with different sizes. Figure S4. SEM and TEM images of T-CMP-Me. Figure S5. SEM images of T-CMP-Me with different sizes. Figure S6. Energy image of T-CMP. Figure S7. Powder X-ray diffraction spectrum of T-CMP and T-CMP-Me. Figure S8. The TG Curves of the molecules TBPT and TEPT, also compared with the porous materials T-CMP and T-CMP-Me. Figure S9. The pore size distribution obtained by the NLDFT method of T-CMP. Figure S10. The pore size distribution obtained by the NLDFT method of T-CMP-Me. Figure S11. H_2 adsorption–desorption isotherms (77 K) of T-CMP. Figure S12. Time-dependent adsorption study of 20 mg/L^{-1} for each kind of MB^+ dye in 20 mL aqueous solution for T-CMP-Me (3 mg). Figure S13. Time-dependent adsorption study of 20 mg/L^{-1} for each kind of CV^+ dye in 20 mL aqueous solution for T-CMP-Me (3 mg). Figure S14. Time-dependent adsorption study of 20 mg/L^{-1} for each kind of RB^+ dye in 20 mL aqueous solution for T-CMP-Me (3 mg). Figure S15. Time-dependent adsorption study of (20 mg/L^{-1} for each kind of MO dye in 20 mL aqueous solution for T-CMP-Me (3 mg).

Figure S16. Time-dependent adsorption study of 20 mg/L^{-1} for each kind of AO dye in 20 mL aqueous solution for T-CMP-Me (3 mg). Figure S17. The removal efficiency of each aqueous mixture: (a) CV^+, (b) MB^+, (c) RB^+, and (d) BR^+ (removal efficiency = $\frac{C0-Ct}{C0} \times 100\%$; C_0: the original concentration; C_t: the concentration at the moment t). Figure S18. Time-dependent adsorption study of 20 mg/L−1 for anionic dyes in 20 mL aqueous solution: (a) MO− and (b) AO−. The removal efficiency of anionic dyes: (c) MO− and (d) AO− (removal efficiency = (C0 − Ct)/C0 × 100%; C0: the original concentration; Ct: the concentration at the moment t). Figure S19. Time-dependent adsorption study of 20 mg/L−1 for neutral dye in 20 mL aqueous solution: (a) DY. The removal efficiency of anionic dyes: (b) DY (removal efficiency = (C0 − Ct)/C0×100%; C0: the original concentration; Ct: the concentration at the moment t). Figure S20. The contact angle images for T-CMP (left) and T-CMP-Me (right) with water. Figure S21. The maximum adsorption of dyes (CV^+, MB^+, RB^+, BR^+, and DY) of T-CMP. Figure S22. Time-dependent UV adsorption rates of MB+ dyes T-CMP and T-CMP-Me, respectively. Figure S23. Pseudo-second-order kinetics for adsorption of MO (a) and AO (b) in the presence of T-CMP-Me. Table S1. Data summary of energy spectrum analysis for T-CMP. Table S2. Data summary of the different kinds of dyes (label, structure, and charge). Table S3. Comparison of constants calculated based on respective pseudo-second-order kinetics.

Author Contributions: Contributions: B.-N.L., X.-L.Z. and X.-H.B. contributed equally to this work. X.-L.Z., data curation; Z.-J.L., X.-L.Z. and X.-H.B., Z.-J.L., formal analysis; B.-N.L., funding acquisition; X.-H.B. and J.L., investigation; X.-Y.F., methodology; B.-N.L., project administration; B.-N.L., supervision; X.-L.Z., X.-L.Z., X.-Y.F. and J.L., validation; B.-N.L., visualization; B.-N.L., writing—original draft; B.-N.L. and X.-L.Z., writing—review and editing. All authors have read and agreed to the published version of the manuscript.

Funding: This work was supported by The Young Talent Promotion Program of Yulin (20220403), the Research Foundation of Yulin University (2023GK05), and the Basic Science Research Project of Shaanxi Province (22JHQ034).

Institutional Review Board Statement: Not applicable.

Informed Consent Statement: Not applicable.

Data Availability Statement: All data are included in the article.

Conflicts of Interest: The authors declare no conflict of interest.

Sample Availability: Samples of Conjugated Microporous Polymers (CMPs) are available from the authors.

References

1. Lee, J.; Kim, J.G.; Chang, J.Y. Fabrication of a conjugated microporous polymer membrane and its application for membrane catalysis. *Sci. Rep.* **2017**, *7*, 13568. [CrossRef]
2. Luo, S.; Almatrafi, E.; Tang, L.; Song, B.; Zhou, C.; Zeng, Y.; Zeng, G.; Liu, Z. Processable Conjugated Microporous Polymer Gels and Monoliths: Fundamentals and Versatile Applications. *ACS Appl. Mater. Interfaces* **2022**, *14*, 39701. [CrossRef]
3. Lee, J.-S.M.; Wu, T.-H.; Alston, B.M.; Briggs, M.E.; Hasell, T.; Hu, C.-C.; Cooper, A.I. Porosity-engineered carbons for supercapacitive energy storage using conjugated microporous polymer precursors. *J. Mater. Chem. A* **2016**, *4*, 7665. [CrossRef]
4. Jiao, Y.; Wu, F.; Xie, A.; Wu, L.; Zhao, W.; Zhu, X.; Qi, X. Electrically conductive conjugate microporous polymers (CMPs) via confined polymerization of pyrrole for electromagnetic wave absorption. *Chem. Eng. J.* **2020**, *398*, 125591. [CrossRef]
5. Haleem, A.; Shafiq, A.; Chen, S.-Q.; Nazar, M. A Comprehensive Review on Adsorption, Photocatalytic and Chemical Degradation of Dyes and Nitro-Compounds over Different Kinds of Porous and Composite Materials. *Molecules* **2023**, *28*, 1081. [CrossRef]
6. Li, Q.; Fan, Z.-L.; Xue, D.-X.; Zhang, Y.-F.; Zhang, Z.-H.; Wang, Q.; Sun, H.-M.; Gao, Z.; Bai, J. A multi-dye@MOF composite boosts highly efficient photodegradation of an ultra-stubborn dye reactive blue 21 under visible-light irradiation. *J. Mater. Chem. A* **2018**, *6*, 2148. [CrossRef]
7. Amin, K.; Ashraf, N.; Mao, L.; Faul, C.F.J.; Wei, Z. Conjugated microporous polymers for energy storage: Recent progress and challenges. *Nano Energy* **2021**, *85*, 105958. [CrossRef]
8. Holst, J.R.; Trewin, A.; Cooper, A.I. Porous organic molecules. *Nat. Chem.* **2010**, *2*, 915. [CrossRef]
9. Jiang, J.-X.; Laybourn, A.; Clowes, R.; Khimyak, Y.Z.; Bacsa, J.; Higgins, S.J.; Adams, D.J.; Cooper, A.I. High Surface Area Contorted Conjugated Microporous Polymers Based on Spiro-Bipropylenedioxythiophene. *Macromolecules* **2010**, *43*, 7577. [CrossRef]
10. Lee, J.M.; Cooper, A.I. Advances in Conjugated Microporous Polymers. *Chem. Rev.* **2020**, *120*, 2171. [CrossRef] [PubMed]
11. Chang, Z.; Zhang, D.S.; Chen, Q.; Bu, X.H. Microporous organic polymers for gas storage and separation applications. *Phys. Chem. Chem. Phys.* **2013**, *15*, 5430. [CrossRef]

12. Liu, Z.; Yin, Y.; Eginligil, M.; Wang, L.; Liu, J.; Huang, W. Two-dimensional conjugated microporous polymer films: Fabrication strategies and potential applications. *Polym. Chem.* **2021**, *12*, 807. [CrossRef]
13. Chen, L.; Honsho, Y.; Seki, S.; Jiang, D. Light-Harvesting Conjugated Microporous Polymers: Rapid and Highly Efficient Flow of Light Energy with a Porous Polyphenylene Framework as Antenna. *J. Am. Chem. Soc.* **2010**, *132*, 6742. [CrossRef]
14. Sun, Q.; Aguila, B.; Song, Y.; Ma, S. Tailored Porous Organic Polymers for Task-Specific Water Purification. *Acc. Chem. Res.* **2020**, *53*, 812. [CrossRef]
15. Tan, Z.; Su, H.; Guo, Y.; Liu, H.; Liao, B.; Amin, A.M.; Liu, Q. Ferrocene-Based Conjugated Microporous Polymers Derived from Yamamoto Coupling for Gas Storage and Dye Removal. *Polymers* **2020**, *12*, 719. [CrossRef]
16. Wang, F.; Sun, W.; Pan, W.; Xu, N. Adsorption of sulfamethoxazole and 17β-estradiol by carbon nanotubes/$CoFe_2O_4$ composites. *Chem. Eng. J.* **2015**, *274*, 17. [CrossRef]
17. Li, Y.; Liu, M.; Chen, L. Polyoxometalate built-in conjugated microporous polymers for visible-light heterogeneous photocatalysis. *J. Mater. Chem. A* **2017**, *5*, 13757. [CrossRef]
18. Tian, S.; Zhang, J.; Chen, J.; Kong, L.; Lu, J.; Ding, F.; Xiong, Y. $Fe_2(MoO_4)_3$ as an Effective Photo-Fenton-like Catalyst for the Degradation of Anionic and Cationic Dyes in a Wide pH Range. *Ind. Eng. Chem. Res.* **2013**, *52*, 13333. [CrossRef]
19. Wang, Z.; Zhu, C.-Y.; Zhao, H.-S.; Yin, S.-Y.; Wang, S.-J.; Zhang, J.-H.; Jiang, J.-J.; Pan, M.; Su, C.-Y. Record high cationic dye separation performance for water sanitation using a neutral coordination framework. *J. Mater. Chem. A* **2019**, *7*, 4751. [CrossRef]
20. Levard, C.; Hotze, E.M.; Lowry, G.V.; Brown, G.E. Environmental Transformations of Silver Nanoparticles: Impact on Stability and Toxicity. *Environ. Sci. Technol.* **2012**, *46*, 6900. [CrossRef] [PubMed]
21. Patel, S.; Hota, G. Iron oxide nanoparticle-immobilized PAN nanofibers: Synthesis and adsorption studies. *RSC Adv.* **2016**, *6*, 15402. [CrossRef]
22. Jiang, J.X.; Su, F.; Trewin, A.; Wood, C.D.; Campbell, N.L.; Niu, H.; Dickinson, C.; Ganin, A.Y.; Rosseinsky, M.J.; Khimyak, Y.Z.; et al. Conjugated microporous poly(aryleneethynylene) networks. *Angew. Chem. Int. Ed.* **2007**, *46*, 8574. [CrossRef]
23. Cheng, G.; Hasell, T.; Trewin, A.; Adams, D.J.; Cooper, A.I. Soluble conjugated microporous polymers. *Angew. Chem. Int. Ed.* **2012**, *51*, 12727. [CrossRef]
24. Luo, L.; Wu, Z.; Wu, Z.; Liu, Y.; Huang, X.; Ling, R.; Ye, L.; Luo, X.; Wang, C. Role of Structure in the Ammonia Uptake of Porous Polyionic Liquids. *ACS Sustain. Chem. Eng.* **2022**, *10*, 4094. [CrossRef]
25. Haleem, A.; Pan, J.-M.; Shah, A.; Hussain, H.; He, W. A systematic review on new advancement and assessment of emerging polymeric cryogels for environmental sustainability and energy production. *Sep. Purif. Technol.* **2023**, *316*, 123678. [CrossRef]
26. Hu, G.; Wu, T.; Liu, Z.; Gao, S.; Hao, J. Application of Molecular Imprinting Technology Based on New Nanomaterials in Adsorption and Detection of Fluoroquinolones. *Anal. Methods* **2023**, *15*, 2467. [CrossRef] [PubMed]
27. Mohanan, M.; Ahmad, H.; Ajayan, P.; Pandey, P.K.; Calvert, B.M.; Zhang, X.; Chen, F.; Kim, S.J.; Kundu, S.; Gavvalapalli, N. Using molecular straps to engineer conjugated porous polymer growth, chemical doping, and conductivity. *Chem. Sci.* **2023**, *14*, 5510. [CrossRef] [PubMed]
28. Segura, J.L.; Mancheno, M.J.; Zamora, F. Covalent organic frameworks based on Schiff-base chemistry: Synthesis, properties and potential applications. *Chem. Soc. Rev.* **2016**, *45*, 5635. [CrossRef] [PubMed]
29. Schwarz, D.; Acharjya, A.; Ichangi, A.; Kochergin, Y.S.; Lyu, P.; Opanasenko, M.V.; Tarabek, J.; Vacek Chocholousova, J.; Vacek, J.; Schmidt, J.; et al. Tuning the Porosity and Photocatalytic Performance of Triazine-Based Graphdiyne Polymers through Polymorphism. *ChemSusChem* **2019**, *12*, 194. [CrossRef]
30. Luo, D.; Li, M.; Ma, Q.; Wen, G.; Dou, H.; Ren, B.; Liu, Y.; Wang, X.; Shui, L.; Chen, Z. Porous organic polymers for Li-chemistry-based batteries: Functionalities and characterization studies. *Chem. Soc. Rev.* **2022**, *51*, 2917. [CrossRef]
31. Zhou, Z.; Shinde, D.B.; Guo, D.; Cao, L.; Nuaimi, R.A.; Zhang, Y.; Enakonda, L.R.; Lai, Z. Flexible Ionic Conjugated Microporous Polymer Membranes for Fast and Selective Ion Transport. *Adv. Funct. Mater.* **2021**, *32*, 2108672. [CrossRef]
32. Zou, Y.; Wang, D.; Guo, J.; Yang, J.; Pu, Y.; Chen, J.-F. Synthesis of poly(9,9-dioctylfluorene) in a rotating packed bed with enhanced performance for polymer light-emitting diodes. *Polym. Chem.* **2022**, *13*, 3506. [CrossRef]
33. Liu, Y.; Cui, Y.; Zhang, C.; Du, J.; Wang, S.; Bai, Y.; Liang, Z.; Song, X. Post-cationic Modification of a Pyrimidine-Based Conjugated Microporous Polymer for Enhancing the Removal Performance of Anionic Dyes in Water. *Chem. Eur. J.* **2018**, *24*, 7480. [CrossRef] [PubMed]
34. Yuan, Y.; Huang, H.; Chen, L.; Chen, Y. N,N′-Bicarbazole: A Versatile Building Block toward the Construction of Conjugated Porous Polymers for CO_2 Capture and Dyes Adsorption. *Macromolecules* **2017**, *50*, 4993. [CrossRef]
35. Tantisriyanurak, S.; Duguid, H.N.; Peattie, L.; Dawson, R. Acid Functionalized Conjugated Microporous Polymers as a Reusable Catalyst for Biodiesel Production. *ACS Appl. Poly. Mater.* **2020**, *2*, 3908. [CrossRef]
36. Zou, L.H.; Johansson, A.J.; Zuidema, E.; Bolm, C. Mechanistic insights into copper-catalyzed Sonogashira-Hagihara-type cross-coupling reactions: Sub-mol% catalyst loadings and ligand effects. *Chem. Eur. J.* **2013**, *19*, 8144. [CrossRef]
37. Wang, Y.; Zhong, H.; Li, L.; Wang, R. Facile Synthesis and Tunable Porosities of Imidazolium-Based Ionic Polymers that Contain In Situ Formed Palladium Nanoparticles. *ChemCatChem* **2016**, *8*, 2234. [CrossRef]
38. Soliman, A.B.; Haikal, R.R.; Hassan, Y.S.; Alkordi, M.H. The potential of a graphene-supported porous-organic polymer (POP) for CO_2 electrocatalytic reduction. *Chem. Commun.* **2016**, *52*, 12032. [CrossRef]

39. Alkordi, M.H.; Haikal, R.R.; Hassan, Y.S.; Emwas, A.-H.; Belmabkhout, Y. Poly-functional porous-organic polymers to access functionality—CO_2 sorption energetic relationships. *J. Mater. Chem. A* **2015**, *3*, 22584. [CrossRef]
40. Li, Q.; Xue, D.-X.; Zhang, Y.-F.; Zhang, Z.-H.; Gao, Z.; Bai, J. A dual-functional indium–organic framework towards organic pollutant decontamination via physically selective adsorption and chemical photodegradation. *J. Mater. Chem. A* **2017**, *5*, 14182. [CrossRef]
41. Liu, Y. Is the Free Energy Change of Adsorption Correctly Calculated? *J. Chem. Eng. Data* **2009**, *54*, 1981. [CrossRef]
42. Ho, Y.S.; McKay, G. Pseudo-second order model for sorption processes. *Process Biochem.* **1999**, *34*, 451. [CrossRef]
43. Wang, S.; Meng, X.; Luo, H.; Yao, L.; Song, X.; Liang, Z. Post-synthetic modification of conjugated microporous polymer with imidazolium for highly efficient anionic dyes removal from water. *Sep. Purif. Technol.* **2022**, *284*, 120245. [CrossRef]

molecules

MDPI

Article

A Carboxyl Group-Functionalized Ionic Liquid Hybrid Adsorbent for Solid-Phase Extraction and Determination of Trace Diclofenac Sodium in Milk Samples

Hongrui Yang [1,2], Chen Wang [1], Wenjuan Zhu [1], Pingning Jin [1], Fei Li [1] and Jing Fan [1,*]

[1] School of Environment, Key Laboratory of Yellow River and Huai River Water Environment and Pollution Control, Ministry of Education, Henan Key Laboratory for Environmental Pollution Control, Henan Normal University, Xinxiang 453007, China; yangyangyang321200@163.com (H.Y.); wang2119084002@163.com (C.W.); wenjuanz87@126.com (W.Z.); jinpingning@126.com (P.J.); lifei40351991@163.com (F.L.)

[2] College of Chemical and Environmental Engineering, Xinjiang Institute of Engineering, Urumqi 830023, China

* Correspondence: fanjing@htu.cn

Abstract: A simple and efficient sample pretreatment technology is very important for the accurate determination of trace drug residues in foods to ensure food safety. Herein, we report a new carboxyl group-functionalized ionic liquid hybrid solid- phase adsorbent (PS-IL-COOH) for the highly efficient extraction and quantitative determination of diclofenac sodium (DS) residue in milk samples. It was found that the adsorption efficiency of PS-IL-COOH for the ppb level of DS was greater than 93.0%, the adsorption capacity was 934.1 mg/g, and the enrichment factor was 620.0, which surpass most of the previously reported values for DS adsorbents. The high concentration of salts did not interfere with the adsorption of DS. Importantly, the recovery of DS was above 90% after 16 adsorption—regeneration cycles. The synergistic effect of the multiple interactions was found to be the main factor for the high efficiency of DS adsorption. The proposed method was applied to the extraction and detection of DS in milk samples, with the relative recovery ranging from 88.2 to 103.0%.

Keywords: carboxyl group-functionalized ionic liquid; solid-phase extraction; diclofenac sodium; milk samples

1. Introduction

Diclofenac sodium (DS) is a nonsteroidal anti-inflammatory drug that is widely used in humans, livestock, and fisheries [1] for the treatment of mammalian mastitis, arthritis, and sports injuries [2], because of its anti-inflammatory, antipyretic, and analgesic properties [3,4]. However, the overuse of DS may lead to unsafe residues in different animal tissues and animal-derived foods, posing a potential risk to consumers [5]. Even at low concentrations, long-term exposure can pose a major threat to human health, such as drug resistance, gastric ulcers, the inhibition of platelet aggregation, and an increased risk of heart attack [4]. Currently, residues of DS have been found in environmental water [6], the muscles of fish and shrimp [7], the livers of cattle and sheep [8], chicken wings, and milk [2,9]. Therefore, developing effective techniques for the accurate determination of trace DS in food samples is of great importance to ensure food safety.

As a drug residue, samples involving DS have the characteristics of complex composition and low content. In order to ensure the reliability of the determination results, sample pretreatment is an essential step. Currently, several methods have been reported for the pretreatment of samples involving DS. Among them, solid-phase extraction has attracted remarkable attention due to its simple operation, low usage amount of organic solvent, and high recovery efficiency [10–12]. In this context, Soheili-Azad et al. [13] reported a zinc/aluminum layered double hydroxide adsorbent (Zn/Al LDH) for the extraction and HPLC determination of trace DS from human serum and hospital wastewater. The DS adsorbed on the Zn/Al LDH was then eluted off with a 2.0 mol/L NaOH solution, and the enrichment factors were 37 (human serum) and 41 (hospital wastewater). Prasetya et al. [14] prepared a hollow fiber of a MOF-808 adsorbent for extracting DS from wastewater; the adsorption equilibrium time was 300 min with an adsorption capacity of up to 833 mg/g, and the adsorbent could withstand four adsorption–regeneration cycles with an eluent of methanol and acetic acid mixture (V:V = 9:1). Gao et al. [7] reported on a UIO-66-NH_2 modified cotton fiber to extract DS from fish and shrimp muscle tissues. Coupled with ultra-high performance liquid chromatography detection, the detection limit was 0.12 ng/mL, and the enrichment factor was 32 with a 2% formic acid and methanol mixture (V:V = 1:1) as the eluent. The performance of the solid-phase extraction material is the key factor affecting the extraction efficiency [15]. Therefore, the design and synthesis of new, functionalized adsorbent materials with a short adsorption equilibrium time, large adsorption capacity, high enrichment factor, and green eluent are still the focus of researchers.

Ionic liquids (ILs) have attracted remarkable interest due to their extraordinary properties, such as low volatility, strong solubility, and controllable structure [16]. Therefore, they are considered to be satisfactory substitutes for traditional volatile organic solvents [17], and have been successfully applied to extraction [18], material preparation [19], organic synthesis [20], and other fields [21]. However, ILs are expensive, hard to regenerate, and easy to lose during use, which greatly limits their practical applications. In order to solve these problems, supported ionic liquid materials have attracted increasing attention because they have the unique properties of ionic liquids in solid supports [22].

In this work, a carboxyl group-functionalized ionic liquid (IL) hybrid solid adsorbent (which is named PS-IL-COOH because it has the properties of the carrier, and is supported by the functional group -COOH and the ionic liquid) is designed and synthesized in one step by chemically grafting N, N-dimethylglycine onto the surface of chloromethyl polystyrene (PS-Cl). It is found that the hybrid adsorbent can be used for the highly efficient separation and enrichment of DS in milk samples. Combined with a simple UV-Vis spectrophotometry, DS at the level of ppb can be directly detected. Moreover, a high concentration of inorganic salts does not affect the adsorption of trace amounts of the targeted drug, which improves the sensitivity and selectivity of the analytical method. In addition, the as-prepared adsorbent also has many advantages, such as a large adsorption capacity, high selectivity, fast adsorption speed, highly efficient regeneration and recycling performance, no need for expensive instruments, and almost no waste production in the whole sample pretreatment. This established separation and analysis method provides a new strategy for the green pretreatment of food samples and DS detection.

2. Results and Discussion

2.1. Characterization of PS-IL-COOH

The morphology of PS-CH_2Cl and PS-IL-COOH are shown in Figure 1a,b,e,f. It is obvious that there are no obvious differences in the morphology between PS-CH_2Cl and PS-IL-COOH. It is also clear that the new characteristic elements of nitrogen and oxygen appear in PS-IL-COOH (Figure 1i,j), which are not displayed in PS-CH_2Cl, indicating that N, N-dimethylglycine successfully bonded onto the surface of the carrier.

Furthermore, as shown in Figure S1A, PS-CH_2Cl remains almost unchanged in weight below 300 °C and gradually decomposes as the temperature rises; a weight loss of ~77% is observed from 300 to 550 °C. In the range of 100–200 °C, the slight weight loss of PS-IL-COOH is due to the evaporation of water, which could be attributed to the hydrophilic structure of the N, N-dimethylglycine that promotes the adsorption of a small amount of water by PS-IL-COOH in air. These results indicate that PS-IL-COOH has good thermal stability below 200 °C. According to the difference between the two thermogravimetric curves in Figure S1A, the grafting percentage of the IL in PS-IL-COOH is calculated to be approximately 17%, demonstrating that N, N-dimethylglycine was successfully bonded onto the PS-CH_2Cl carrier.

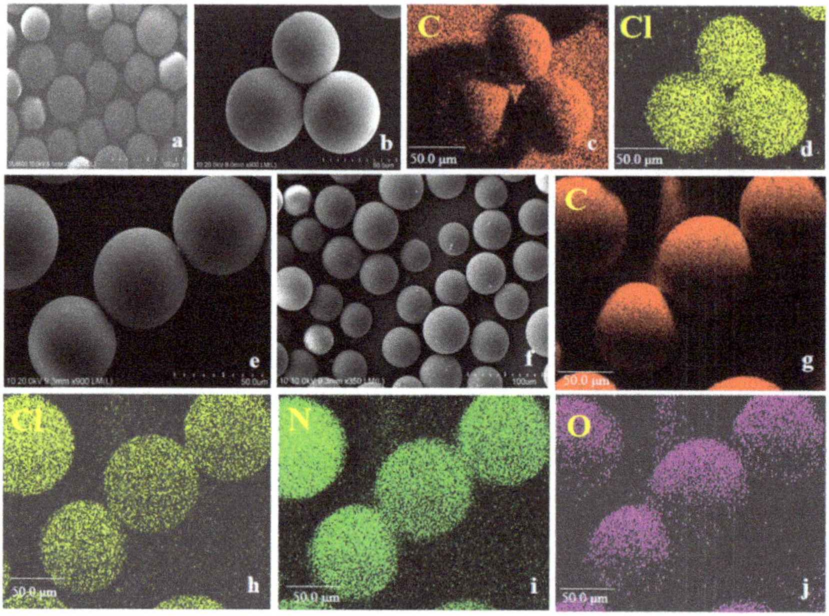

Figure 1. SEM images of PS-CH_2Cl (**a**,**b**) and PS-IL-COOH (**e**,**f**), EDS mapping of PS-CH_2Cl (**c**,**d**) and PS-IL-COOH (**g**–**j**).

Next, the FT-IR spectra in Figure S1B show that the peaks at 1264 cm^{-1} and 671 cm^{-1} disappear in PS-IL-COOH, which is attributed to the C-Cl bending vibration and stretching vibration of PS-CH_2Cl, respectively. However, a new characteristic peak of PS-IL-COOH appears at 1633 cm^{-1} (the COOH bending vibration) [23–25], suggesting that the N, N-dimethylglycine group was successfully grafted onto the surface of the carrier (Scheme 1). On the surface of the PS-CH_2Cl carrier, there is the structure of an ionic liquid composed of carboxyl-functionalized cation and chloride anion. The IL on the surface of PS-IL-COOH can provide more active sites, give full play to the advantages of ionic liquids and, thus, significantly improve its adsorption capacity.

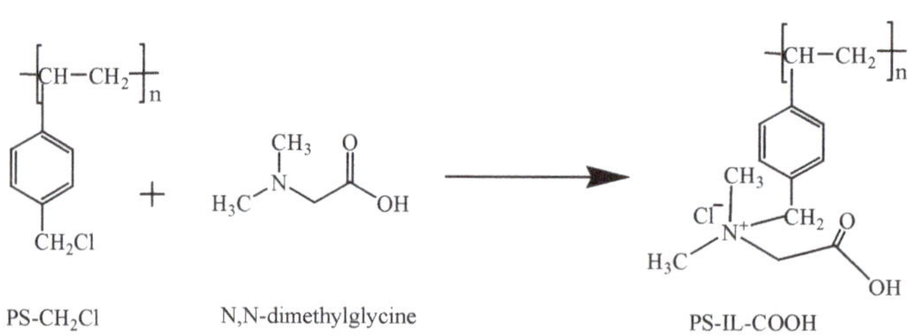

Scheme 1. Preparation of PS-IL-COOH.

2.2. *Optimization of Adsorption Performance*

2.2.1. Effects of pH and Temperature

The pH value would affect not only the existing form of the target analyte but also the charge type of the adsorbent. Here, the effect of pH was investigated from 1 to 12. As shown in Figure 2A, it was found that PS-CH_2Cl and N, N-dimethylglycine hardly adsorbed DS. However, for the as-prepared PS-IL-COOH adsorbent, the adsorption performance was enhanced significantly under a pH range from 1 to 5, and the adsorption capacity was at a maximum in the pH range of 5–11. This could be attributed to the fact that diclofenac (pKa = 4.15) mostly exists as neutral molecules in water in the range of pH < pKa [14]. When pH > pKa, DS mainly exists in the form of anions, as indicated by the positive zeta potential of PS-IL-COOH (Figure S2), and there is an electrostatic interaction between DS and PS-IL-COOH, which is favorable for improving the adsorption performance. All these points indicate that PS-IL-COOH had a good adsorption performance for DS in a wide range of pH. In order to avoid adjusting the pH of the sample solution, the following tests were carried out at a natural pH (pH $\approx$ 6) value.

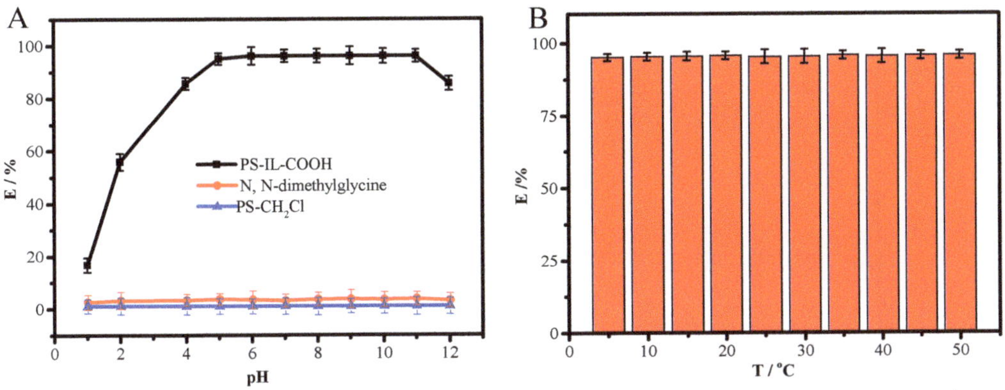

Figure 2. Effects of initial pH (**A**) and temperature (**B**) on the extraction efficiency of DS by PS-IL-COOH. Experimental conditions: $m_{PS\text{-}IL\text{-}COOH}$ = 10 mg; C_{DS} = 20 µg/mL; V_{DS} = 5 mL; time = 30 min; room temperature in (**A**); and pH $\approx$ 6 in (**B**).

The extraction efficiency of PS-IL-COOH towards DS was studied for a temperature range from 5 to 50 °C (Figure 2B). Clearly, temperature has little effect on the extraction efficiency, demonstrating that PS-IL-COOH can be applied at different temperatures without decreasing its adsorption performance. Therefore, room temperature was chosen for the subsequent experiments.

2.2.2. Adsorption Capacity

The adsorption capacity and adsorption equilibrium time are important parameters for assessing the adsorption performance of an adsorbent. As shown in Figure 3A, the adsorption capacity increases quickly with time in the initial absorption stage, and the adsorption equilibrium time is extended with an increase in DS initial concentration. This result can be ascribed to the abundant active sites of PS-IL-COOH in the initial stage. As these active adsorption sites are occupied, the speed of adsorption slows down, and the adsorption equilibrium is achieved in 15–25 min. As shown in Figure 3B, there is a significant rise in the equilibrium adsorption capacity of DS when the equilibrium concentration is smaller than 500 mg/L, then it stabilizes until adsorption reaches saturation with a maximum equilibrium adsorption capacity of 934.1 mg/g. In addition, the extraction efficiency of DS by PS-IL-COOH is 99.8–97.9% at an initial DS concentration range of 50–1000 mg/L, while the $PS\text{-}CH_2Cl$ carrier hardly adsorbs DS (Figure 3C), demonstrating that PS-IL-COOH is an excellent adsorbent for the extraction of DS.

To study the interaction between DS and PS-IL-COOH, the experimental data were fitted using various adsorption isotherm models (Table S1). The results are shown in Table S2. As shown in Figure 3D, the Langmuir model fitted the data best, and the maximum equilibrium adsorption capacity (965.3 mg/g) estimated using the Langmuir model is very close to the experimental value (934.1 mg/g).

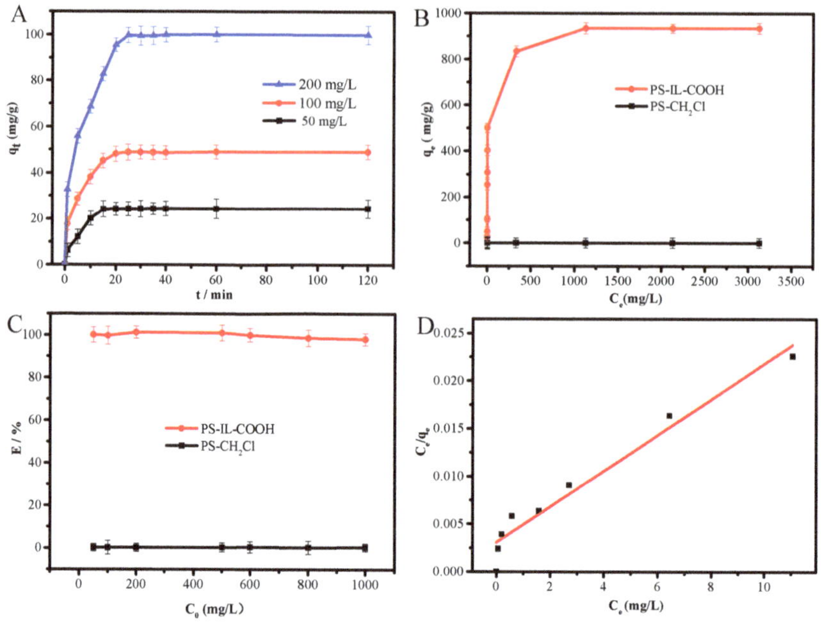

Figure 3. Adsorption capacity as a function of initial DS concentration. (**A**) The adsorption isotherm of PS-IL-COOH and PS-CH_2Cl, (**B**) the extraction efficiency of PS-IL-COOH and PS-CH_2Cl, and (**C**) the Langmuir model for the extraction of DS by PS-IL-COOH (**D**). Experimental conditions: $m_{PS\text{-}IL\text{-}COOH}$ = 10 mg; time = 120 min; V_{DS} = 5 mL; pH ≈ 6; room temperature.

2.2.3. Extraction of Diclofenac Sodium at the Ppb Level

Because of the low concentration of residual DS in the actual samples, the ability of PS-IL-COOH to extract low concentrations of DS was estimated. An amount of 1.0 mL, 5 μg/mL solution of DS was diluted into a series of solutions in the concentration range of 0.0025–0.10 μg/mL, and then flowed through a column containing 100 mg of PS-IL-COOH at the optimized flow speed of 6 mL/min. Finally, the type and volume of eluent were investigated. As shown in Table 1, among aqueous HCl, NaOH, C_2H_5OH, and their mixtures, 3 mL of HCl-C_2H_5OH (V:V = 1:9) was the best eluent, with an elution recovery of 94.8%. Even when the concentration of DS was lower than 2.5 ppb, the elution recovery still reached 93.0%, with an enrichment factor (EF) as high as 620.0 (Table S3), as calculated using the equation EF = C_f/C_i [26], where C_f and C_i are the concentration of eluent and initial sample solution, respectively. All this suggests that PS-IL-COOH is a promising material for the extraction of DS at the ppb level.

Table 1. Elution recovery of DS in different eluents.

Eluent	R%	Eluent	R%
6 mol/L HCl	16.5	6 mol/L HCl-C_2H_5OH (5:5)	61.5
3 mol/L HCl	27.8	1 mol/L NaOH	13.3
C_2H_5OH	49.8	1 mol/L NaOH-C_2H_5OH (1:9)	88.7
6 mol/L HCl-C_2H_5OH (1:9)	94.8	1 mol/L NaOH-C_2H_5OH (2:8)	80.8
6 mol/L HCl-C_2H_5OH (2:8)	91.5	1 mol/L NaOH-C_2H_5OH (3:7)	73.1
6 mol/L HCl-C_2H_5OH (3:7)	80.0	1 mol/L NaOH-C_2H_5OH (4:6)	70.6
6 mol/L HCl-C_2H_5OH (4:6)	71.3	1 mol/L NaOH-C_2H_5OH (5:5)	67.1

Experimental conditions: $m_{PS\text{-}IL\text{-}COOH}$ = 100 mg; pH ≈ 6; room temperature.

2.2.4. Selectivity of the Proposed Extraction Strategy

Coexisting substances in actual samples may affect the extraction and enrichment of DS. Thus, the influence of eight kinds of inorganic ions and seven kinds of organic compounds on the extraction efficiency of DS were studied in detail. For this purpose, different concentrations of the interfering substances were added to a solution of DS at a concentration of 10 μg/mL. After pretreatment with the proposed extraction strategy, the absorbance of DS in the eluent before and after adding the interfering substances was determined. It was considered that the concentration did not interfere with the extraction of DS when the relative standard deviation (RSD) was not greater than ±5%. As can be seen from Table 2, less interference is present from the inorganic ions; even when their concentrations were 1000 times that of DS, they still did not affect the extraction efficiency of PS-IL-COOH towards DS. The effect of divalent ions are greater than those of monovalent ions, supporting the viewpoint that electrostatic interaction is the main factor for the extraction of DS. However, the interference of the organic compounds is larger than that of the inorganic ions, and organic acid anions are more interferential than neutral molecules, suggesting that there may be hydrogen bonding in addition to the electrostatic interaction during the adsorption process. All this demonstrates that PS-IL-COOH has a potential application for the extraction and enrichment of DS from actual samples.

Table 2. The permissible interference ratio of common coexisting substances.

Coexisting Substances	Concentration (μg/mL)	Permit Ratio	RSD (%)
K^+	100,000	10,000	1.6
Na^+	100,000	10,000	2.3
Mg^{2+}	10,000	1000	1.7
Ca^{2+}	10,000	1000	1.5
Cl^-	100,000	10,000	1.3
$H_2PO_4{}^-$	100,000	10,000	1.9
$CO_3{}^{2-}$	10,000	1000	1.7
$SO_4{}^{2-}$	10,000	1000	2.4
glucose	18,000	1800	3.9
starch	3500	350	4.9
acetaminophen	4000	400	4.1
sulfanilamide	2000	200	2.6
citric acid	1000	100	1.9
L-alanine	1000	100	3.8
L-proline	1000	100	4.6

Experimental conditions: $m_{PS\text{-}IL\text{-}COOH}$ = 10 mg; C_{DS} = 10 μg/mL; pH ≈ 6; room temperature.

2.2.5. Reusability of PS-IL-COOH

Regeneration and reuse is one of the most important properties for assessing the actual application of adsorbents [27]. Meanwhile, no waste discharge during sample pretreatment is also a goal of sustainable green sample processing. From the above discussion in Section 2.2.3, HCl-C_2H_5OH (V:V = 1:9) was used to regenerate PS-IL-COOH and recover DS. It was found that the elution recovery of DS was still above 90% even after 16 adsorption—desorption cycles (Figure 4). In addition, PS-IL-COOH could be reused until adsorption saturation was reached. On the other hand, it is clearly seen from Figure 4 that PS-IL-COOH could maintain a high extraction efficiency in the first 14 adsorption cycles without regeneration, but exhibited a significant reduction in adsorption efficiency in the 15th cycle. However, after regeneration by using the eluent, the adsorption capacity of PS-IL-COOH recovered in the 16th cycle. Compared to the first strategy for the regeneration of PS-IL-COOH (regeneration after each recycle), the second strategy (regeneration after the 14th recycle) can greatly shorten the elution and sample- processing time, simplify the operation process, and reduce the amount of elution agent, which has a great actual application value for the recovery of DS. In addition, there was no clear difference in the UV-Vis adsorption of the recovered DS and the original DS under the same conditions (Figure S3), which demonstrates that the recovered DS has high purity. Simultaneously, as shown in Figure S4, there is little variation in the FT-IR spectrum between the regenerated PS-IL-COOH and fresh PS-IL-COOH, suggesting that the structure of the regenerated adsorbent was not destroyed, and the adsorption performance remained good. These results indicate that PS-IL-COOH is a potential material for the extraction of DS in actual samples.

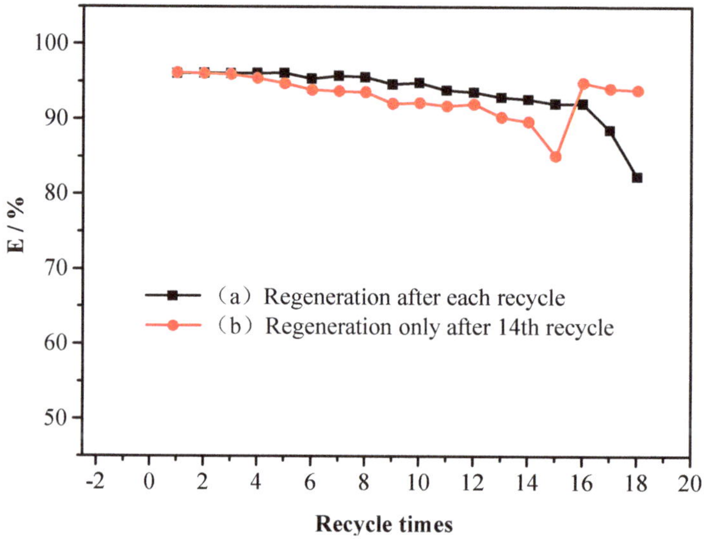

Figure 4. Influence of the recycling number of PS-IL-COOH on the extraction efficiency of DS: regeneration after each cycle (a) and regeneration only after the 14th cycle (b).

2.2.6. Method Validation

In order to evaluate the applicability of the proposed method, the limit of detection (LOD, S/N = 3) and limit of quantification (LOQ, S/N = 10) were determined. The precision was assessed using the relative standard deviations of inter-day (n = 6) and intra-day (n = 6), and the accuracy was validated by the average recovery. The results given in Table S4 indicate that the proposed method has the potential to detect DS in milk samples.

2.2.7. Real Sample Analysis

Due to the complex matrix effects of food and environmental samples, most actual samples need to be pre-treated before instrumental analysis in order to remove matrix effects and enrich trace analytes. To verify the applicability of the established method, PS-IL-COOH was further used to separate and enrich different concentrations of DS (0.01, 1.0, and 10.0 μg/mL) in three milk samples. The milk samples were pretreated and determined by following the procedures described in Sections 3.5 and 3.6. It is evident that the standard addition recovery is in the range of 88.2–103.0%, and the RSD is 2.5–7.3% (Table 3). The same 300 mL of DS (0.01 μg/mL) was flown through the column at 6 mL/min, and then the DS adsorbed on the PS-IL-COOH was eluted off with 3 mL of HCl-C_2H_5OH (V:V = 1:9) at 0.3 mL/min; the DS in the eluent was detected using UV-Vis spectrophotometry and HPLC (Figure 5A,B). It was clearly seen that the adsorbent had a good enrichment effect on trace DS, which was beneficial for the accurate determination of trace DS in actual samples.

Table 3. The recovery of PS-IL-COOH towards DS from milk samples.

Sample	Added Values (μg/mL)	Detected (μg/mL)	Recovery (%)	RSD (n = 3)
Milk sample 1	0.0	ND	ND	ND
	0.01	0.0091	91.0	2.9
	1.0	0.96	96.0	3.7
	10.0	9.41	94.1	5.2
Milk sample 2	0.0	ND	ND	ND
	0.01	0.0103	103.0	2.5
	1.0	1.02	102.0	3.4
	10.0	10.1	101.1	2.9
Milk sample 3	0.0	ND	ND	ND
	0.01	0.0098	98.0	3.1
	1.0	0.89	89.0	5.8
	10.0	8.82	88.2	7.3

ND: not detected.

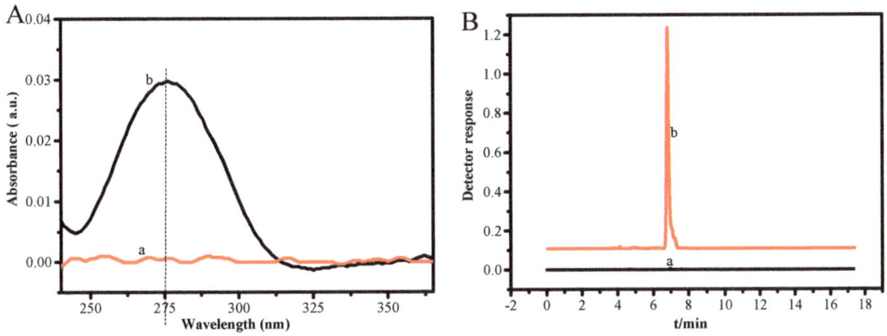

Figure 5. UV-Vis spectra of DS at 0.01 μg/mL before (a) and after (b) enrichment (**A**), and HPLC of DS at 0.01 μg/mL before (a) and after (b) enrichment (**B**). Experimental conditions of (**A**,**B**): pH ≈ 6; $m_{PS\text{-}IL\text{-}COOH}$ = 100 mg; room temperature. Experimental conditions of (**B**): mobile phase= methanol-water (95:5, V:V); column temperature = 30 °C; sample volume = 20 μL; and detection wavelength = 280 nm.

To further evaluate the accuracy of the proposed method of sample pretreatment, the actual samples were separated and enriched with PS-IL-COOH; the determination results of UV-Vis spectrophotometry and HPLC were checked using the significance test (Table 4). It was found that there was no significant difference between UV-Vis and HPLC ($t = 2.92 < t_{critical} = 4.30$), indicating that UV-Vis spectrophotometry is reliable for the determination of trace DS in foods and environmental samples. From Figure 5A,B, it can be seen that the trace DS (0.01 μg/mL) could not be detected before enrichment because the instrument sensitivity was not high, while it could be detected successfully with UV-Vis spectrophotometry and HPLC after enrichment with PS-IL-COOH, with a recovery of 91.3% and 93.3%, respectively. This confirms that trace DS residues in milk samples pretreated with PS-IL-COOH can be accurately determined using UV-Vis spectrophotometry, which is simple, inexpensive, and does not consume large amounts of organic solvents, which is especially useful in economically undeveloped areas [28].

2.2.8. Comparison with Other Adsorbents

To more clearly display the advantages of the PS-IL-COOH materials prepared in this study, the adsorption performance of PS-IL-COOH was compared to that of some previously reported adsorbents used for the adsorption of DS. It is shown in Table 5 that the PS-IL-COOH material has obvious advantages in terms of adsorption capacity, recycling number, and enrichment factor. Simultaneously, compared to commercially available adsorbents, PS-IL-COOH has obvious advantages in the extraction of DS both at low and high amounts under the same adsorption conditions (Table S5). The adsorption capacity of PS-IL-COOH is about 2.4–35.5 times that of commercially available adsorbents, indicating that PS-IL-COOH has significant potential for the extraction of DS from foods and environmental samples.

Table 4. Comparison of the UV-Vis spectrophotometry and HPLC results.

Sample	Spiked (μg/mL)	UV-Vis Detected (μg/mL)	HPLC Detected (μg/mL)	*t*-Value [a]
Milk sample 1	0.01	0.0092	0.0091	2.92
	0.01	0.0089	0.0097	
	0.01	0.0093	0.0092	

Experimental conditions: $m_{PS\text{-}IL\text{-}COOH}$ = 100 mg; pH ≈ 6; room temperature. [a] paired *t*-test, 95% confidence level, $n = 3$, $t_{critical} = 4.30$.

Table 5. Comparison of the adsorption of DS by different adsorbents.

Adsorbent	q_e	t_{eq}	RN	EF	t_{dt}	Ref.
CS@PANI@LDH composite	618.2	120	5	-	-	[29]
Fe_3O_4@MIL-100 (Fe)	400	60	5	-	-	[30]
Amine-functionalized egg albumin hydrogel	232.5	180	4	-	-	[31]
Bilayer amino-functionalized cellulose nanocrystals/chitosan composite	444.4	50	5	-	-	[32]
Amine-functionalized chitosan	469.5	60	4	-	-	[33]
Mg/Al layered double hydroxide-poly(m-phenylenediamine) composite	588	2	3	-	-	[34]
MOF-808 and its hollow fibre	833	300	4	-	-	[14]
Magnetic covalent organic frameworks	-	60	3	37.2	5	[15]
PS-IL-COOH	934.1	30	16	620.0	10	This work

q_e: adsorption capacity (mg/g); t_{eq}: equilibrium time (min); RN: recycling number; EF: enrichment factor; t_{dt}: desorption time (min); t_{et}: enrichment time (min), $t_{et} = t_{eq} + t_{dt}$.

2.2.9. Adsorption Mechanism

The study of the adsorption mechanism of pollutants is very beneficial to the development of an effective treatment technology. First, a comparison experiment was performed to study the main interactions between PS-IL-COOH and DS in an aqueous solution. We found that methyl orange (contains a typical organic anion) could be efficiently extracted, but rhodamine B (contains a typical organic cation) could not. Considering the fact that DS contains a bulk organic anion while PS-IL-COOH contains a bulk organic cation, this result suggests that electrostatic attraction plays an important role in the adsorption process, which was confirmed by the above mentioned pH effect of DS and the zeta potential of PS-IL-COOH.

In addition, it was found that after DS was adsorbed onto the surface of PS-IL-COOH (Figure S5), the stretching vibration peak of C=O of PS-IL-COOH (at 1633 cm^{-1}) and the -N-H stretching vibration of DS (at 1310 cm^{-1}) [35] were, respectively, shifted to 1630 cm^{-1} and 1315 cm^{-1}. All these changes demonstrate that the -COOH on the surface of PS-IL-COOH may form a hydrogen bond with the N-H of DS.

Another hypothesis is that there is an ion exchange between the anion of DS and the Cl- of PS-IL-COOH. To confirm this, 1 mL of supernatant was taken after the DS was adsorbed by the PS-IL-COOH, followed by adding one drop of HNO_3 (0.1 mol/L) and then one drop of $AgNO_3$ (0.1 mol/L), to see whether AgCl precipitate was produced. Subsequently, the result was compared to different blanks (Figure S6a–c) under the same conditions. It was found that AgCl precipitate appeared in the supernatant after the DS was adsorbed by PS-IL-COOH (Figure S6d), demonstrating that there was an ion exchange between the chloride ion of PS-IL-COOH and the anion of DS. And thus, the synergistic effect of electrostatic attraction, hydrogen bonding, and ion exchange plays a vital role in the super-high adsorption capacity of DS by PS-IL-COOH. The possible adsorption mechanism is shown in Figure 6.

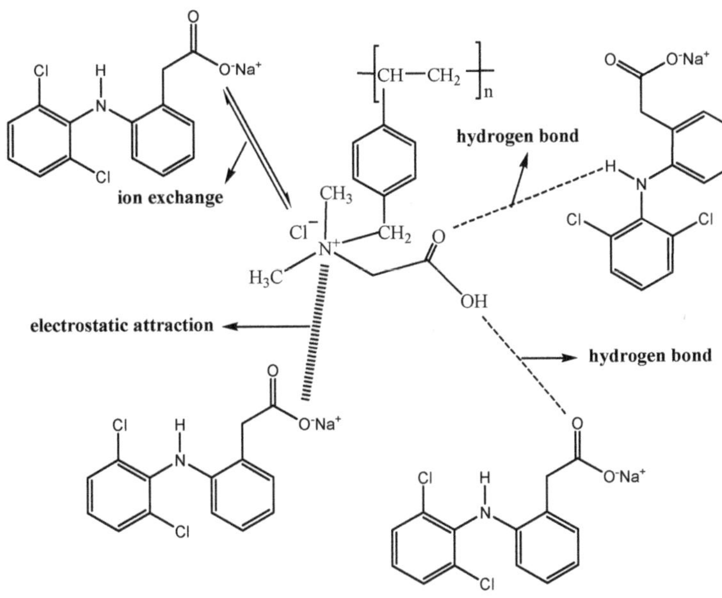

Figure 6. The possible adsorption mechanism of DS by PS-IL-COOH.

3. Experimental Section

3.1. Reagents and Materials

Chloromethyl polystyrene resin (PS-CH_2Cl, 200–400 mesh, chlorinity 3.5 mmol/g) was purchased from Tianjin Nankai Chemical Co. Ltd. (Tianjin, China). N, N-dimethylglycine (97%), N-methyl-2-pyrrolidone (98%), diclofenac sodium (99%), and trifluoroacetic acid (99%) were obtained from Aladdin Biochemical Technology Co., Ltd. (Shanghai, China). The other chemicals were all analytical grade.

3.2. Apparatus

A UV–Vis spectrophotometer (TU-1810, Beijing, China) was used for UV-Vis analysis of diclofenac sodium. The morphology and thermal stability of the materials were characterized using a Hitachi SU8010 scanning electron microscopy (Tokyo, Japan) and a Netzsch STA449C thermal analyzer (Selb, Germany), respectively. A FTS-40 Fourier-transform infrared spectroscopy (Waltham, MA, USA) was used to identify functional groups and interactions between materials and contaminants. A DC101-C oil bath (Zhengzhou, China) was used as the thermostat bath for synthesis purposes. A SPE-12B solid-phase extraction instrument (Tianjin, China) was used for the column adsorption experiments. The chromatographic analysis was performed using a waters 1525 HPLC system (Fleming, NY, USA), which was equipped with a 7725i manual sampler and 2998 ultraviolet detector.

3.3. Preparation of the Hybrid Solid-Phase Adsorbent

The carboxyl group-functionalized ionic liquid hybrid adsorbent (PS-IL-COOH) was synthesized using the following procedures: First, 0.5 g of PS-CH_2Cl and 20 mL of N-methyl-2-pyrrolidone were added to a 50 mL round-bottomed flask and swelled the PS-CH_2Cl for 12 h. Then, 1.084 g N, N-dimethylglycine was added with a speed of 700 rpm for a reaction time of 12 h at 80 °C in the oil bath. After the above mixtures were cooled to room temperature, they were transferred to a 50 mL centrifuge tube, and cleaned with deionized water and ethanol, respectively, and then centrifuged at 8000 rpm for 5 min. And then the supernatant was discarded, the mixture alternately cleaned with ethanol and deionized water until the N-methyl-2-pyrrolidone and the residue N, N-dimethylglycine were washed out. Finally, the target product was collected and then dried under vacuum at 60 °C for 12 h.

3.4. Preparation of the Solid-Phase Extraction Column Filled with PS-IL-COOH

The purchased polyethylene extraction column (10.0 mm in diameter and 62 mm in length) and several sieve pieces were soaked in ethanol for 24 h, cleaned with deionized water, and dried for use. A certain amount of PS-IL-COOH was weighed accurately and soaked in deionized water for 24 h as a backup. A sieve sheet was placed at the bottom of the extraction column, and the column was filled with deionized water. The column was loaded with the PS-IL-COOH using the overflow method, then a sieve sheet was placed on the column, and the PS-IL-COOH was compacted to make a miniature solid-phase extraction separation column for use.

3.5. Solid-Phase Extraction and Determination of Diclofenac Sodium

Adsorption experiments on the carboxyl group-functionalized ionic liquid hybrid adsorbent (PS-IL-COOH) towards diclofenac sodium (DS) were carried out using batch adsorption and column adsorption experiments. The batch adsorption experiment was performed at room temperature, where 5 mL sample of DS at different concentrations and 10 mg of PS-IL-COOH were added to a series of 20 mL sample bottles. After magnetic stirring for 30 min at 500 rpm to ensure the equilibrium of adsorption, the mixtures were transferred to a 10 mL centrifuge tube and then centrifuged at 8000 rpm for 5 min. A series of DS solutions with different concentrations were prepared with deionized water to make a standard curve, and then the concentration of the DS before and after adsorption could be determined using UV-Vis spectrophotometry at 275 nm. Finally, the concentration of DS adsorbed by the PS-IL-COOH could be quantitatively calculated according to the concentration difference before and after adsorption. The pH, temperature, adsorption equilibrium time, adsorption capacity, and anti-interference capability were investigated using batch experiments. The equilibrium adsorption capacity (q_e) and the extraction efficiency (E%) of PS-IL-COOH towards DS were, respectively, calculated using Equations (1) and (2):

$$q_e = \frac{(C_0 - C_e)V}{m} \tag{1}$$

$$E\% = \frac{(C_0 - C_e)}{C_0} \times 100\% \tag{2}$$

where C_0 (mg/L) and C_e (mg/L) are the concentrations of DS at the initial and the equilibrium (supernatant) concentration after adsorption, m (mg) is the mass of PS-IL-COOH, V (mL) is the volume of the solution of DS. In addition, we use q_t (mg/g) to present the absorption capacity at a time, t, and use R% to indicate the relative recovery of DS by PS-IL-COOH. All the experimental data were measured in triplicate.

The enrichment factor, real sample analysis, and regenerative capacity of the adsorbent were estimated with column adsorption experiments. Here, 100 mg of PS-IL-COOH was placed in a solid-phase extraction column and the DS solution was passed through the column at 6 mL/min. The DS adsorbed by the column was eluted with eluent at a flow speed of 0.3 mL/min, and then a series of DS solutions with different concentrations as reference eluents were used to make standard curves; the DS in the eluent was detected using UV-Vis spectrophotometry and confirmed using HPLC (mobile phase: methanol/water mixture (95: 5 = V:V), temperature of column: 30 °C, volume of sample: 20 μL, detection wavelength: 280 nm). The relative recovery (R%) was calculated according to Equation (3):

$$R\% = \frac{q_{eluted}}{q_{added}} \times 100\% \tag{3}$$

where q_{eluted} and q_{added} are, respectively, the mass of eluted DS from the adsorbent and the added DS in the actual sample.

3.6. Real Sample Pretreatment and Analysis

The sample preparation procedure for three different brands of milk samples purchased at a local supermarket was based on a reported method [36]. A mixed solution of 60 mL of milk sample and 3 mL of trifluoroacetic acid (TFA) was centrifuged at 500 rpm for 5 min to separate the protein, the supernatant was filtered using a 0.22 μm membrane, and then the experiments of solid-phase extraction and determination were performed in accordance with Section 3.5.

4. Conclusions

In this work, a carboxyl group-functionalized ionic liquid hybrid adsorbent (PS-IL-COOH) was prepared using a one-step grafting reaction. The systematic studies show that PS-IL-COOH is a chemically stable solid phase adsorbent with an excellent adsorption performance for DS. The extraction efficiency still exceeded 93.0% even when the DS content was lower than 2.5 ppb. The enrichment factor and adsorption capacity reached up to 620.0 and 934.1 mg/g, respectively, which exceeds most of the previously reported values for DS adsorbents. In particular, PS-IL-COOH exhibited good salt resistance, and no effect was observed on the adsorption performance when the concentration of common inorganic salts was 1000–10,000 times that of DS. Moreover, under a wide range of pH and temperature, the recovery of DS was still above 90% after 16 adsorption–regeneration cycles, demonstrating its excellent environmental adaptability. These excellent adsorption properties may be attributed to the synergistic effects of electrostatic attraction, ion exchange, and hydrogen bonding of PS-IL-COOH with DS. All these results suggest that the adsorbent reported in the present work is outstanding and has great potential as an appealing sample pretreatment material to monitor trace pharmaceutical residuals in real food samples.

Supplementary Materials: The following supporting information can be downloaded at: https://www.mdpi.com/article/10.3390/molecules28176216/s1. Table S1. The isotherm model and its equations; Table S2. The related parameters of various adsorption isotherm models; Table S3. The recovery and enrichment factor of PS-IL-COOH at different concentrations of DS; Table S4. Analytical performance of the proposed method on milk samples; Table S5. Comparison of the extraction of DS using PS-IL-COOH and commercial adsorbents; Figure S1. (A): The TGA curves of PS-CH_2Cl (a) and PS-IL-COOH (b); (B): FT-TR spectra of PS-CH_2Cl (a) and PS-IL-COOH (b); Figure S2. Zeta potential of PS-IL-COOH; Figure S3. UV-visible adsorption spectra of original DS (a) and the recovered DS (b); Figure S4. FT-IR spectra of the fresh PS-IL-COOH (a) and the regenerated PS-IL-COOH (b); Figure S5. FT-IR spectra of the PS-IL-COOH (a), PS-IL-COOH-DS (b), and DS (c); Figure S6. The photographs of different blank solutions and the adsorbed DS solution after addition of aqueous $AgNO_3$: (a) deionized water, (b) aqueous DS solution, (c) supernatant after soaking of PS-IL-COOH, and (d) supernatant after DS was adsorbed by PS-IL-COOH. Experimental conditions: C_{DS} = 100 mg/L; V_{DS} = 5 mL; $m_{PS\text{-}IL\text{-}COOH}$ = 10 mg; time = 30 min; pH $\approx$ 6; room temperature.

Author Contributions: H.Y.: investigation, methodology, formal analysis, data curation, visualization, writing—original Draft, writing—review and editing; C.W.: formal analysis; W.Z.: methodology; P.J.: methodology, formal analysis; F.L.: methodology, formal analysis; J.F.: conceptualization, methodology, validation, writing—review and editing, visualization, supervision, funding acquisition. All authors have read and agreed to the published version of the manuscript.

Funding: This work was financially supported by funding from the National Natural Science Foundation of China (No. 21777038).

Institutional Review Board Statement: Not applicable.

Informed Consent Statement: Not applicable.

Data Availability Statement: This study presents novel concepts and did not report any data.

Conflicts of Interest: The authors declare no conflict of interest.

Sample Availability: Samples of the compounds are available from the authors.

References

1. Jedziniak, P.; Szprengier-Juszkiewicz, T.; Pietruk, K.; Sledzinska, E.; Zmudzki, J. Determination of non-steroidal anti-inflammatory drugs and their metabolites in milk by liquid chromatography-tandem mass spectrometry. *Anal. Bioanal. Chem.* **2012**, *403*, 2955–2963. [CrossRef] [PubMed]
2. Wang, Y.; Ou, Y.; Xie, S.; Chen, D.; Wang, X.; Pan, Y.; Wang, Y.; Huang, L.; Cheng, G.; Qu, W.; et al. Magnetic Graphene Solid-Phase Extraction for the Determination of 47 Kinds of Non-steroidal Anti-inflammatory Drug Residues in Animal Food with Liquid Chromatography Tandem Mass Spectrometry. *Food Anal. Methods* **2019**, *12*, 1346–1368. [CrossRef]
3. Oral, O.; Kantar, C. Diclofenac removal by pyrite-Fenton process: Performance in batch and fixed-bed continuous flow systems. *Sci. Total Environ.* **2019**, *664*, 817–823. [CrossRef]
4. Liu, D.; Wang, J.; Zhou, J.; Xi, Q.; Li, X.; Nie, E.; Piao, X.; Sun, Z. Fabricating I doped TiO_2 photoelectrode for the degradation of diclofenac: Performance and mechanism study. *Chem. Eng. J.* **2019**, *369*, 968–978. [CrossRef]
5. Dowling, G.; Malone, E.; Harbison, T.; Martin, S. Analytical strategy for the determination of non-steroidal anti-inflammatory drugs in plasma and improved analytical strategy for the determination of authorized and non-authorized non-steroidal anti-inflammatory drugs in milk by LC-MS/MS. *Food Addit. Contam.* **2010**, *27*, 962–982. [CrossRef]
6. Zhou, Y.; Xu, J.; Lu, N.; Wu, X.; Zhang, Y.; Hou, X. Development and application of metal-organic framework@GA based on solid-phase extraction coupling with UPLC-MS/MS for the determination of five NSAIDs in water. *Talanta* **2021**, *225*, 121846. [CrossRef] [PubMed]
7. Gao, Y.; Wang, S.; Zhang, N.; Xu, X.; Bao, T. Novel solid-phase extraction filter based on a zirconium meta-organic framework for determination of non-steroidal anti-inflammatory drugs residues. *J. Chromatogr. A* **2021**, *1652*, 462349. [CrossRef]
8. Shishov, A.; Gerasimov, A.; Nechaeva, D.; Volodina, N.; Bessonova, E.; Bulatov, A. An effervescence-assisted dispersive liquid–liquid microextraction based on deep eutectic solvent decomposition: Determination of ketoprofen and diclofenac in liver. *Microchem. J.* **2020**, *156*, 104837. [CrossRef]
9. Zhang, N.; Gao, Y.; Xu, X.; Bao, T.; Wang, S. Hydrophilic carboxyl supported immobilization of UiO-66 for novel bar sorptive extraction of non-steroidal anti-inflammatory drugs in food samples. *Food Chem.* **2021**, *355*, 129623. [CrossRef]
10. Liang, X.; Zou, Y.; Liu, S.; Chen, C.; Wang, J.; Hu, H.; Yao, S. Facile and robust dual interaction modification of hexadecyldimethyl amine magnetic nanoparticles for the ultrasensitive analysis of perfluorinated compounds in environmental water. *J. Sep. Sci.* **2015**, *38*, 1394–1401. [CrossRef]
11. Qin, S.B.; Fan, Y.H.; Li, X.S.; Zhang, Y.; Qi, S.H. Rapid preparation of methyltrimethoxy-modified magnetic mesoporous silica as an effective solid-phase extraction adsorbent. *J. Sep. Sci.* **2018**, *41*, 669–677. [CrossRef] [PubMed]
12. Wang, M.; Han, Y.; Qiao, F.; Yan, H. Improved Solid-Phase Extraction for Simple, Sensitive, and Efficient Determination of Trace Plant Growth Regulators in Cherry Tomatoes by High-Performance Liquid Chromatography. *J. Agric. Food Chem.* **2020**, *68*, 8447–8454. [CrossRef] [PubMed]
13. Soheili-Azad, P.; Yaftian, M.R.; Dorraji, M.S.S. Application of zinc/aluminum layered double hydroxide nanosorbent in a fixed-bed column for SPE-preconcentration followed by HPLC determination of diclofenac in biological and hospital wastewater samples. *Microchem. J.* **2019**, *148*, 270–276. [CrossRef]
14. Prasetya, N.; Li, K. MOF-808 and its hollow fibre adsorbents for efficient diclofenac removal. *Chem. Eng. J.* **2021**, *417*, 129216. [CrossRef]
15. Huang, L.; Shen, R.; Liu, R.; Xu, S.; Shuai, Q. Facile fabrication of magnetic covalent organic frameworks for magnetic solid-phase extraction of diclofenac sodium in milk. *Food Chem.* **2021**, *347*, 129002. [CrossRef]
16. Dai, C.; Zhang, J.; Huang, C.; Lei, Z. Ionic Liquids in Selective Oxidation: Catalysts and Solvents. *Chem. Rev.* **2017**, *117*, 6929–6983. [CrossRef]
17. Cheng, M.; Jiang, J.; Wang, J.; Fan, J. Highly Salt Resistant Polymer Supported Ionic Liquid Adsorbent for Ultrahigh Capacity Removal of p-Nitrophenol from Water. *ACS Sustain. Chem. Eng.* **2019**, *7*, 8195–8205. [CrossRef]
18. Lu, Y.; Li, P.; Yan, H.; Shen, S. Ionic Liquid Modified Porous Polymer as a Dispersive Filter Extraction Adsorbent for Simple, Sensitive, and Efficient Determination of Chlorotriazine Herbicides in Irrigation Water. *J. Agric. Food Chem.* **2022**, *70*, 1327–1334. [CrossRef]
19. Wang, Y.; Yang, C.; Sun, Y.; Qiu, F.; Xiang, Y.; Fu, G. Synthesis of thermo-responsive bovine hemoglobin imprinted nanoparticles by combining ionic liquid immobilization with aqueous precipitation polymerization. *J. Sep. Sci.* **2018**, *41*, 765–773. [CrossRef]
20. Durga, G.; Kalra, P.; Kumar Verma, V.; Wangdi, K.; Mishra, A. Ionic liquids: From a solvent for polymeric reactions to the monomers for poly(ionic liquids). *J. Mol. Liq.* **2021**, *335*, 116540. [CrossRef]
21. Valverde, D.; Porcar, R.; Lozano, P.; García-Verdugo, E.; Luis, S.V. Multifunctional Polymers Based on Ionic Liquid and Rose Bengal Fragments for the Conversion of CO_2 to Carbonates. *ACS Sustain. Chem. Eng.* **2021**, *9*, 2309–2318. [CrossRef]
22. Zhang, H.; Wu, X.; Yuan, Y.; Han, D.; Qiao, F.; Yan, H. An ionic liquid functionalized graphene adsorbent with multiple adsorption mechanisms for pipette-tip solid-phase extraction of auxins in soybean sprouts. *Food Chem.* **2018**, *265*, 290–297. [CrossRef] [PubMed]
23. Guo, Y.; Huang, W.; Chen, B.; Zhao, Y.; Liu, D.; Sun, Y.; Gong, B. Removal of tetracycline from aqueous solution by MCM-41-zeolite A loaded nano zero valent iron: Synthesis, characteristic, adsorption performance and mechanism. *J. Hazard. Mater.* **2017**, *339*, 22–32. [CrossRef] [PubMed]

24. Fu, J.; Chen, Z.; Wang, M.; Liu, S.; Zhang, J.; Zhang, J.; Han, R.; Xu, Q. Adsorption of methylene blue by a high-efficiency adsorbent (polydopamine microspheres): Kinetics, isotherm, thermodynamics and mechanism analysis. *Chem. Eng. J.* **2015**, *259*, 53–61. [CrossRef]
25. Goyne, K.W.; Chorover, J.; Kubicki, J.D.; Zimmerman, A.R.; Brantley, S.L. Sorption of the antibiotic ofloxacin to mesoporous and nonporous alumina and silica. *J. Colloid Interface Sci.* **2005**, *283*, 160–170. [CrossRef]
26. Zheng, X.; He, L.; Duan, Y.; Jiang, X.; Xiang, G.; Zhao, W.; Zhang, S. Poly(ionic liquid) immobilized magnetic nanoparticles as new adsorbent for extraction and enrichment of organophosphorus pesticides from tea drinks. *J. Chromatogr. A* **2014**, *1358*, 39–45. [CrossRef]
27. Zhao, G.; Huang, X.; Tang, Z.; Huang, Q.; Niu, F.; Wang, X. Polymer-based nanocomposites for heavy metal ions removal from aqueous solution: A review. *Polym. Chem.* **2018**, *9*, 3562–3582. [CrossRef]
28. Zhang, K.G.R.; Wang, Y.; Nie, Q.; Zhu, G. One-step derivatization and temperature-controlled vortex-assisted liquid-liquid microextraction based on the solidification of floating deep eutectic solvents coupled to UV–Vis spectrophotometry for the rapid determination of total iron in water and food samples. *Food Chem.* **2022**, *384*, 132414. [CrossRef]
29. Xu, H.; Zhu, S.; Xia, M.; Wang, F. Rapid and efficient removal of diclofenac sodium from aqueous solution via ternary core-shell CS@PANI@LDH composite: Experimental and adsorption mechanism study. *J. Hazard. Mater.* **2021**, *402*, 123815. [CrossRef]
30. Li, S.; Cui, J.; Wu, X.; Zhang, X.; Hu, Q.; Hou, X. Rapid in situ microwave synthesis of Fe_3O_4@MIL-100(Fe) for aqueous diclofenac sodium removal through integrated adsorption and photodegradation. *J. Hazard. Mater.* **2019**, *373*, 408–416. [CrossRef]
31. Godiya, C.B.; Kumar, S.; Xiao, Y. Amine functionalized egg albumin hydrogel with enhanced adsorption potential for diclofenac sodium in water. *J. Hazard. Mater.* **2020**, *393*, 122417. [CrossRef] [PubMed]
32. Hu, D.; Huang, H.; Jiang, R.; Wang, N.; Xu, H.; Wang, Y.G.; Ouyang, X.K. Adsorption of diclofenac sodium on bilayer amino-functionalized cellulose nanocrystals/chitosan composite. *J. Hazard. Mater.* **2019**, *369*, 483–493. [CrossRef] [PubMed]
33. Liang, X.X.; Omer, A.M.; Hu, Z.H.; Wang, Y.G.; Yu, D.; Ouyang, X.K. Efficient adsorption of diclofenac sodium from aqueous solutions using magnetic amine-functionalized chitosan. *Chemosphere* **2019**, *217*, 270–278. [CrossRef] [PubMed]
34. Xiong, T.; Yuan, X.; Wang, H.; Wu, Z.; Jiang, L.; Leng, L.; Xi, K.; Cao, X.; Zeng, G. Highly efficient removal of diclofenac sodium from medical wastewater by Mg/Al layered double hydroxide-poly(m-phenylenediamine) composite. *Chem. Eng. J.* **2019**, *366*, 83–91. [CrossRef]
35. Zhao, R.; Wang, Y.; Li, X.; Sun, B.; Wang, C. Synthesis of beta-Cyclodextrin-Based Electrospun Nanofiber Membranes for Highly Efficient Adsorption and Separation of Methylene Blue. *ACS Appl. Mater. Interfaces* **2015**, *7*, 26649–26657. [CrossRef]
36. Ye, Z.; Huang, Y.; Luo, Q.; Wang, L.; Huang, X. Preparation of highly fluorinated and boron-rich adsorbent for magnetic solid-phase extraction of fluoroquinolones in water and milk samples. *J. Chromatogr. A* **2019**, *1601*, 86–94. [CrossRef]

molecules

MDPI

Article

Construction of Ketoenamine-Based Covalent Organic Frameworks with Electron-Rich Sites for Efficient and Rapid Removal of Iodine from Solution

Qi Tao [1,*], Xiao Zhang [2], Liping Jing [3,*], Lu Sun [4] and Peipei Dang [5]

1 College of Energy and Electrical Engineering, Hohai University, Nanjing 211100, China
2 Key Laboratory of Applied Surface and Colloid Chemistry Ministry of Education (MOE), School of Chemistry and Chemical Engineering, Shaanxi Normal University, Xi'an 710062, China
3 College of Chemistry and Chemical Engineering, Qiqihar University, Qiqihar 161006, China
4 Department of Chemistry, Northeast Normal University, Changchun 130024, China
5 State Key Laboratory of Rare Earth Resource Utilization, Changchun Institute of Applied Chemistry, Chinese Academy of Sciences, Changchun 130022, China
* Correspondence: 20220924@hhu.edu.cn (Q.T.); 03707@qqhru.edu.cn (L.J.)

Abstract: Porous covalent organic frameworks (COFs) have been widely used for the efficient removal of iodine from solution due to their abundance of electron-rich sites. In this study, two kinds of ketoenamine-based COFs, TpBD-$(OMe)_2$ and TpBD-Me_2, are successfully synthesized via Schiff base reaction under solvothermal conditions using 1, 3, 5-triformylphoroglucinol as aldehyde monomer, o-tolidine and o-dianisidine as amino monomers. The ability of TpBD-$(OMe)_2$ and TpBD-Me_2 to adsorb iodine in cyclohexane or aqueous solutions has been quantitatively analyzed and interpreted in terms of adsorption sites. TpBD-Me_2 possesses two adsorption sites, -NH- and -C=O, and exhibits an adsorption capacity of 681.67 mg/g in cyclohexane, with an initial adsorption rate of 0.6 g/mol/min with respect to COF unit cell. The adsorption capacity of TpBD-$(OMe)_2$ can be as high as 728.77 mg/g, and the initial adsorption rate of TpBD-$(OMe)_2$ can reach 1.2 g/mol/min in the presence of oxygen atoms between the methyl group and the benzene ring. Compared with TpBD-Me_2, the higher adsorption capacity and adsorption rate of TpBD-$(OMe)_2$ towards iodine are not only reflected in organic solvents, but also in aqueous solutions. It is proven through X-ray photoelectron spectroscopy and Raman spectroscopy that iodine exists in the form of I_2, I_3^-, and I_5^- within TpBD-$(OMe)_2$ and TpBD-Me_2 after adsorption. This work not only expands the application of COFs in the field of iodine adsorption, but also provides research ideas and important an experimental basis for the optimization of iodine adsorption sites.

Keywords: ketoenamine-based; covalent organic frameworks; adsorption; iodine

1. Introduction

The rapid development of the nuclear industry has gravely harmed the environment and posed a hazard to human health due to nuclear waste, making its removal an urgent issue that needs to be addressed [1–3]. The nuclear reaction process generates the radioactive isotopes ^{129}I and ^{131}I, ^{129}I having a significantly longer half-life [2,4]. Iodine isotopes entering the human body without the right safeguards in place may lead to serious health conditions, such as acute radiation sickness and tumor formation [5,6]. In addition, iodine leakage into water bodies in the form of ions causes severe environmental pollution, while its solubility in organic solvents provides a considerable obstacle to industrial production and solvent recycling [7,8]. Therefore, the efficient removal of iodine from solution has emerged as a matter of widespread concern.

The commonly used iodine adsorbent consists of activated carbon impregnated with triethylenediamine (TEDA) or KI [9–11]. However, the accumulation of TEDA or KI within the adsorbents could lead to pore blockage in the activated carbon, thereby reducing their adsorption capacity over time [12]. Ag-based adsorbents include Ag-exchanged zeolites [13,14], titanosilicates [15], silver-doped porous oxide materials [16–18], silver-functionalized silica aerogels [19,20], and so on. The adsorption of iodine by silver-based adsorbents results in the formation of stable AgI [21,22], which exhibits low desorption efficiency when using suitable solvents for iodine, thereby limiting the recyclability of Ag-based adsorbents. The functional macroreticular resin exhibits a high specific surface area and exceptional stability [23,24]. Moreover, the amino, amide, and cationic components grafted onto the resin effectively enhance its adsorption capacity for iodine. Importantly, the absorbed iodine can be desorbed using an appropriate solvent, enabling the reuse of the resin material [24,25]. Newly developed metal organic frameworks (MOFs) [26–29] and porous organic polymers (POPs) [2,30–33] have gained significant attention due to their exceptional designability as monomers, high specific surface area, adjustable pore size, and pore volume, making them highly suitable for iodine adsorption in both vapor and solution systems. Covalent organic frameworks (COFs) connected by robust covalent bonds using multidentate organic building units emerge as significant adsorbents for iodine in solution owing to their excellent crystallinity, high specific surface area, one-dimensional open pores, low density, and exceptional thermal stability [34,35].

Considering the facile modifiability of COF monomers, electron-rich non-metallic heteroatoms, such as P, N, O, and S, were commonly incorporated into the skeleton of COFs. The incorporation of these elements effectively enhances the adsorption capacity of COFs towards iodine by leveraging the interaction between lone-pair electrons and iodine molecules [36–40]. The enol form of COFs, derived from the Schiff base reaction between 1, 3, 5-triformylphoroglucinol (Tp) and binary amino monomers, can undergo a conversion into a more thermodynamically stable ketone form, resulting in the formation of ketoenamine-based COFs [34]. The oxygen atom, as an electron-rich atom, is typically found in ketoenamine-based COFs in the form of a carbonyl or ether bond, which serve as important binding sites for iodine molecules. Since the oxygen in the ether bond possesses more lone electron pairs, its role in iodine adsorption cannot be overlooked. Current studies have primarily focused on investigating the influence of COFs' pore size and specific surface area on their iodine adsorption performance [41], while there is a lack of clear experimental result to elucidate the impact of electron-rich sites on the adsorption performance. In this study, two types of COFs, namely TpBD-Me_2 and TpBD-$(OMe)_2$, were synthesized through the Schiff base reaction between Tp and o-tolidine (BD-Me_2) and o-dianisidine (BD-$(OMe)_2$), respectively. These COFs exhibit excellent crystallinity, high specific surface area, and remarkable thermal stability. We investigated their adsorption capacity and rate towards iodine in solution. It was observed that the presence of ether oxygen atoms in COFs led to enhanced adsorption capacity and faster adsorption kinetics for iodine uptake. This study provides new insights into understanding how TpBD-Me_2 and TpBD-$(OMe)_2$ interact with iodine at an atomic level while offering two novel materials for the efficient removal of iodine from solutions.

2. Results and Discussion

2.1. The Structure of TpBD-Me_2 and TpBD-$(OMe)_2$

To study the iodine adsorption performance of COFs from the perspective of adsorption sites, it is necessary to select suitable monomers to construct COFs in order to facilitate comparing the effects of differences in molecular structure on iodine adsorption. The structural differences between the already-reported TpBD-Me_2 and TpBD-$(OMe)_2$, which only exist in the oxygen atoms, allow us to simultaneously reveal the influence of both the ketoenamine-based skeletons and the ether oxygen atoms on iodine adsorption properties. The synthesis of TpBD-Me_2 and TpBD-$(OMe)_2$ was achieved by reacting Tp with BD-Me_2 and BD-$(OMe)_2$ (Figure 1a), respectively, under solvothermal conditions, as previously reported in the literature [34,38]. The molar ratio of Tp to BD-Me_2 or BD-$(OMe)_2$ in the reaction was 2:3, indicating that two molecules of Tp reacted with three molecules of BD-Me_2

or BD-(OMe)$_2$ to form a COF unit cell. TpBD-Me$_2$ possesses two electron-rich functional groups consisting of secondary amine (-NH-) and carbyl oxygen (C=O), which serve as binding sites for iodine adsorption in solution. The incorporation of a methoxy component into the COF structure for TpBD-(OMe)$_2$ additionally introduces ether oxygen sites (-O-), which could effectively enhance its adsorption capacity and rate of iodine (Scheme 1).

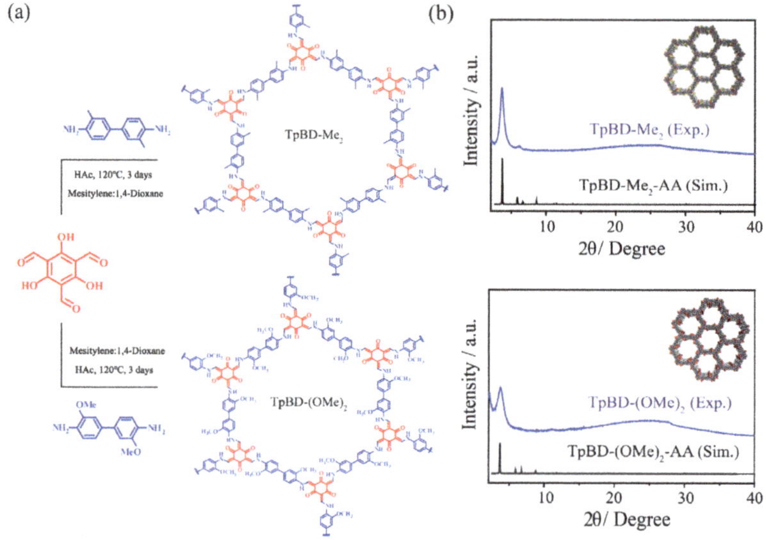

Figure 1. (**a**) Synthetic conditions and molecular structure of TpBD-Me$_2$ and TpBD-(OMe)$_2$; (**b**) PXRD results obtained from the experiments and simulations of TpBD-Me$_2$ (**top**) and TpBD-(OMe)$_2$ (**bottom**). Inset: AA stacking structural models of TpBD-Me$_2$ and TpBD-(OMe)$_2$.

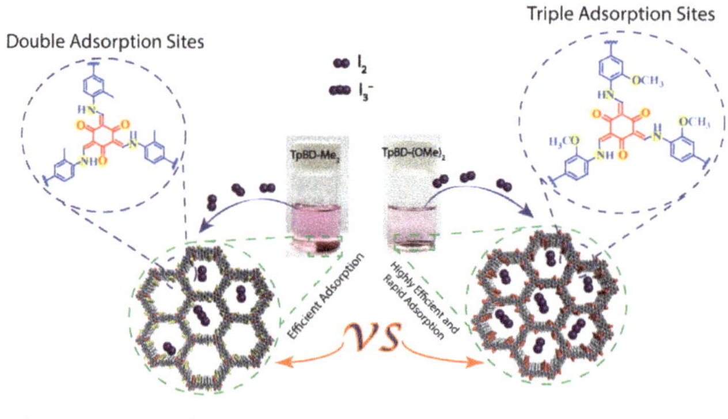

Scheme 1. Two kinds of ketoenamine-based covalent organic frameworks (COFs) with electron-rich sites, namely TpBD-Me$_2$ (double adsorption sites) and TpBD-(OMe)$_2$ (triple adsorption sites) for iodine adsorption in solution.

Before studying the iodine absorption property, we characterized the as-synthesized TpBD-Me$_2$ and TpBD-(OMe)$_2$ and compared the results with those in the literature to confirm their successful preparation. As shown in Figure 1b, the experimental powder X-ray diffraction (PXRD) results of TpBD-Me$_2$ and TpBD-(OMe)$_2$ are consistent with the theoretical simulations based on the AA stacking model using the Materials Studio 2020 software package, which can prove that they have good crystallinity. Fourier transform infrared spectroscopy (FT-IR) was employed for the characterization of the functional groups present in as-prepared TpBD-Me$_2$ and TpBD-(OMe)$_2$ (Figure S1). We carried out ^{13}C solid-state nuclear magnetic resonance (NMR) spectroscopy to verify the atomic-level construction of as-prepared COFs (Figure S2). The thermal stability of the two types of COFs were assessed via thermogravimetric analysis in ambient air, revealing that TpBD-Me_2 and TpBD-$(OMe)_2$ exhibited notable thermal stability up to 300 °C without any discernible mass loss, which is comparable to that of other reported imine-based COFs [34,42] (Figure S3). The X-ray photoelectron spectroscopy (XPS) analysis of both types of COFs reveals the presence of C, N, and O elements (Figure S4, Table S1). The high-resolution XPS analysis of C element in TpBD-Me_2 and TpBD-$(OMe)_2$ reveals that the C1s signals can be categorized into three distinct carbon environments (Figure S5). The surface irregular flocculation morphology of TpBD-Me_2 and TpBD-$(OMe)_2$ was observed via scanning electron microscopy (SEM) and transmission electron microscopy (TEM) (Figures S6 and S7). In addition, it was demonstrated using PXRD that TpBD-Me_2 and TpBD-$(OMe)_2$ remained well stabilised in different solvents (Figure S8).

According to nitrogen adsorption-to-desorption isotherms (Figure S9a), both COFs exhibit a typical type I reversible isotherm, indicating significant nitrogen adsorption at low relative pressure and the existence of micropores in these materials. The N_2 desorption curve of TpBD-$(OMe)_2$ shows a slight hysteresis compared to TpBD-Me_2, suggesting the presence of narrowly distributed mesopores. The Brunauer–Emmet–Teller (BET) plots to calculate surface area are shown in Figure S10. The surface area of TpBD-Me_2 is 942.81 g/m^2, surpassing that of TpBD-$(OMe)_2$ (539.30 g/m^2) (Table S2). The surface area of TpBD-Me_2 is approximately twice as high as that of TpBD-$(OMe)_2$, because integrating the ether oxygen atoms in the pore walls of the COFs further blocks the pores. The primary pore size of TpBD-Me_2 is 1.83 nm, while that for TpBD-$(OMe)_2$ is 1.88 nm, with a discernible presence of mesoporous regions (Figure S9b), which is consistent with the N_2 desorption profile. Considering that the pore size of both COFs exceeds the molecular diameter of I_2 (0.53 nm), the removal of I_2 in solutions by TpBD-Me_2 and TpBD-$(OMe)_2$ could be attributed to chemisorption, while the influence of pore size on adsorption can be disregarded.

2.2. Adsorption of Iodine by TPBD-Me_2 and TPBD-$(OMe)_2$ in Cyclohexane

To facilitate the separation and collection process, TpBD-Me_2 or TpBD-$(OMe)_2$ was compressed into a flaky shape for efficient iodine adsorption, and their morphologies are shown in Figure S11. The PXRD of the COFs in flake form is consistent with that of powders, indicating that they still maintain their original ordered structure (Figure S12). The thermogravimetric analysis of the flaky TpBD-Me and TpBD-$(OMe)_2$ reveals that their decomposition temperatures remain above 300 °C, indicating that the applied pressure has not compromised their skeleton structure (Figure S13). The flecks of TpBD-Me_2 and TpBD-$(OMe)_2$ were immersed in a 100 mg/L iodine cyclohexane solution, respectively. Over time, the color intensity of the solutions gradually decreased. Notably, after 48 h of adsorption, it is evident that TpBD-$(OMe)_2$ exhibits a significantly higher iodine absorption capacity compared to TpBD-Me_2 (Figure 2a,b). UV-vis spectroscopy was used to monitor the concentration of iodine in cyclohexane to quantify the amount of iodine absorbed by COFs (Figure S14). A standard equation, based on UV-vis absorption values at various iodine concentrations, was employed to determine the concentration of iodine in cyclohexane solution during COF-mediated iodine uptake (Figure S15). The time-dependent change in adsorption capacity per unit mass (q_t) of 10 mg TpBD-Me_2 and TpBD-$(OMe)_2$ in a 100mg/L iodine cyclohexane solution is depicted in Figure 2c (Table S3). Adsorption equilibrium is achieved within approximately 30 h, with the adsorption capacity of TpBD-$(OMe)_2$ for iodine being significantly higher compared to that of TpBD-Me_2. The iodine absorption capacity of TpBD-$(OMe)_2$ with respect to unit cell (q_t^{uc}) is significantly higher than that of TpBD-Me_2, suggesting the crucial role played by the ether oxygen atoms in the process of iodine adsorption (Table S3). Based on the molar ratio of iodine molecule to COF unit cell (q_t^{Mol}), TpBD-$(OMe)_2$ exhibits an additional adsorption capacity of 0.8 mol I_2 per unit cell compared to TpBD-Me_2 at equilibrium. The maximum adsorption capacities (q_e and q_e^{uc}) at equilibrium were determined by simulating the values of q_t and q_t^{uc} using both pseudo-first-order and pseudo-second-order kinetic models (Tables S4 and S5). It is observed that the temporal changes in q_t and q_t^{uc} were better described by the pseudo-second-order kinetic model, suggesting a multi-site adsorption mechanism for iodine on the two COF materials, with charge transfer occurring between iodine molecules and adsorption sites. By calculating the adsorption rate of COFs of iodine ($\Delta q_t^{uc}/\Delta t$), it is evident that the initial adsorption rate of TpBD-(OMe)$_2$ (1.2 g/mol/min) exhibits a twofold increase compared to that of TpBD-Me$_2$ (0.6 g/mol/min), while maintaining a consistent adsorption rate even after 100 min (Figure 2d).

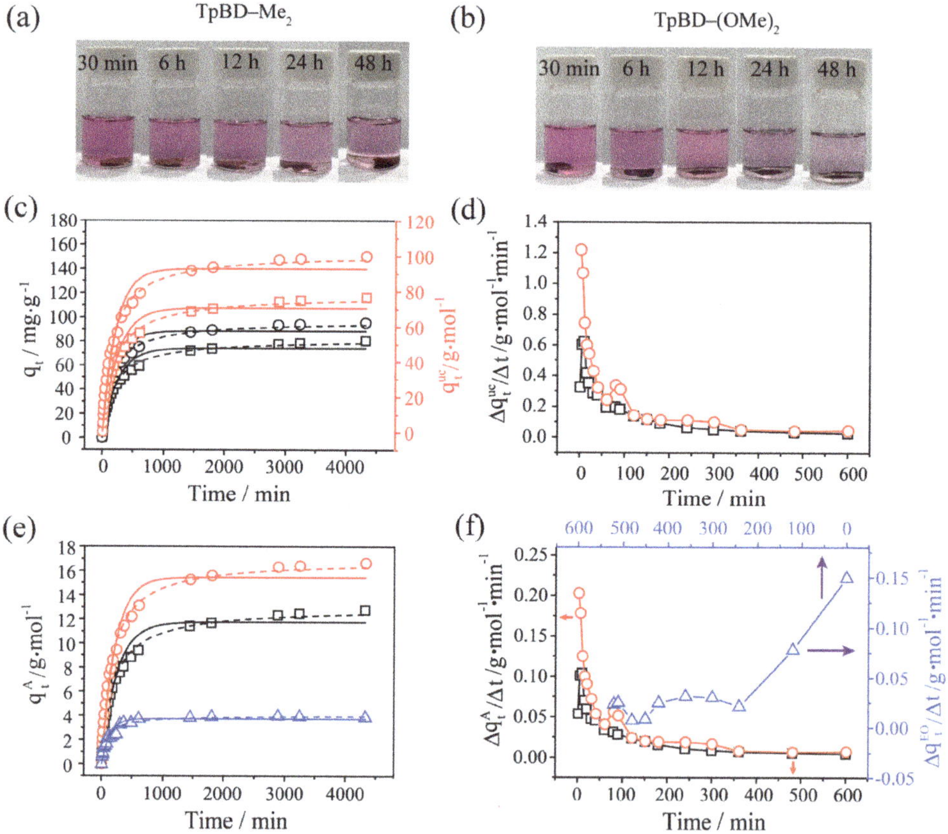

Figure 2. (**a**,**b**) The photographs of the process of I_2 uptake after 10 mg TpBD-Me$_2$ (**a**) and TpBD-(OMe)$_2$ (**b**) were immersed in 10 mL of cyclohexane solution of I_2 at the initial concentration of 100 mg·L^{-1}; the incubation times are marked on the photos. (**c**) The values of q_t (black) and q_t^{uc} (red) were plotted against time for the adsorption kinetics of TpBD-Me$_2$ (square) and TpBD-(OMe)$_2$ (circle), which were fitted using pseudo-first-order (solid curves) and pseudo-second-order (dash curves) kinetic models. (**d**) Plots of the values of $\Delta q_t^{uc}/\Delta t$ for TpBD-Me$_2$ (black square) and TpBD-(OMe)$_2$ (red circle) in cyclohexane solution versus adsorption time. (**e**) Plots of the values of q_t^{N+CO} (black square) for TpBD-Me$_2$, q_t^{N+O} (red circle) and q_t^{EO} (blue triangle) for TpBD-(OMe)$_2$ in cyclohexane solution versus adsorption time, which were fitted by pseudo-first-order (solid curves) and pseudo-second-order (dash curves) kinetic models. (**f**) Plots of the values of $\Delta q_t^{N+CO}/\Delta t$ (black square) for TpBD-Me$_2$, $\Delta q_t^{N+O}/\Delta t$ (red circle), and $\Delta q_t^{EO}/\Delta t$ (blue triangle) for TpBD-(OMe)$_2$ in cyclohexane solution versus adsorption time. The arrows are used to mark the horizontal and vertical coordinates of the corresponding curves.

By analyzing the iodine molecule binding sites within the crystal structure of both COFs, it is found that TpBD-Me_2 possesses two distinct binding sites for iodine molecules (-NH-(N) and -C=O(CO)), while TpBD-$(OMe)_2$ exhibits three binding sites (-NH-(N), -C=O(CO), and -O-Me (EO)). Each unit cell of the COFs contains six electron-rich sites of the same atoms for iodine adsorption. The value of q_t^{uc} is divided by six to obtain the mass of adsorbed iodine per mole of adsorption sites (q_t^A) (Figure 2e). For TpBD-Me_2, the adsorption capacity per mole of iodine binding site is represented by q_t^{N+CO}, whereas for TpBD-$(OMe)_2$, it is denoted as q_t^{N+O}. By calculating the difference between q_t^{N+O} and q_t^{N+CO}, the quantity of iodine adsorbed by each mole of oxygen atom in the ether bond (q_t^{EO}) could be determined (Table S6). According to experimental results and kinetic fitting, the maximum value of q_e^{N+O} is greater than that of q_e^{N+CO}, indicating that the oxygen atoms in the methoxyl group play a crucial role in the iodine adsorption process (Table S7). The kinetic fitting results of q_t^{N+O} and q_t^{N+CO} are in better agreement with the pseudo-second-order kinetic model, providing further evidence for the multi-site adsorption nature of iodine molecules on COFs. Time-dependent calculation of the value of initial adsorption rate as for adsorption sites ($\Delta q_t^A / \Delta t$) reveals that $\Delta q_t^{N+O} / \Delta t$ for TpBD-$(OMe)_2$ at the three adsorption sites surpass that of TpBD-Me_2 at the two sites, while the initial adsorption rate for oxygen atoms alone ($\Delta q_t^{EO} / \Delta t$) exceeded that of TpBD-$Me_2$ (Figure 2f). The above results demonstrate that the introduction of an oxygen atom not only enhances the iodine adsorption capacity, but also accelerates the adsorption rate. Although TpBD-Me_2 has a higher surface area, its adsorption capacity and rate of iodine are lower than those of TpBD-$(OMe)_2$, indicating that the adsorption site is the decisive factor in the process of the removal of iodine in solution.

To further assess the iodine adsorption capacity of COFs, the iodine adsorption isotherm was determined by measuring the maximum amount of iodine adsorption at various equilibrium concentrations (Table S8, Figure 3a). Here, both the adsorption capacity with respect to the unit mass (q_e) and unit cell (q_e^{uc}) of these two COFs were compared, and Langmuir and Freundlich models were used to fit them (Tables S9 and S10). The maximum value of q_e obtained by fitting the Langmuir model for TpBD-Me_2 and TpBD-$(OMe)_2$ are 681.67 mg/g and 728.77 mg/g, respectively. Additionally, the maximum values of q_e^{uc} are determined to be 646.99 g·mol^{-1} and 761.68 g·mol^{-1}, corresponding to 2.5 and 3.0 I_2 per unit cell for TpBD-Me_2 and TpBD-$(OMe)_2$. The adsorption curve was fitted using the Langmuir equation with a correlation coefficient (R^2) of 0.982 for TpBD-Me_2 and 0.921 for TpBD-$(OMe)_2$, while it was fitted using the Freundlich equation with a correlation coefficient (R^2) of 0.992 for TpBD-Me_2 and 0.998 for TpBD-$(OMe)_2$. The adsorption data for the COF materials exhibit a higher degree of conformity with the Freundlich model rather than the Langmuir model, implying that iodine adsorption takes place in multiple layers within the pores of the COF materials. The value of $1/n$ fell within the range of 0.1 to 0.5 for Freundlich model, indicating a pronounced affinity between iodine molecules and the sites within the COF channels. The cyclability of COFs was evaluated by subjecting them to three cycles of elution using ethanol to achieve the complete desorption of the iodine molecule, followed by subsequent iodine adsorption in cyclohexane. The adsorption capacity of iodine in cyclohexane for TpBD-Me_2 and TpBD-$(OMe)_2$, as shown in Figure 3b, remains unchanged compared to the freshly prepared sample after three cycles, indicating their excellent cyclability and reusability.

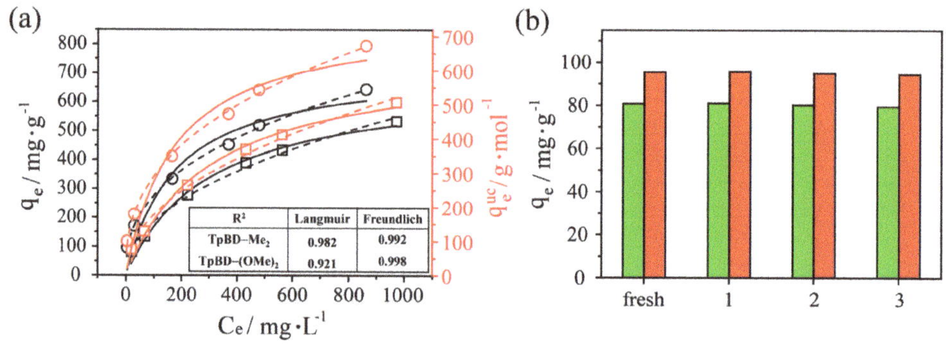

Figure 3. (**a**) Adsorption isotherms of TpBD-Me$_2$ (square) and TpBD-(OMe)$_2$ (circle) towards I_2 in cyclohexane obtained by the values of q_e (black) and q_e^{uc} (red) at different equilibrium I_2 concentration (C_e), which were fitted by Langmuir (solid curves) and Freundlich models (dashed curves). Inset: correlation coefficient of adsorption isotherms fitted by Langmuir and Freundlich models for TpBD-Me$_2$ and TpBD-(OMe)$_2$. (**b**) Recyclability of TpBD-Me$_2$ (green) and TpBD-(OMe)$_2$ (red) for I_2 adsorption in the cyclohexane solution at the initial concentration of 100 mg·L^{-1}.

The surface morphology of TpBD-Me$_2$ and TpBD-(OMe)$_2$ after iodine adsorption in cyclohexane (referred to as TpBD-Me$_2$@I$_2$ and TpBD-(OMe)$_2$@I$_2$) was observed using SEM, and their morphologies remain consistent with freshly prepared samples (Figure 4a,b). The PXRD patterns of TpBD-Me$_2$@I$_2$ and TpBD-(OMe)$_2$@I$_2$ show that they still have good crystallinity, indicating that the COF materials do not change their ordered structure after adsorption of iodine (Figure S16). The elemental distribution of TpBD-Me$_2$@I$_2$ and TpBD-(OMe)$_2$@I$_2$ was confirmed using energy dispersive spectroscopy (EDS). Notably, the variations in C, N, and O contents originate from the COFs themselves. Moreover, the higher iodine content observed in TpBD-(OMe)$_2$@I$_2$ compared to that in TpBD-Me$_2$@I$_2$ further substantiates the superior iodine absorption capacity of TpBD-(OMe)$_2$ (Figure 4a,b). Raman spectroscopy was employed for the investigation of iodine molecule presence within COF channels (Figure 4c). The vibration peaks at 108 cm^{-1} and 140 cm^{-1} correspond to the symmetric and asymmetric stretching vibrations of I_3^- ions, respectively. Additionally, a prominent peak at 165 cm^{-1} is observed, which corresponds to the stretching vibrations of I_5^- (I_2 + I_3^-) ions. The chemical state of iodine in COF material after iodine adsorption was further confirmed via X-ray photoelectron spectroscopy (XPS). The comparison of XPS spectra of TpBD-Me$_2$ and TpBD-(OMe)$_2$ before and after iodine uptake reveals that TpBD-Me$_2$@I$_2$ and TpBD-(OMe)$_2$@I$_2$ exhibit I 3d peaks at 631 eV and 620 eV, corresponding to I 3d$_{3/2}$ and I 3d$_{5/2}$, respectively, indicating the successful adsorption of iodine molecules (Figure S17). The atomic concentrations of various elements are presented in Table S11. Notably, the atomic concentration of iodine in TpBD-(OMe)$_2$@I$_2$ surpasses that of TpBD-Me$_2$@I$_2$, thereby reaffirming the superior adsorption capacity of TpBD-(OMe)$_2$ towards iodine. The high-resolution XPS spectrum for I 3d reveals two splitting peaks in I 3d$_{3/2}$ and I 3d$_{5/2}$. The peaks with binding energies at 618.84 eV and 630.86 eV correspond to I_3^-, while the peaks at 620.39 eV and 632.40 eV represent I_2 (Figure 4d). The XPS results for the analysis of chemical state of iodine in COF are consistent with those obtained from Raman spectroscopy.

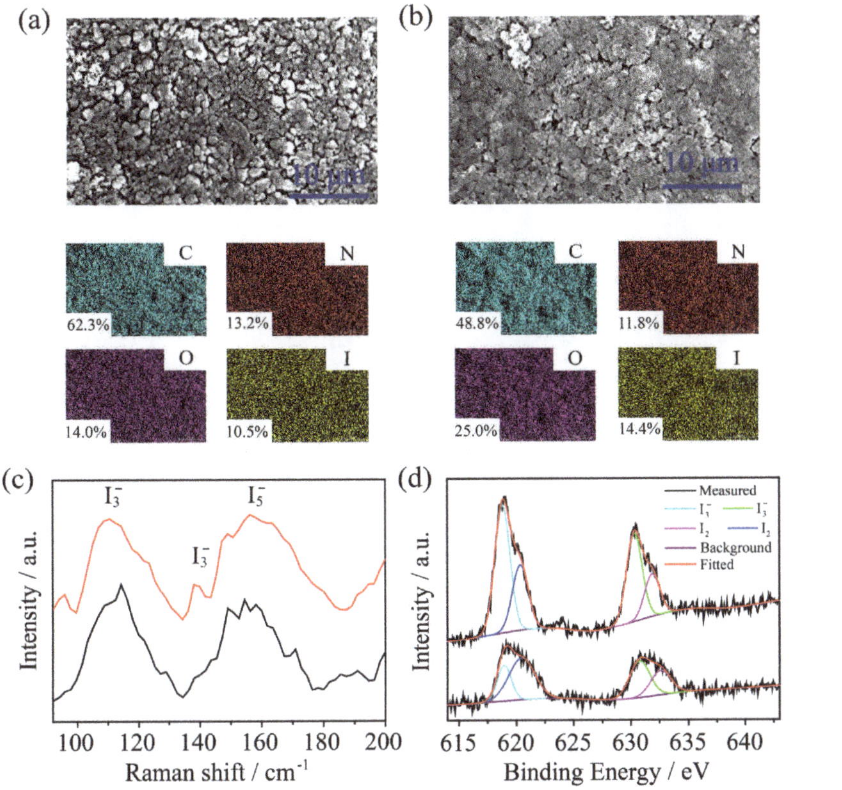

Figure 4. (**a**,**b**) SEM images and corresponding elemental mapping images of TpBD-Me$_2$@I$_2$ (**a**) and TpBD-(OMe)$_2$@I$_2$ (**b**) and atomic concentrations of different elements were marked on the photos. (**c**) Raman spectra of TpBD-Me$_2$@I$_2$ (black) and TpBD-(OMe)$_2$@I$_2$ (red) after adsorption equilibrium. (**d**) High-resolution XPS I3d spectrum of TpBD-Me$_2$@I$_2$ (bottom) andTpBD-(OMe)$_2$@I$_2$ (top). In (**a**–**d**), the I$_2$ adsorption was carried out at the initial concentration of 500 mg·L^{-1} in cyclohexane solution.

The iodine adsorption capacity of TpBD-Me$_2$ and TpBD-(OMe)$_2$ in cyclohexane is compared with the those reported in the literature for COFs [36,39,40,43–47], revealing that while their adsorption capacity does not reach the level of P-COF, they surpass most other COFs (Figure 5). The high adsorption capacity of TpBD-Me$_2$ and TpBD-(OMe)$_2$ could be attributed to their multiple binding sites for iodine molecules, as well as their large pore size and volume. Comparing TpBD-Me$_2$ and TpBD-(OMe)$_2$ with other porous materials, it is found that their iodine absorption capacity in organic solutions exceeds that of most MOFs and POPs (Table S12). Due to their inherent stability and recyclability as adsorption materials, COFs have been widely used in the field of iodine adsorption.

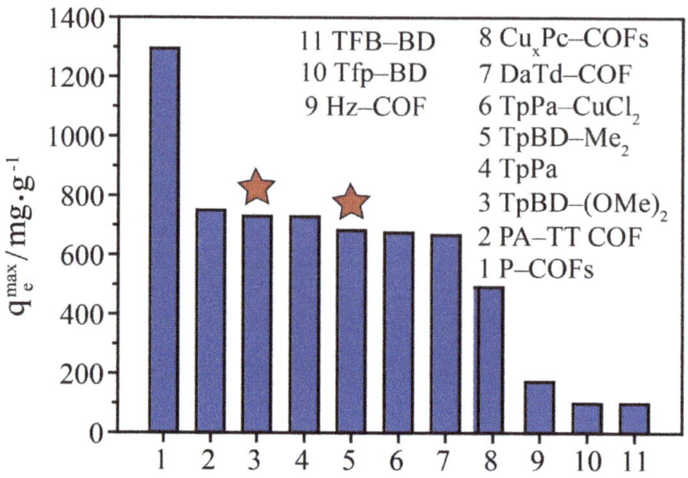

Figure 5. Comparison of maximum adsorption capacity (q_e^{max}) towards iodine for different COFs in organic solvent. Red stars are used to label TpBD-Me_2 and TpBD-$(OMe)_2$ prepared in this paper [36,39,40,43–46].

2.3. *Adsorption of* I_3^- *by TPBD-*Me_2 *and TPBD-*$(OMe)_2$ *in Aqueous Solution*

Although I_2 is poorly soluble in water, it can form I_3^- with I^- ions, which in turn dissolves in water and becomes one of the major pollutants in water bodies. To expand the application scope of COF materials for solvent-based iodine adsorption, the iodine adsorption capacity of TpBD-Me_2 and TpBD-$(OMe)_2$ in aqueous solution was further investigated here. To enhance the solubility of iodine in water, I_2 was dissolved in an aqueous solution containing KI (with a molar ratio of I_2 to I^- as 1:5) to generate I_3^- for subsequent adsorption. The flakes of TpBD-Me_2 and TpBD-$(OMe)_2$ were placed into the I_3^- solution, and it was found that the solution became basically colorless after 24 h, indicating that both COFs possessed high removal efficiency for I_3^- ions (Figure 6a,b). Depending on the concentration of iodine dissolved in the KI solution, the UV-vis spectrum of the I_3^- solution was acquired and a standard curve was employed to quantify the iodine content in the solution (Figure S18). According to the standard curve, the quantity of iodine adsorbed by COF was determined based on the UV-vis absorbance of the solution (Figure S19). The temporal evolution of adsorption capacity per unit mass of COFs (Q_t) and per COF unit cells (Q_t^{uc}) towards I_2 are depicted in Figure 6c (Table S13). The Q_t and Q_t^{uc} values of TpBD-Me_2 and TpBD-$(OMe)_2$ show a time-dependent increase, eventually reaching equilibrium. Notably, the Q_t value of TpBD-$(OMe)_2$ is slightly greater than that of TpBD-Me_2, while the Q_t^{uc} values display more pronounced differences between the two COFs. These findings suggest that TpBD-$(OMe)_2$ offers enhanced advantages in adsorbing iodine from aqueous solution. The temporal evolutions of Q_t and Q_t^{uc} values exhibit greater conformity with the pseudo-second-order kinetic model, thereby further substantiating the existence of multiple iodine molecule binding sites within COFs and highlighting the occurrence of charge sharing and transfer between iodine molecules and these binding sites (Tables S14 and S15). The time-dependent iodine adsorption rate with respect to COF unit cell ($\Delta Q_t^{uc}/\Delta t$) is presented in Figure 6d, where the initial adsorption rate of TpBD-$(OMe)_2$ is twice as high as that of TpBD-Me_2, and both rates reach equilibrium rapidly. It is noteworthy that both COFs exhibit a higher rate of iodine adsorption in water compared to cyclohexane, attributed to the poor solubility of iodine molecules in aqueous solutions.

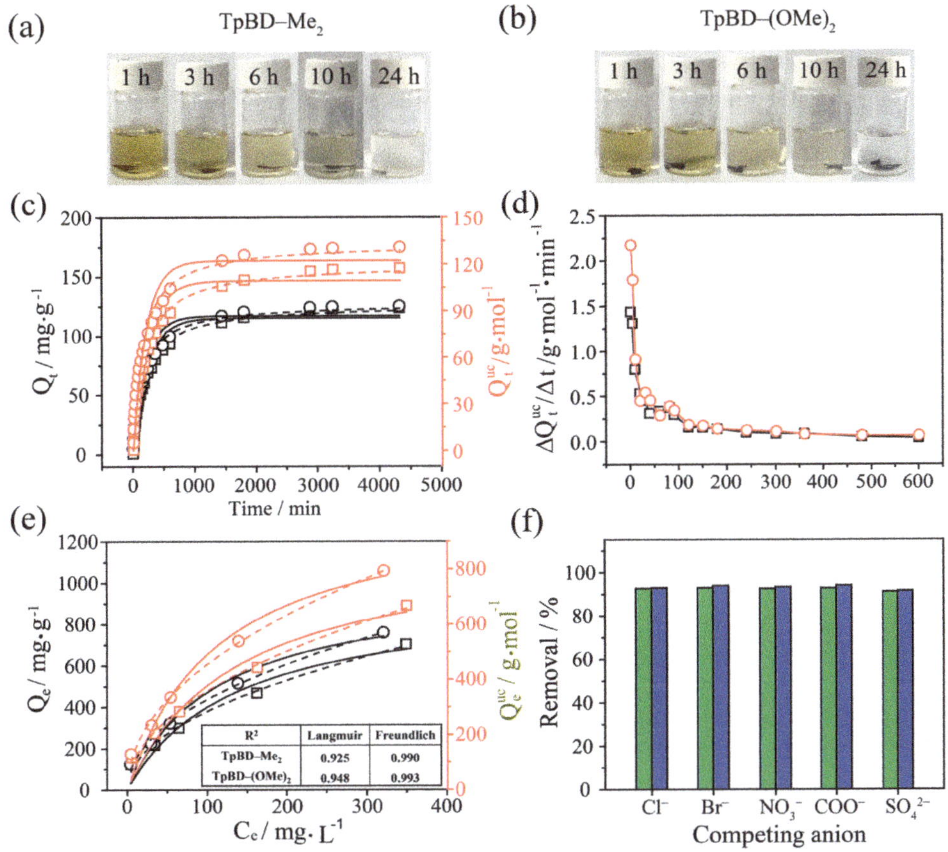

Figure 6. (**a**,**b**) The photographs of the process of I_3^- uptake after 5 mg TpBD-Me_2 (**a**) and TpBD-$(OMe)_2$ (**b**) were immersed in 10 mL of aqueous solution of I_2/KI at the initial concentration of 67 mg·L^{-1} of I_2; the incubation times were marked on the photos. (**c**) The values of Q_t (black) and Q_t^{uc} (red) were plotted against time for the adsorption kinetics of TpBD-Me_2 (square) and TpBD-$(OMe)_2$ (circle), which were fitted by pseudo-first-order (solid curves) and pseudo-second-order (dash curves) kinetics models. (**d**) Plots of the values of $\Delta Q_t^{uc}/\Delta t$ for TpBD-Me_2 (black square) and TpBD-$(OMe)_2$ (red circle) in aqueous solution versus adsorption time. (**e**) Adsorption isotherms of TpBD-Me_2 (square) and TpBD-$(OMe)_2$ (circle) towards I_2 in KI/I_2 aqueous solution obtained by the values of Q_e (black) and Q_e^{uc} (red) at different equilibrium I_2 concentration (C_e), which were fitted by Langmuir (solid curves) and Freundlich models (dashed curves). Inset: correlation coefficient of adsorption isotherms fitted by Langmuir and Freundlich models for TpBD-Me_2 and TpBD-$(OMe)_2$. (**f**) Effect of typical competing anions on the removal percentage of iodine for TpBD-Me_2 (green) and TpBD-$(OMe)_2$ (blue) at an initial I_2 concentration of 67 mg·L^{-1}, in which the concentration of competing anions is 0.1 M.

The adsorption isotherms for iodine adsorption in aqueous solution by TpBD-Me_2 and TpBD-$(OMe)_2$ were determined based on the maximum values of Q_e and Q_e^{uc} at various equilibrium concentrations (Table S16, Figure 6e). The maximum iodine uptake of TpBD-Me_2 and TpBD-$(OMe)_2$, as determined by the Langmuir model, are 939.56 mg/g and 997.13 mg/g, respectively (Table S17). These values significantly exceed the iodine uptake in cyclohexane, indicating that water, despite being considered an undesirable solvent for iodine, exhibits a more favorable propensity for iodine adsorption. The maximum values of Q_e^{uc} obtained from Langmuir model simulations for TpBD-Me_2 and TpBD-$(OMe)_2$ are 891.77 g/mol and 1042.15 g/mol, respectively, and TpBD-$(OMe)_2$ exhibits a superior adsorption capability (Table S18). Based on the correlation coefficient (R^2) obtained from the fitting analysis (Figure 6e), it is evident that the adsorption of iodine in water by COFs aligns more closely with the Freundlich model, implying a multilayered adsorption behavior of iodine molecules within the pores of COFs. Freundlich model simulations reveal that TpBD-Me_2 and TpBD-$(OMe)_2$ exhibit adsorption strengths of 1/n less than 0.5, thereby demonstrating the favorable affinity between the active adsorption sites in the COFs and iodine. Different types of ions tend to dissolve in water when COFs were employed for iodine uptake. However, the efficiency of iodine removal by the COFs in aqueous solutions is found to be unaffected by different anions, indicating that the presence of anionic species does not interfere with the iodine uptake ability of TpBD-Me_2 and TpBD-$(OMe)_2$ (Figure 6f and Figure S20).

Raman spectroscopy and X-ray photoelectron spectroscopy were employed to confirm the presence of iodine species in TpBD-Me_2 and TpBD-$(OMe)_2$ after the adsorption of I_3^- ions (designated as TpBD-Me_2@I_3^- and TpBD-$(OMe)_2$@I_3^-) in an aqueous medium. The Raman spectrum reveals that the peaks observed at 108 cm^{-1} and 140 cm^{-1} correspond to the symmetric and asymmetric stretching vibrations of I_3^-, respectively, while the peaks detected at 160 cm^{-1} are attributed to the vibrations of I_5^- (I_2 + I_3^-) (Figure 7a). The two peaks observed at 632 eV and 620 eV in the XPS of TpBD-Me_2@I_3^- and TpBD-$(OMe)_2$@I_3^- correspond to I $3d_{3/2}$ and I $3d_{5/2}$, respectively (Figure S21). High-resolution XPS is able to separate I $3d_{3/2}$ and I $3d_{5/2}$ into two characteristic peaks, respectively, with the peaks at binding energies of 618.79 eV and 630.28 eV corresponding to I_3^- and the peaks at binding energies of 620.67 eV and 632.28 eV corresponding to I_2 (Figure 7b). The atomic concentrations of TpBD-Me_2@I_3^- and TpBD-$(OMe)_2$@I_3^- indicate that TpBD-$(OMe)_2$ still exhibits a higher affinity for adsorbing I_3^- ions (Table S19). The adsorption capacities of TpBD-Me_2 and TpBD-$(OMe)_2$ exhibit variations in different solvents, while the chemical state of iodine within the COF channels remains consistent.

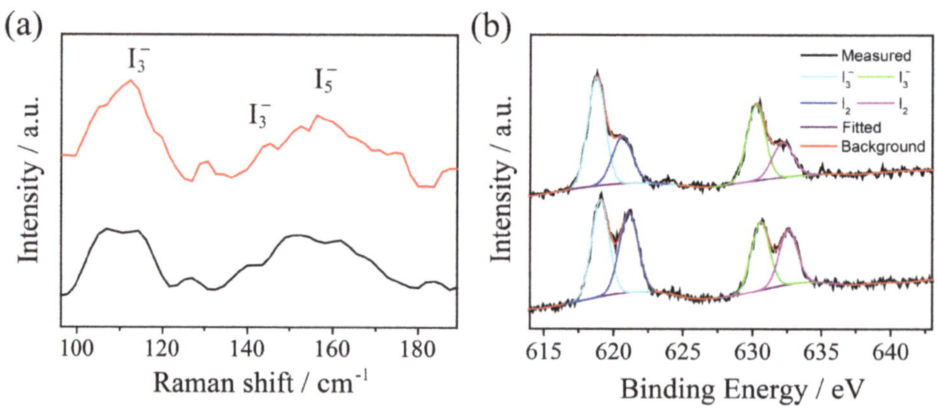

Figure 7. (**a**) Raman spectra of TpBD-Me_2@I_3^- (black) and TpBD-$(OMe)_2$@I_3^- (red) after adsorption equilibrium. (**b**) The high-resolution XPS spectrum of I 3d for TpBD-Me_2@I_3^- (bottom) and TpBD-$(OMe)_2$@I_3^- (top).

3. Materials and Methods

3.1. Materials

2,4,6-Trihydroxylbenzene-1,3,5-tricarbaldehyde (Tp), o-Tolidine (BD-Me_2), and o-Dianisidine (BD-$(OMe)_2$) were purchased from Jilin Province Yanshen Technology Co., Ltd, Changchun, China. Iodine, Mesitylene, 1,4-dioxane, NaCl, NaBr, $NaNO_3$, CH_3COONa, and Na_2SO_4 were purchased from Sigma-Aldrich, Shanghai, China. EtOH, THF, DCM, Acetone, Cyclohexane, and CH_3COOH were purchased from Beijing Chemical Industry

Group Co. Ltd., Beijing, China. All the chemicals and materials were used without further purification. Water with resistivity of 18 MΩ·cm^{-1} was prepared using a Millipore Milli-Q system and used in all experiments.

3.2. Characterization

Powder X-ray diffraction (PXRD) was performed using D8 Advance X-ray Diffractometer (Bruker AXS, Karlsruhe, Germany) using Cu-Kα radiation (λ = 1.5418 Å), operating at 40 kV and 40 mA with a scanning rate of 1°·min^{-1} (2θ). FT-IR spectra was recorded on a Bruker IFS 66v/S Fourier transform infrared spectrometer. ^{13}C NMR spectra were obtained on a Bruker Avance III model 400 MHz solid-state nuclear magnetic resonance (NMR) spectrometer at a magic angle spinning (MAS) rate of 5 kHz. X-ray photoelectron spectroscopy (XPS) was performed using a Thermo Scientific K-Alpha XPS (XPS, ESCALAB 250, Thermofisher, Waltham, U.S.) with a characteristic energy of 1486.7 eV. UV-vis absorption spectroscopy was performed by a UV-2450 spectrophotometer (HITACHI, Tokyo, Japan). Thermogravimetric analysis (TGA) was performed on a Netzch Sta 449c thermal analyzer system in the temperature range from 30 °C to 800 °C at a heating rate of 10 °C·min^{-1} in an air atmosphere. The N_2 adsorption–desorption isotherms were measured on a Quantachrome Autosorb-iQ2 analyzer. For N_2 adsorption measurements, the masses of TpBD-Me_2 and TpBD-$(OMe)_2$ used were 69.2 mg and 61.8 mg, respectively. Prior to measurements, the COFs were degassed under vacuum at 120 °C for 8 h in order to remove the gas adsorbed on the surface of the materials. Raman spectroscopy was performed on a Horiba LabRAM HR evolution instrument in the wavenumber range of 50–1000 cm^{-1} by using 633 nm laser (50 mW). The accumulation time of 600 s was applied to collect all the spectroscopic data. Scanning electron microscopy (SEM) imaging and energy dispersive spectroscopic (EDS) elementary mapping were performed on a Hitachi FE-SEM S-4800 scanning electron microscope with an acceleration voltage of 3 kV. The TEM images were observed on JEM-2100F, JEOL transmission electron microscopy.

3.3. Synthesis of TpBD-Me_2 and TpBD-$(OMe)_2$ Powders

In the typical synthesis, a Pyrex tube (o.d. × i.d. = 10 × 8 mm^2 and length 18 cm) was charged with 0.2 mM 1,3,5-triformylphloroglucinol (Tp, 42.0 mg), 0.3 mM o-tolidine (BD-Me_2, 63.7 mg), or o-dianisidine (BD-$(OMe)_2$, 73.3 mg), 1.5 mL of mesitylene, 1.5 mL of 1,4-dioxane and 0.2 mL of 6 M aqueous acetic acid. This reaction mixture was sonicated for 10 min to obtain a homogeneous dispersion. After being frozen at 77 K (liquid N_2 bath) and degassed by three freeze–pump–thaw cycles, the reaction tube was flame-sealed, reducing the total length by ca. 8.0 cm. Upon warming to room temperature, the reaction tube was then heated to 120 °C for 3 days. After cooling down to room temperature, the precipitates were isolated through filtration and washed with THF/DCM/acetone thrice. The products obtained were further purified through Soxhlet extraction using THF as the solvent for two days. After that, the products were dried in vacuo at 120 °C for 24 h to produce TpBD-Me_2 in 90.4% yield and TpBD-$(OMe)_2$ in 91.2% yield.

3.4. Study of Iodine Adsorption Property of TpBD-Me_2 and TpBD-$(OMe)_2$ in Cyclohexane

To avoid iodine sublimation, all iodine adsorption experiments in different solutions were carried out in dark brown bottles at room temperature. The absorption intensity of I_2 cyclohexane solution with different concentrations of 5, 10, 25, 50, 100, 250, 500 mg/L was measured using UV-vis absorption spectroscopy, and the absorption intensity was plotted against the concentration to obtain the standard curve of I_2 cyclohexane solution. After as-prepared TpBD-Me_2 and TpBD-$(OMe)_2$ were placed into the I_2 cyclohexane solutions, the I_2 concentrations in cyclohexane were monitored by means of UV-vis absorption spectroscopy. The I_2 concentration in cyclohexane solutions at a given adsorption time (t, min) was calculated from the standard curve. The I_2 adsorption capacity with respect to the mass of COFs (q_t, mg·g^{-1}) and molar number of COF unit cells (q_t^{uc}, g·mol^{-1}) and the ratio of the molar number of adsorbed I_2 molecules to that of COF unit cells (q_t^{mol}) were calculated using Equations (1)–(3):

$$q_t = ((C_0 - C_t)V)/W_{COF} \tag{1}$$

$$q_t^{uc} = ((C_0 - C_t)V)/(W_{COF}/M_{COF}) \tag{2}$$

$$q_t^{mol} = ((C_0 - C_t)V/M_{I_2})/(W_{COF}/M_{COF}) \tag{3}$$

where C_0 is the initial concentration of I_2 cyclohexane solution, C_t is the concentration of the solution at any time during the adsorption process, V is the volume of solution used for adsorption, W_{COF} is the mass of the COFs and M_{COF} is the molar mass of COF unit cell, which are 949.06 g/mol and 1045.06 g/mol, respectively, for TpBD-Me_2 and TpBD-$(OMe)_2$ according to the molar ratio of the reactive monomers. M_{I_2} is the molar mass of the I_2 molecule, the value of which is 253.8 g/mol.

The measured adsorption kinetics data of TpBD-Me_2 and TpBD-$(OMe)_2$ were fitted using pseudo-first-order and pseudo-second-order models, as shown in Equations (4) and (5), respectively.

$$ln(q_e - q_t) = lnq_e - k_1t \tag{4}$$

$$t/q_t = 1/k_2(q_e)^2 + t/q_e \tag{5}$$

In these equations, q_t and q_e are the adsorption capacity at a given time (t, min) and at equilibrium state. When q_t is the adsorption capacity with respect to COF unit cell (q_t^{uc}), q_e is defined as adsorption capacity per molar number of COF unit cell (q_e^{uc}), and k_1 and k_2 are the pseudo-first-order rate constant (min^{-1}) and the pseudo-second-order rate constant ($g{\cdot}mg^{-1}{\cdot}min^{-1}$), respectively.

When adsorption equilibrium is reached, adsorption capacity with respect to the mass of COFs (q_e) and the molar number of COF unit cell (q_e^{uc}), and the ratio of the molar number of adsorbed I_2 molecules to that of COF unit cells (q_e^{mol}) at different initial concentrations (C_0) can be calculated using Equations (1)–(3) according to different equilibrium concentrations (C_e). The Langmuir isotherm is used to study the adsorption process occurring on the surface of the adsorbent (Equation (6)). In this equation, K_L is the Langmuir constant, q_m ($mg{\cdot}g^{-1}$, $g{\cdot}mol^{-1}$) is the maximum adsorption capacity, and C_e ($mg{\cdot}L^{-1}$) is the iodine concentration at equilibrium state.

$$q_e = q_mK_LC_e/(1 + K_LC_e) \tag{6}$$

The Freundlich isotherm is often used to study multi-layer adsorptions, as described in Equation (7),

$$lnq_e = 1/nlnC_e + lnK_F \tag{7}$$

where $1/n$ represents the adsorption intensity ($mg{\cdot}L^{-1}$), K_F is the Freundlich constant ($mg{\cdot}g^{-1}$, $g{\cdot}mol^{-1}$), and C_e ($mg{\cdot}L^{-1}$) is the equilibrated concentration of iodine.

3.5. Study of Iodine Adsorption Property of TpBD-Me_2 and TpBD-$(OMe)_2$ in Aqueous Solution

The absorption intensity of KI/I_2 aqueous solutions with different concentrations of I_2 at 10, to 20, 30, 40, 50, 60, and 80 mg/L was measured using UV-vis absorption spectroscopy, and the absorption intensity was plotted against the concentration to obtain the standard curve. After as-prepared TpBD-Me_2 and TpBD-$(OMe)_2$ were placed into the KI/I_2 aqueous solutions, the I_2 concentration in the solution was monitored by means of UV-vis absorption spectroscopy. The I_2 concentration in aqueous solutions at a given adsorption time (t, min) was calculated from the standard curve. The I_2 adsorption capacity in KI/I_2 aqueous solution with respect to the mass of COFs (Q_t, $mg{\cdot}g^{-1}$) and molar number of COF unit cells (Q_t^{uc}, $g{\cdot}mol^{-1}$) was calculated using Equations (1)–(3) using I_2 concentration in solution at any time (C_t). The adsorption kinetics of I_2 in aqueous solution can also be fitted through the pseudo-first-order and pseudo-second-order models (Equations (4) and (5)).

The maximum adsorption capacity towards I_2 with respect to the mass of COFs (Q_e) and the molar number of COF unit cell (Q_e^{uc}) in KI/I_2 aqueous solutions can also be calculated using Equations (1) and (2) according different I_2 concentrations at equilibrium states (C_e). The adsorption isotherms could be simulated by Langmuir and Freundlich models to obtain the maximum adsorption capacity (Q_m) and the adsorption intensity ($1/n$) in aqueous solutions (Equations (6) and (7)).

3.6. Study of the Reusability of Iodine Uptake by TpBD-Me$_2$ and TpBD-(OMe)$_2$ in Cyclohexane

Freshly prepared TpBD-Me$_2$ and TpBD-(OMe)$_2$ were immersed in a cyclohexane solution containing 100 mg/L of I_2 for three days to achieve adsorption equilibrium. Afterwards, TpBD-Me$_2$ and TpBD-(OMe)$_2$, which were saturated with adsorbed iodine, were immersed in ethanol for desorption. The ethanol was replaced every 12 h for a total of five cycles to ensure complete desorption of iodine molecules existed in the pores of COF materials. Next, TpBD-Me$_2$ and TpBD-(OMe)$_2$ were again immersed in 100 mg/L of I_2 cyclohexane solution for I_2 adsorption. This cycle was repeated 3 times, and the amount of iodine adsorbed by TpBD-Me$_2$ and TpBD-(OMe)$_2$ each time was compared with that of fresh preparation.

3.7. Evaluation of the Resistance to Anionic Interference on the Iodine Absorption Property in Aqueous Solutions

The KI/I_2 solution, containing an I_2 concentration of 67 mg/L, was supplemented with 0.1 M NaCl, NaBr, $NaNO_3$, CH_3COONa, and Na_2SO_4, respectively. Subsequently, 5 mg of TpBD-Me$_2$ and TpBD-(OMe)$_2$ were immersed in the aforementioned mixed solution. After a period of 3 days, the adsorption equilibrium was established, and the I_2 concentration in the solution was determined using UV-vis spectroscopy. Subsequently, the adsorption capacity of iodine (Q_e) by COFs after mixing other anions in the KI/I_2 solution was calculated based on Equation (1). The removal efficiency of iodine can be determined using Equation (8).

$$Removal\ \% = Q_e/(C_0 V) \tag{8}$$

4. Conclusions

The present study successfully synthesized two COFs, namely TpBD-Me$_2$ and TpBD-(OMe)$_2$, through the Schiff base reaction under solvothermal conditions. PXRD analysis, thermogravimetry, and N_2 adsorption–desorption confirmed their high crystallinity, excellent thermal stability, large specific surface area, and presence of open pores. Considering the abundance of electron-rich N and O atoms in both COFs, their potential for efficient iodine absorption in cyclohexane and aqueous solution was explored. Both in water and in cyclohexane, TpBD-OMe$_2$ exhibited higher iodine uptake and faster iodine uptake rates relative to TpBD-Me$_2$, which was attributed to the existence of ether oxygen atoms increasing the number of its adsorption sites. The present study confirmed the crucial role of ether oxygen atoms in multiple COF monomers for iodine adsorption. Comparing the iodine adsorption of TpBD-(OMe)$_2$ with other porous materials, it was found that its adsorption capacity was able to exceed that of most of the POPs (including COFs) and MOFs, making it a preferred material for iodine adsorption. Raman spectroscopy and XPS analysis revealed that iodine predominantly existed as monomeric species and their aggregates (I_2, I_3^-, I_5^-) within the pores of the COFs. In this work, the adsorption capacities of TpBD-Me$_2$ and TpBD-(OMe)$_2$ in solutions were investigated for the first time from the perspective of adsorption sites, which improved the experimental basis and research ideas for the design of COF materials for iodine adsorption and provided new materials for the efficient removal of iodine from solutions.

Supplementary Materials: The following supporting information can be downloaded at: https://www.mdpi.com/article/10.3390/molecules28248151/s1, Figure S1: FT-IR spectra of COFs; Figure S2: Solid-state ^{13}C CP-MAS NMR spectra of COFs; Figure S3: TGA profiles of COFs; Figure S4: XPS spectra of COFs; Figure S5: High-resolution XPS spectra of C1s; Figure S6: SEM images of COFs; Figure S7: TEM images of COFs; Figure S8: PXRD patterns of COFs in different solvents; Figure S9: Nitrogen adsorption-desorption isotherms curves and the pore size distribution of COFs; Figure S10: Plots of BET surface area of COFs; Figure S11: Photos of disk-shaped sheet of COFs; Figure S12: PXRD patterns of COF sheets; Figure S13: TGA profiles of COF sheets; Figure S14: Temporal absorption spectral evolution of cyclohexane solution of I_2 during the adsorption process; Figure S15: UV-Vis spectra of cyclohexane solutions of I_2 and plot of the absorption intensity versus the I_2 concentration; Figure S16: PXRD patterns of COFs@I_2; Figure S17: XPS spectra of COFs@I_2; Figure S18: UV-vis spectra of I_3^- aqueous solution and plot of the absorbance intensity of I_3^- aqueous solution versus I_2 concentration; Figure S19: Temporal absorption spectral evolution of aqueous solution of I_2/KI during the adsorption process; Figure S20: Plots of the values of Q_e in the presence of different kinds of competing anions; Figure S21: XPS spectra of COFs@I_3^-; Table S1: Summary of C, N, O atomic concentration of COFs; Table S2: Summary of the BET fitting results of nitrogen adsorption isotherms of COFs; Table S3: Summary of the values of q_t, q_t^{uc} and q_t^{Mol}; Table S4: Summary of the experimental values of q_e and calculated value of q_e from fitting of their adsorption kinetics profiles; Table S5: Summary of the experimental values of q_e^{uc} and the calculated value of q_e^{uc} from fitting of their adsorption kinetics profiles; Table S6: Summary of the values of q_t^{N+CO} of TpBD-Me$_2$ and q_t^{N+O}, q_t^{EO} of TpBD-(OMe)$_2$; Table S7: Summary of the values of q_e^{N+CO} of TpBD-Me$_2$ and the values of q_e^{N+O} and q_e^{EO} of TpBD-(OMe)$_2$ and the calculated value of q_e^{N+CO}, q_e^{N+O} and q_e^{EO} from fitting of their adsorption kinetics profiles; Table S8: Summary of the q_e, q_e^{uc} and q_e^{Mol} values at different initial I_2 concentrations; Table S9: Summary of the fitting results of I_2 adsorption isotherms in terms of q_e; Table S10: Summary of the fitting results of I_2 adsorption isotherms in terms of q_e^{uc}; Table S11: Summary of the element atomic concentration of COFs@I_2; Table S12: Summary of I_2 adsorption capacity of reported porous materials in organic solvent; Table S13: Summary of the values of Q_t, Q_t^{uc}; Table S14: Summary of the experimental values of Q_e and the calculated value of Q_e from fitting of their adsorption kinetics profiles; Table S15: Summary of the experimental values of Q_e^{uc} and the calculated value of Q_e^{uc} from fitting of their adsorption kinetics profiles; Table S16: Summary of the Q_e and Q_e^{uc} values; Table S17: Summary of the fitting results of I_2 adsorption isotherms in terms of Q_e; Table S18: Summary of the fitting results of I_2 adsorption isotherms in terms of Q_e^{uc}; Table S19: Summary of the element atomic concentration of COFs@I_3^-. References [48–76] are cited in the supplementary materials.

Author Contributions: Q.T. conceived the idea, designed, and carried out the experiments, and performed the data interpretation. X.Z. carried out XPS spectra measurements. L.J. designed and carried the experiments. L.S. performed the data interpretation. P.D. carried the XRD measurements. Q.T. wrote the manuscript with assistance of all other authors. All authors contributed to the general discussion. All authors have read and agreed to the published version of the manuscript.

Funding: This research received no external funding.

Institutional Review Board Statement: Not applicable.

Informed Consent Statement: Not applicable.

Data Availability Statement: The data presented in this study are available on request from the corresponding author.

Acknowledgments: We are grateful to Jialu Li (Department of Chemistry, Tsinghua University, Beijing, China) for COF model construction and PXRD simulation.

Conflicts of Interest: The authors declare no conflict of interest.

References

1. Bates, J.K.; Bradley, J.P.; Teetsov, A.; Bradley, C.R.; Buchholtz Ten Brink, M. Colloid Formation During Waste Form Reaction: Implications for Nuclear Waste Disposal. *Science* **1992**, *256*, 649–651. [CrossRef] [PubMed]
2. Wang, C.; Jiang, S.; Ma, W.; Liu, Z.; Liu, L.; Zou, Y.; Xu, B.; Tian, W. Polymorphic Covalent Organic Frameworks: Molecularly Defined Pore Structures and Iodine Adsorption Property. *Molecules* **2023**, *28*, 449. [CrossRef] [PubMed]
3. Dehner, G.; McBeth, M.K.; Moss, R.; Woerden, I. A Zero-Carbon Nuclear Energy Future? Lessons Learned from Perceptions of Climate Change and Nuclear Waste. *Energies* **2023**, *16*, 2025. [CrossRef]

4. Fujiwara, H. Observation of Radioactive Iodine (I-131, I-129) in Cropland Soil after the Fukushima Nuclear Accident. *Sci. Total Environ.* **2016**, *566*, 1432–1439. [CrossRef] [PubMed]
5. Vienna, J.D. Nuclear Waste Vitrification in the United States: Recent Developments and Future Options. *Int. J. Appl. Glass. Sci.* **2010**, *1*, 309–312. [CrossRef]
6. Ewing, R.C.; Hippel, F.N. Nuclear Waste Management in the United States-Starting Over. *Science* **2009**, *325*, 151–152. [CrossRef] [PubMed]
7. Teien, H.C.; Wada, T.; Kashparov, V.; Lopez-Gutierrez, J.M.; Garcia-Tenorio, R.; Hinton, T.G.; Salbu, B. Transfer of 129 I to Freshwater Fish Species within Fukushima and Chernobyl Exclusion Zones. *J. Environ. Radioact.* **2023**, *207*, 107269. [CrossRef]
8. Zhou, W.; Li, A.; Zhou, M.; Xu, Y.; Zhang, Y.; He, Q. Nonporous Amorphous Superadsorbents for Highly Effective and Selective Adsorption of Iodine in Water. *Nat. Commun.* **2023**, *14*, 5388. [CrossRef]
9. Nandanwar, S.U.; Coldsnow, K.; Utgikar, V.; Sabharwall, P.; Aston, D.E. Capture of Harmful Radioactive Contaminants from Off-gas Stream Using Porous Solid Sorbents for Clean Environment—A Review. *Chem. Eng. J.* **2016**, *306*, 369–381. [CrossRef]
10. Herdes, C.; Prosenjak, C.; Román, S.; Müller, E.A. Fundamental Studies of Methyl Iodide Adsorption in DABCO Impregnated Activated Carbons. *Langmuir* **2013**, *29*, 6849–6855. [CrossRef]
11. González-García, C.M.; Román, S.; González, J.F.; Sabio, E.; Ledesma, B. Surface Free Energy Analysis of Adsorbents Used for Radioiodine Adsorption. *Appl. Surf. Sci.* **2013**, *282*, 714–717. [CrossRef]
12. Huve, J.; Ryzhikov, A.; Nouali, H.; Lalia, V.; Auged, G.; Daou, T.J. Porous Sorbents for the Capture of Radioactive Iodine Compounds: A Review. *RSC Adv.* **2018**, *8*, 29248–29273. [CrossRef]
13. Jabraoui, H.; Hessou, E.P.; Chibani, S.; Cantrel, L.; Lebegue, S.; Badawi, M. Adsorption of Volatile Organic and Iodine Compounds over Silver-Exchanged Mordenites: A Comparative Periodic DFT Study for Several Silver Loadings. *Appl. Surf. Sci.* **2019**, *485*, 56–63. [CrossRef]
14. Yadav, A.; Chong, S.; Riley, B.J.; MoCloy, J.S.; Goel, A. Iodine Capture by Ag-Loaded Solid Sorbents Followed by Ag Recycling and Iodine Immobilization: An End-to-End Process. *Ind. Eng. Chem. Res.* **2023**, *62*, 3635–3646. [CrossRef]
15. Wu, L.; Sawada, J.A.; Kuznicki, D.B.; Kuznicki, T.; Kuznicki, S.M. Iodine Adsorption on Silver-Exchanged Titania-Derived Adsorbents. *J. Radioanal. Nucl. Chem.* **2014**, *302*, 527–532. [CrossRef]
16. Muhire, C.; Reda, A.T.; Zhang, D.; Xu, Y.; Cui, C. An Overview on Metal Oxide-Based Materials for Iodine Capture and Storage. *Chem. Eng. J.* **2022**, *431*, 133816. [CrossRef]
17. Liu, S.; Kang, S.; Wang, H.; Wang, G.; Zhao, H.; Cai, W. Nanosheets-Built Flowerlike Micro/Nanostructured $Bi_2O_{2.33}$ and Its Highly Efficient Iodine Removal Performances. *Chem. Eng. J.* **2016**, *289*, 219–230. [CrossRef]
18. Bo, A.; Sarina, S.; Zheng, Z.; Yang, D.; Liu, H.; Zhu, H. Removal of Radioactive Iodine from Water Using Ag_2O Grafted Titanate Nanolamina as Efficient Adsorbent. *J. Hazard. Mater.* **2013**, *246*, 199–205. [CrossRef]
19. Asmussen, R.M.; Matyas, J.; Qafoku, N.P.; Kruger, A.A. Silver-Functionalized Silica Aerogels and Their Application in the Removal of Iodine from Aqueous Environments. *J. Hazard. Mater.* **2019**, *379*, 119364. [CrossRef] [PubMed]
20. Tang, S.; Choi, S.; Nan, Y.; Tavlarides, L.L. Adsorption of Methyl Iodide on Reduced Silver-Functionalized Silica Aerogel: Kinetics and Modeling. *AIChE J.* **2021**, *67*, e17137. [CrossRef]
21. Buck, T.; Chibani, S.; Paul, J.-F.; Cantrel, L.; Badawi, M. Dissociative Iodomethane Adsorption on Ag-MOR and the Formation of Agl Clusters: An &ITab Initio&IT Molecular Dynamics Study. *Phys. Chem. Chem. Phys.* **2017**, *19*, 27530–27543.
22. Wang, B.; Zhu, H.; Duan, T.; He, G.; Wei, Y.; Zhou, J. Multi-Layer Active Interface Construction with Polyphenols and Nano-Silver on Nano Collagen Fiber for Efficient Capturing Iodine Vapor. *Appl. Surf. Sci.* **2022**, *596*, 153585. [CrossRef]
23. Yu, X.; Cui, W.; Zhang, F.; Guo, Y.; Deng, T. Removal of Iodine from the Salt Water Used for Caustic Soda Production by Ion-Exchange Resin Adsorption. *Desalination* **2019**, *458*, 76–83. [CrossRef]
24. Chen, D.; Ma, T.; Zhao, X.; Jing, X.; Zhao, R.; Zhu, G. Multi-Functionalization Integration into the Electrospun Nanofibers Exhibiting Effective Iodine Capture from Water. *ACS Appl. Mater. Interfaces* **2022**, *14*, 47126–47135. [CrossRef]
25. Xiang, Y.; Wang, Y.; Chen, H.; Fu, J.; Zhu, Q.; Yang, X.; Xu, X.; Qin, S.; He, L.; Tao, G. Efficient Enrichment of Iodine by Supported Ionic Liquid with Three Effective Adsorption Sites: Heteroatoms, Fused Aromatic Rings and Ionic Bond. *Chem. Eng. J.* **2023**, *456*, 140979. [CrossRef]
26. Zhang, X.; Maddock, J.; Nenoff, T.M.; Denecke, M.A.; Yang, S.; Schroder, M. Adsorption of Iodine in Metal–Organic Framework Materials. *Chem. Soc. Rev.* **2022**, *51*, 3243. [CrossRef]
27. Yu, R.; Li, Q.; Li, Z.; Wang, X.; Xia, L. Analysis of Radioactive Iodine Trapping Mechanism by Zinc-Based Metal−Organic Frameworks with Various N-Containing Carboxylate Ligands. *ACS Appl. Mater. Interfaces* **2023**, *15*, 35082–35091. [CrossRef]
28. Zhang, Y.; Cui, X.; Xing, H. Recent Advances in the Capture and Abatement of Toxic Gases and Vapors by Metal–Organic Frameworks. *Mater. Chem. Front.* **2021**, *5*, 5970. [CrossRef]
29. Li, Z.; Ju, Y.; Yu, B.; Wu, X.; Lu, H.; Li, Y.; Zhou, J.; Guo, X.; Zhang, Z.; Lin, J.; et al. Modulated Synthesis and Isoreticular Expansion of Th-MOFs with Record High Pore Volume and Surface Area for Iodine Adsorption. *Chem. Commun.* **2020**, *56*, 6715. [CrossRef] [PubMed]
30. Du, W.; Qin, Y.; Ni, C.; Dai, W.; Zou, J. Efficient Capture of Volatile Iodine by Thiophene-Containing Porous Organic Polymers. *ACS Appl. Polym. Mater.* **2020**, *2*, 5121–5128. [CrossRef]
31. Kurisingal, J.F.; Yun, H.; Hong, C.S. Porous Organic Materials for Iodine Adsorption. *J. Hazard. Mater.* **2023**, *458*, 131835. [CrossRef] [PubMed]

32. Wang, H.; Qiu, N.; Kong, X.; Hu, Z.; Zhong, F.; Li, Y.; Tan, H. Novel Carbazole-Based Porous Organic Polymer for Efficient Iodine Capture and Rhodamine B Adsorption. *ACS Appl. Mater. Interfaces* **2023**, *15*, 14846–14853. [CrossRef] [PubMed]
33. Yan, Z.; Yuan, Y.; Tian, Y.; Zhang, D.; Zhu, G. Highly Efficient Enrichment of Volatile Iodine by Charged Porous Aromatic Frameworks with Three Sorption Sites. *Angew. Chem. Int. Ed.* **2015**, *54*, 12733–12737. [CrossRef] [PubMed]
34. Chandra, S.; Kandambeth, S.; Biswal, B.P.; Lukose, B.; Kunjir, S.M.; Chaudhary, M.; Babarao, R.; Heine, T.; Banerjee, R. Chemically Stable Multilayered Covalent Organic Nanosheets from Covalent Organic Frameworks via Mechanical Delamination. *J. Am. Chem. Soc.* **2013**, *135*, 17853–17861. [CrossRef]
35. Liu, T.; Zhao, Y.; Song, M.; Pang, X.; Shi, X.; Jia, J.; Chi, L.; Lu, G. Ordered Macro−Microporous Single Crystals of Covalent Organic Frameworks with Efficient Sorption of Iodine. *J. Am. Chem. Soc.* **2023**, *145*, 2544–2552. [CrossRef] [PubMed]
36. Song, S.; Shi, Y.; Liu, N.; Liu, F. Theoretical Screening and Experimental Synthesis of Ultrahigh-Iodine Capture Covalent Organic Frameworks. *ACS Appl. Mater. Interfaces* **2021**, *13*, 10513–10523. [CrossRef] [PubMed]
37. Zhang, Z.; Dong, X.; Yin, J.; Li, Z.; Li, X.; Zhang, D.; Pan, T.; Lei, Q.; Liu, X.; Xie, Y.; et al. Chemically Stable Guanidinium Covalent Organic Framework for the Efficient Capture of Low-Concentration Iodine at High Temperatures. *J. Am. Chem. Soc.* **2022**, *144*, 6821–6829. [CrossRef]
38. Xie, Y.; Pan, T.; Lei, Q.; Chen, C.; Dong, X.; Yuan, Y.; Shen, J.; Cai, Y.; Zhou, C.; Pinnau, I.; et al. Ionic Functionalization of Multivariate Covalent Organic Frameworks to Achieve an Exceptionally High Iodine-Capture Capacity. *Angew. Chem. Int. Ed.* **2021**, *60*, 22432–22440. [CrossRef]
39. Li, Y.; Li, X.; Li, J.; Liu, W.; Cheng, G.; Ke, H. Phosphine-Based Covalent Organic Framework for Highly Efficient Iodine Capture. *Micropor. Mesopor. Mater.* **2021**, *325*, 111351. [CrossRef]
40. Yan, X.; Yang, Y.; Li, G.; Zhang, J.; He, Y.; Wang, R.; Lin, Z.; Cai, Z. Thiophene-Based Covalent Organic Frameworks for Highly Efficient Iodine Capture. *Chin. Chem. Lett.* **2023**, *34*, 107201. [CrossRef]
41. Sun, Y.; Song, S.; Xiao, D.; Gan, L.; Wang, Y. Easily Constructed Imine-Bonded COFs for Iodine Capture at Ambient Temperature. *ACS Omega* **2020**, *5*, 24262–24271. [CrossRef] [PubMed]
42. Ding, S.; Gao, J.; Wang, Q.; Zhang, Y.; Song, W.; Su, C.; Wang, W. Construction of Covalent Organic Framework for Catalysis: Pd/COF-LZU1 in Suzuki-Miyaura Coupling Reaction. *J. Am. Chem. Soc.* **2011**, *133*, 19816–19822. [CrossRef] [PubMed]
43. Jing, L.; Cheng, C.; Wang, B.; Wang, S.; Xie, R.; Xia, H.; Wang, D. Controlled Iodine Phase Transfer of Covalent Organic Framework Membranes for Instant but Sustained Disinfection. *Langmuir* **2023**, *39*, 597–609. [CrossRef] [PubMed]
44. Wu, Z.; Wei, W.; Ma, J.; Luo, J.; Zhou, Y.; Zhou, Z.; Liu, S. Adsorption of Iodine on Adamantane-Based Covalent Organic Frameworks. *Chem. Select* **2021**, *6*, 10141–10148. [CrossRef]
45. Liu, X.; Zhang, A.; Ma, R.; Wu, B.; Wen, T.; Ai, Y.; Sun, M.; Jin, J.; Wang, S.; Wang, X. Experimental and Theoretical Insights into Copper Phthalocyanine-Based Covalent Organic Frameworks for Highly Efficient Radioactive Iodine Capture. *Chin. Chem. Lett.* **2022**, *33*, 3549–3555. [CrossRef]
46. Mokhtari, N.; Dinari, M. Developing Novel Amine-Linked Covalent Organic Frameworks towards Reversible Iodine Capture. *Sep. Purif. Technol.* **2022**, *301*, 121948. [CrossRef]
47. Song, S.; Shi, Y.; Liu, N.; Liu, F. C=N Linked Covalent Organic Framework for the Efficient Adsorption of Iodine in Vapor and Solution. *RSC Adv.* **2021**, *11*, 10512. [CrossRef]
48. Wang, Z.; Huang, Y.; Yang, J.; Li, Y.; Zhuang, Q.; Gu, J. The Water-Based Synthesis of Chemically Stable Zr-Based MOFsUsing Pyridine-Containing Ligands and Their Exceptionally High Adsorption Capacity for Iodine. *Dalton Trans.* **2017**, *46*, 7412–7420. [CrossRef]
49. Xu, T.; Li, J.; Jia, M.; Li, G.; Liu, Y. Contiguous Layer Based Metal–Organic Framework with Conjugated π-Electron Ligand for High Iodine Capture. *Dalton Trans.* **2021**, *50*, 13096–13102. [CrossRef]
50. Ju, Y.; Li, Z.; Qiu, J.; Li, X.; Yang, J.; Zhang, Z.; He, M.; Wang, J.; Lin, J. Adsorption and Detection of Iodine Species by a Thorium-Based Metal−Organic Framework. *Inorg. Chem.* **2023**, *62*, 8158–8165. [CrossRef]
51. Zhao, Y.; Zhang, N.; Wang, Y.; Bai, F.; Xing, Y.; Sun, L. Ln-MOFs with Window-Shaped Channels Based on Triazine Tricarboxylic Acid as a Linker for the Highly Efficient Capture of Cationic Dyes and Iodine. *Inorg. Chem. Front.* **2021**, *8*, 1736–1746. [CrossRef]
52. Xu, L.; Zheng, Q.; Wang, Y.; Jiang, L.; Jiang, J.; Qiu, J. A Pillared Double-Wall Metal-Organic Framework Adsorption Membrane for the Efficient Removal of Iodine from Solution. *Sep. Purif. Technol.* **2021**, *274*, 118436. [CrossRef]
53. Zhang, N.; Sun, L.; Xing, Y.; Bai, F. A Double-Walled Thorium-Based Metal−Organic Framework as a Promising Dual-Function Absorbent for Efficiently Capturing Iodine and Dyes. *Cryst. Growth Des.* **2019**, *19*, 5686–5695. [CrossRef]
54. Cai, H.; Zeng, G.; You, Z.; Wang, C.; Sun, L.; Bai, F.; Xing, Y. Cu(II) and Zn(II) Frameworks Constructed by Directional Tuning of Diverse Substituted Groups on a Triazine Skeleton and Supermassive Adsorption Behavior for Iodine and Dyes. *Dalton Trans.* **2022**, *51*, 5457–5470. [CrossRef] [PubMed]
55. Zhang, N.; Sun, L.; Bai, F.; Xing, Y. Thorium-Organic Framework Constructed with a Semirigid Triazine Hexacarboxylic Acid Ligand: Unique Structure with Thorium Oxide Wheel Clusters and Iodine Adsorption Behavior. *Inorg. Chem.* **2020**, *59*, 3964–3973. [CrossRef]
56. Liu, M.; Cai, H.; Jiang, S.; Xing, Y.; Bai, F. Construction of a Triazine Polycarboxylate Co-MOF with Flexible and Rigid Coordination Arms as well as Excellent Catalytic Reduction and Adsorption Properties. *Dalton Trans.* **2023**, *52*, 6773–6781. [CrossRef]
57. Li, Z.; Yue, Z.; Ju, Y.; Wu, X.; Ren, Y.; Wang, S.; Li, Y.; Zhang, Z.; Guo, X.; Lin, J. Ultrastable Thorium Metal−Organic Frameworks for Efficient Iodine Adsorption. *Inorg. Chem.* **2020**, *59*, 4435–4442.

58. Huang, J.; Hu, H.; Deng, S.; Cai, S.; Fan, J.; Zhang, W.; Zheng, S. A Ni(II) metal–organic framework with helical channels for the capture of iodine via guest exchange induced amorphization. *New J Chem.* **2022**, *46*, 7144–7152. [CrossRef]
59. Hu, H.; Chen, F.; Zhang, Z.; Liu, D.; Liang, Y.; Chen, Z. Heterometallic Metal-Organic Framework Based on $[Cu_4I_4]$ and $[Hf_6O_8]$ Clusters for Adsorption of Iodine. *Front. Chem.* **2022**, *10*, 864131. [CrossRef]
60. Falaise, C.; Volkringer, C.; Facqueur, J.; Bousquet, T.; Gasnot, L.; Loiseau, T. Capture of Iodine in Highly Stable Metal–Organic Frameworks: A Systematic Study. *Chem. Commun.* **2013**, *49*, 10320–10322. [CrossRef]
61. Li, Z.; Ju, Y.; Lu, H.; Wu, X.; Yu, X.; Li, Y.; Wu, X.; Zhang, Z.; Lin, J.; Qian, Y. Boosting the Iodine Adsorption and Radioresistance of Th-UiO-66 MOFs via Aromatic Substitution. *Chem. Eur. J.* **2020**, *27*, 1286–1291. [CrossRef]
62. Wang, G.; Huang, J.; Huang, X.; Deng, S.; Zheng, S.; Cai, S.; Fan, J.; Zhang, W. A Hydrolytically Stable Cage-Based Metal–Organic Framework Containing two Types of Building Blocks for the Adsorption of Iodine and Dyes. *Inorg. Chem. Front.* **2021**, *8*, 1083–1092. [CrossRef]
63. Chen, H.; Fan, L.; Zhang, X.; Ma, L. Robust Heterometallic Tb^{III}/Mn^{II}−Organic Framework for CO_2/CH_4 Separation and I_2 Adsorption. *ACS Appl. Nano Mater.* **2020**, *3*, 2680–2686. [CrossRef]
64. Yu, R.; Li, Q.; Zhang, T.; Li, Z.; Xia, L. Zn, O Co-Adsorption Based on MOF-5 for Efficient Capture of Radioactive Iodine. *Process Saf. Environ.* **2023**, *174*, 770–777. [CrossRef]
65. Yang, Y.; Tu, C.; Shi, J.; Zhao, T.; Liu, Z.; Cheng, F.; Luo, F. Highly Stable Cd(II)-MOFs Based on 2,6-Naphthanlenedisulfonate and Bisimidazole Ligands: A New Platform for Selective Detection of Cu^{2+} and Efficient Removal of Iodine. *J. Solid State Chem.* **2021**, *302*, 122439. [CrossRef]
66. Zhang, N.; Xing, Y.; Bai, F. A Uranyl-Organic Framework Featuring Two-Dimensional Graphene-like Layered Topology for Efficient Iodine and Dyes Capture. *Inorg. Chem.* **2019**, *58*, 6866–6876. [CrossRef]
67. Li, M.; Yuan, G.; Zeng, Y.; Yang, Y.; Liao, J.; Yang, J.; Liu, N. Flexible surface-supported MOF membrane via a convenient approach for efficient iodine adsorption. *J. Radioanal. Nucl. Ch.* **2020**, *324*, 1167–1177. [CrossRef]
68. Al Lafi, A.G.; Assfour, B.; Assaad, T. Metal Organic Framework MIL-101(Cr): Spectroscopic Investigations to Reveal Iodine Capture Mechanism. *J. Inorg. Organomet. P.* **2020**, *30*, 1218–1230. [CrossRef]
69. Li, L.; Sun, Z.; Yang, A.; Zhang, X.; Zhu, X.; Li, W.; Liu, Y.; Luan, J. A Novel Azo-Linked Polymer Bearing Trifluoromethyl Groups for I_2 Capture. *Macromol. Rapid Comm.* **2023**, *44*, 2200982. [CrossRef] [PubMed]
70. Thurakkal, L.; Cheekatla, S.; Porel, M. Superfast Capture of Iodine from Air, Water, and Organic Solvent by Potential Dithiocarbamate-Based Organic Polymer. *Int. J. Mol. Sci.* **2023**, *24*, 1466. [CrossRef] [PubMed]
71. Pan, X.; Ding, C.; Zhang, Z.; Ke, H.; Cheng, G. Functional Porous Organic Polymer with High S and N for Reversible Iodine Capture. *Micropor. Mesopor. Mat.* **2020**, *300*, 110161. [CrossRef]
72. Yang, X.; Duan, L.; Ran, X.; Ran, B.; Yi, S. Two Nitrogen-Rich Conjugated Microporous Polymers for Efficient Iodine Sequestration and Removal. *J. Polym. Res.* **2022**, *29*, 499. [CrossRef]
73. Sigen, A.; Zhang, Y.; Li, Z.; Xia, H.; Xue, M.; Liu, X.; Mu, Y. Highly Efficient and Reversible Iodine Capture Using a Metalloporphyrin-Based Conjugated Microporous Polymer. *Chem. Commun.* **2014**, *50*, 8495–8498.
74. Liu, L.; Song, C.; Kong, A. Nitrogen and Sulfur-Enriched Porous Bithiophene-Melamine Covalent Organic Polymers for Effective Capture of CO_2 and Iodine. *Mater. Lett.* **2020**, *277*, 128291. [CrossRef]
75. Mohan, A.; Al-Sayah, M.H.; Ahmed, A.; El-Kadri, O.M. Triazine-Based Porous Organic Polymers for Reversible Capture of Iodine and Utilization in Antibacterial Application. *Sci. Rep.* **2022**, *12*, 2638. [CrossRef]
76. Yang, M.; Shi, W.; Liu, S.; Xu, K. Multifunctional Diphenyl Ether-Based, Cross-Linked Polyisocyanide for Efficient Iodine Capture and NO_2^-/SO_3^{2-} Electrochemical Probing. *Colloid Surface A* **2022**, *642*, 128680. [CrossRef]

molecules

MDPI

Review

Effective Removal of Boron from Aqueous Solutions by Inorganic Adsorbents: A Review

Xiang-Yang Lou [1,†], Lucia Yohai [1,2,†], Roberto Boada [1,*], Montserrat Resina-Gallego [1], Dong Han [1] and Manuel Valiente [1]

1 Grup de Tècniques de Separació en Química (GTS-UAB Research Group), Department of Chemistry, Facultat de Ciències, Universitat Autònoma de Barcelona, 08193 Bellaterra, Spain; xiangyang.lou@uab.cat (X.-Y.L.); yohai@fi.mdp.edu.ar (L.Y.); montserrat.resina@uab.cat (M.R.-G.); dong.han@uab.cat (D.H.); manuel.valiente@uab.cat (M.V.)

2 Instituto de Investigaciones en Ciencia y Tecnología de Materiales (INTEMA), Universidad Nacional de Mar del Plata-Consejo Nacional de Investigaciones Científicas y Técnicas (UNMdP-CONICET), Mar del Plata B7608FDQ, Argentina

* Correspondence: roberto.boada@uab.cat; Tel.: +34-935814638

† These authors contributed equally to this work.

Abstract: Increasing levels of boron in water exceeding acceptable thresholds have triggered concerns regarding environmental pollution and adverse health effects. In response, significant efforts are being made to develop new adsorbents for the removal of boron from contaminated water. Among the various materials proposed, inorganic adsorbents have emerged as promising materials due to their chemical, thermal, and mechanical stability. This review aims to comprehensively examine recent advances made in the development of inorganic adsorbents for the efficient removal of boron from water. Firstly, the adsorption performance of the most used adsorbents, such as magnesium, iron, aluminum, and individual and mixed oxides, are summarized. Subsequently, diverse functionalization methods aimed at enhancing boron adsorption capacity and selectivity are carefully analyzed. Lastly, challenges and future perspectives in this field are highlighted to guide the development of innovative high-performance adsorbents and adsorption systems, ultimately leading to a reduction in boron pollution.

Keywords: boron; adsorption; inorganic adsorbents; surface modification; metal (hydr)oxides

1. Introduction

Boron (B) is a non-metallic element that is widely found on the surface of the earth and in water in various forms, such as boric acid and borates [1]. Its average concentration in the environment varies across different media. In soil, the average concentration of boron is approximately 30 mg/kg; in seawater, it is approximately 4.5 mg/L; in groundwater, it can range from 0.3 to 100 mg/L [2]. This widespread presence of boron underscores its significance and ubiquity in the natural environment.

1.1. Boron Sources and Related Problems

Boron is introduced into the environment through a combination of natural phenomena and human activities [3,4]. Naturally occurring boron compounds originate from volcanic gases and hot springs near volcanic active areas. These compounds include borosilicate, boroaluminosilicate, and borate minerals [5]. In addition, boron is released into the environment through the weathering of rocks and leaching of salt deposits [6]. Furthermore, it is noteworthy that boron can be found naturally in seawater [7]. The concentration of boron in seawater varies significantly depending on the geography and location of ocean bodies. For instance, in the Mediterranean Sea, boron concentrations can reach as high as 9.6 mg/L [8]. Because of the geochemical nature of the drainage area, boron is also found in rainfall in coastal areas [9]. Industrial activity is another important source of boron released into the environment [10,11]. Boric acid and boron salts are widely utilized for manufacturing [12,13], with the glass industry alone consuming over half of the total boron compounds produced [14].

The boron released into the environment exhibits high volatility. It can evaporate, forming acid rain that falls back to the ground, contaminating drinking water, accumulating in the soil, and being taken up by plants, resulting in a series of environmental and health issues [15]. Indeed, the concentration range between boron deficiency and toxicity is extremely narrow, and this means that even a slight increase in boron levels can be hazardous to certain organisms. Unfortunately, the prevalence of toxic effects due to excess boron is more common than boron deficiency in the environment [16,17]. Hence, despite its essential role in plant growth, it is necessary to control the boron concentration in both drinking and irrigation water. The World Health Organization (WHO) established a guideline for boron, limiting it to 2.4 mg/L in drinking water and 1.0 mg/L for irrigation water to avoid damaging sensitive crops [18]. In addition, many countries have established regulations. For example, the boron threshold for drinking water in the European Union was recently set to 1.5 mg/L, whereas 1.0 mg/L was established in the United Kingdom, South Korea, and Japan, and the limit in the United States ranges between 0.6 and 1.0 mg/L depending on the state [19].

1.2. Chemistry of Boron in Aqueous Solution

In aqueous solutions, boron usually exists in the form of boric acid, $B(OH)_3$, and various kinds of borates, depending on the pH and the concentration [20]. The distribution of borate species in solution under different pH values and at different initial concentrations was performed using Hydra-Medusa software v1.0 [21,22]. $B(OH)_3$ and $B(OH)_4^-$ are mainly present at low boron concentrations (<216 mg/L), as shown in Figure 1. At high boron concentrations (>270 mg/L), and with an increase in pH value from 6–10, water-soluble polyborate ions such as $B_2(OH)_5^-$, $B_3O_3(OH)_4^-$, and $B_4O_5(OH)_4^{2-}$ are formed [23]. The formation of these polynuclear ions is attributed to the interaction of boric acid and borate ions in solution. In general, regardless of the concentration, boric acid dominates at low pH, while borate ions dominate at high pH.

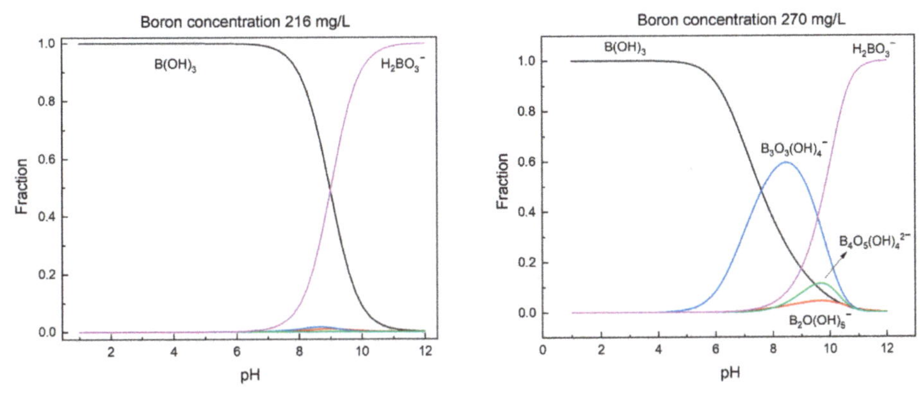

(a) (b)

Figure 1. Distribution of borate species in solution as a function of pH, depicted for concentrations of 216 mg/L (**a**) and 270 mg/L (**b**).

1.3. Boron Removal Methods

The development and investigation of efficient boron removal processes from aqueous solutions have become a crucial task due to the increasing boron concentration in surface and groundwater, as well as the need to treat seawater in desalination plants and wastewater from industries using boric acid as raw material [24–26]. The major challenge in selectively removing boron lies in the fact that it exists in water in the forms of various chemical species and that their concentrations vary from one region to another [27].

In past decades, a variety of methodologies such as precipitation [28], extraction [29], ion exchange [30], adsorption [31], and membrane [32] processes have been studied for boron removal [33]. The reported results indicate that boron is much more difficult to remove than many metalloids. For instance, the coagulation, flocculation, and filtration techniques that are frequently used for water purification offer little effectiveness in boron removal [24]. Reverse osmosis (RO) exhibits poor boron rejection primarily because boron, being a small species and undissociated acid, cannot be effectively rejected by the membrane owing to its lack of charge [34]. Many advances have been introduced to membrane processes to reach a removal rate higher than 90%; however, this also increased the operational time, cost, and system complexity [14]. Commercial chelating resins with functional groups have also been reported for effective boron removal through ion exchange with high selectivity [30,35]. The boron adsorption capacities of some commercial resins, modified resins, and other modified adsorbents are displayed in Table 1. However, their poor chemical and thermal stability, as well as their high price, easy fouling, and high recycling cost, severely limit their practical applications.

Table 1. Boron adsorption capacities and particle size of some commercial resins, modified resins, and other modified commercial adsorbents.

Adsorbents	Adsorption Capacity (mg/g)	Particle Size (μm)	Reference
CL-RESIN	8.4	722–855	[36]
NCL-RESIN	8.6	710–845	[36]
IRA 743-RESIN	10.9	550–700	[36]
P(GMA-co-TRIM)-EN-PG	29.2	<241	[37]
P(GMA-co-TRIM)-TETA-PG	23.3	<273	[37]
CL-MCM-41	19.5	-	[38]
NCL-MCM-41	16.7	-	[38]
T-RESIN	21.3	-	[39]

In contrast, adsorption offers several advantages over the aforementioned techniques for removing boron from water [40]. It is easy to implement, requires no additional chemicals, and is relatively inexpensive to operate. Some adsorbents can be regenerated multiple times with proper treatment, significantly reducing the cost per unit of treated water compared to other processes. Additionally, adsorption is an efficient process capable of removing substantial quantities of boron in a short time. Among the traditional adsorbents, activated carbon has been widely used in many applications; however, it has not been extensively used for boron removal because of its poorly performance for adsorbing polar adsorbates, such as boron, which confers a low effective surface active area, causing low boron adsorption selectivity unless modified [19,41,42]. In contrast, nanostructured and porous inorganic materials have emerged as the most promising materials due to their large specific surface areas and stable structures [43,44]. Recent efforts have been devoted to the development of novel low-cost inorganic adsorbents with high adsorption capacities. These research findings have significantly promoted the development of adsorption techniques for effective boron removal from water.

Herein, a comprehensive analysis of the recent advances in boron adsorption from aqueous solutions using inorganic adsorbents is provided. The review begins by presenting an overview of numerous adsorbents reported in recent years. Subsequently, a careful review of various functionalization strategies aimed at enhancing both boron adsorption capacity and selectivity is presented. Lastly, the current challenges and future perspectives in this field have been outlined to guide the development of next-generation, high-performance adsorbent materials and adsorption systems for boron removal.

2. Inorganic Adsorbents for Boron Removal

Among all the adsorbents used for boron adsorption, inorganic adsorbents like aluminum, iron, and magnesium oxides are of great interest because of their stability in a wide range of temperatures and times of exposure; in addition, they have low cost, are easy to synthesize, and could provide a moderate to large specific surface area depending on the porosity or surface to volume ratio of the nanostructured materials. Table 2 shows a summary of the boron adsorption performance of some inorganic adsorbents. It is worth mentioning that numerous synthesis methodologies for obtaining inorganic adsorbents such as sol-gel, gas phase, coprecipitation, hydrothermal synthesis, and microemulsion reaction, among others, have been reported [45]; however, they will not be extensively discussed here.

Table 2. Summary of boron adsorption capacities of different inorganic adsorbents.

Adsorbent	Surface Area (m^2/g)	Adsorption Capacity (mg/g)	Temperature (°C)	Contact Time (h)	pH	Ref.
Al_2O_3	150	0.35 at 15 mg/L B concentration	25	72	8	[46]
	-	0.65 at 50 mg/L B concentration	25	0.5	8–8.5	[47]
	250	138.50 *	25	2	8.0	[48]
	169	0.43 *	22 ± 3		8.5	[49]
	-	6.38 *	20	72	8	[50]
	150	25.86 *	35	24	10	[51]
Fe(O)OH	-	0.32 *	22	72	8	[52]
Fe_3O_4	-	0.26 at 21,622 mg/L B concentration	22	48	6	[53]
	107	49.41 at 7567.7 mg/L B concentration	45	1.5	7	[54]
MgO	4.8	232.44 *	60	24	<8	[55]
	168	87.03 *	90	3	10	[56]
	15	5.4 *	RT	0.5	<8	[57]
	-	54.2 *	RT	48	9.5–10.5	[58]
	165	27.3 at 500 mg/L B concentration	-	1	7	[59]
	280	202.43 *	25	6	9	[60]
Mixed oxides	213	77.84 *	25	-	7	[61]
	-	33 at 21,622 mg/L B concentration	90	3	9	[62]
	10	37.90 *	RT	4	9	[63]
	-	25 *	-	-	10	[64]
	-	41.08 *	30	2	10	[65]
	40.5	0.98 *				
	34.6	0.70 *	RT	24	8.2–8.5	[66]
	14.5	0.19 *				

* Maximum adsorption capacity obtained from fitting the adsorption isotherm.

2.1. Aluminum Oxide

Aluminum oxide (Al_2O_3) is a versatile and widely used material due to its good mechanical properties, physicochemical stability, corrosion resistance, and high specific surface area (150–280 m^2/g) when presented as a nanostructured porous material. Its great adsorption affinity towards inorganic compounds has made alumina one of the most studied materials in water treatment technologies for wastewater treatment or groundwater remediation [67].

The optimum pH for the removal of boron by alumina is found in the range of 7 to 9, normally with maximum adsorption observed at pH 8 for both amorphous and crystalline alumina [46,47,68]. The pH level aligns closely with the pH_{pzc} value of alumina (8.0–9.0), slightly lower than the pKa of boric acid (9.2). This indicates that the most favorable conditions for efficient boron removal occur when the adsorbent surfaces are uncharged and boric acid is the predominant boron species in the solution. In most of the reported studies, researchers worked with initial concentrations lower than 270 mg/L to avoid the appearance of polyborates such as $B_2O(OH)_6^{2-}$, $B_3O_3(OH)_4^{-}$, $B_4O_5(OH)_4^{2-}$, and $B_5O_6(OH)_4^{-}$ [69,70]. Additionally, some studies have demonstrated that ionic strength has a negligible effect on the adsorption process, suggesting that electrostatic interactions are of minor importance for alumina and that adsorption occurs mainly via inner-sphere complex formation [48].

Boron chemistry and reported studies suggest that boron adsorption onto alumina prefers to be carried out at pH 8 and room temperature (RT). This pH is not only the one with maximum adsorption capacity but is also highly representative of surface and underground waters, which is an advantage when designing a purification or desalination process. In this way, Bouguerra et al. [47] used activated alumina, for which they found that for initial concentrations of 5 and 50 mg/L, the uptake of boron reaches approximately 40% and 65%, respectively, with an adsorbent dose of 0.8 and 5.0 g per 100 mL, and a contact time of 0.5 h. In terms of adsorption capacities, these values are 0.25 and 0.65 mg/g, respectively. In the case of Konstantinou et al. [46], where the adsorption mechanism is based on inner-sphere complexation, the maximum adsorption capacity for alumina was 0.4 mg/g. The maximum removal efficiency for alumina was obtained at a relatively low boron concentration and with a high contact time of 72 h. Similar results were obtained by Demetriou et al. [49], with a maximum adsorption capacity of 0.43 mg/g, represented via inner-sphere complex formation. However, in the study of Demey et al. [50], the adsorption capacity reached 4.41 mg/g at pH 8 after 72 h of contact time; this capacity is around 10 times that of previous works. One of the latest studies with nano-γ-Al_2O_3, prepared from aluminum foils, reached a boron adsorption capacity of 25.86 mg/g at 35 °C, 24 h of contact time, and pH 10 [51]. In addition, Lou et al. [48] reported a hierarchical alumina microspheres sorbent synthesized via a microwave-assisted coprecipitation method. The adsorption capacity was 51.60 mg/g at an initial boron concentration of 800 mg/L, while the theoretically calculated maximum adsorption capacity using the Langmuir model reached 138.50 mg/g. This outstanding performance can be primarily attributed to the microporosity of the hierarchical dandelion-like structure of the adsorbent.

2.2. Iron Oxide

Iron oxide materials have been intensively studied in recent decades because of their properties to be used in fields like catalysis, biomedical applications, gas sensing materials, electromagnetism, and water purification [71]. Some materials can adsorb contaminants in water, such as heavy metal ions, organic compounds, and other contaminants like arsenic and boron [72]. However, the applications of iron oxides strongly depend on the type of oxide and the stability of the particles under specific conditions. Sixteen pure phases of iron (hydr)oxides, i.e., oxides, hydroxides, and oxyhydroxides, are known to date [73]. Common synthetic methods for magnetic nanoparticles (MNPs) encompass coprecipitation, microemulsion, thermal decomposition, sol-gel, and solvothermal/hydrothermal processes [74]. Among all, coprecipitation is currently the most widely used method, and its main advantage is that a single synthesis can produce large quantities of products, and the reaction time is short, which is suitable for large-scale industrial production. However, it is still a challenge to regulate particle size and avoid agglomeration [72]. Another approach proposed by Glasgow et al. [75] is the synthesis of crystalline and uniform Fe_3O_4 through the thermal decomposition of iron precursors in a continuous flow method. Nevertheless, the Fe_2O_3/γ-Fe_2O_3 phase might be present together with Fe_3O_4.

The most common magnetic nanoparticles used in boron removal are magnetite (Fe_3O_4), maghemite (γ-Fe_2O_3), and some iron oxide-hydroxides such as Fe(O)OH. γ-Fe_2O_3 has been studied for boron adsorption, demonstrating peak adsorption capacity at pH levels ranging from 8 to 9. On the other hand, Fe_3O_4 possesses strong magnetic characteristics, enabling straightforward separation from the reaction medium after adsorption via a simple magnet [72]. Demetriou et al. [52] studied boron adsorption using iron(III) oxide hydroxide

(Fe(O)OH). The maximum adsorption capacity was found to be 0.32 mg/g at pH 8 after 72 h of contact time. Thermodynamic values and a lack of ionic strength dependence suggest that adsorption is based on inner-sphere complexation, as found for alumina adsorbents. Liu et al. [53] studied Fe_3O_4 and two different surface modification adsorbents derived from it. They found that, for all MNPs, boron is adsorbed in the forms of both H_3BO_3 and $B(OH)_4^-$. Furthermore, it was observed that in all cases, the highest level of boron adsorption occurred in neutral solutions, likely attributed to the presence of hydrogen bonding, electrostatic attractions, and hydrophobic interactions. Conversely, the lowest adsorption was observed in basic solutions due to electrostatic repulsion. Additionally, the study demonstrated a decrease in adsorption as ionic strength increased. In another study, Chen et al. [54] reached equilibrium rapidly in 1.5 h, showing a high capacity of 49.41 mg/g at pH 7 and 45 °C. Surface analysis indicated the formation of new Fe-O-B bonds in this case.

Despite their versatility and wide range of applications, iron oxides are frequently used after surface modification and/or functionalization to increase their affinity and selectivity towards boron [76]. Moreover, the surface modification of iron oxide magnetic nanoparticles serves to decrease surface energy, reduce agglomeration in solution, and enhance hydrophobic characteristics. The various surface modifications applied to boron adsorbents will be discussed in detail in the following section.

2.3. *Magnesium Oxide*

Magnesium oxide (MgO) has various applications in industry. The physical and chemical properties of magnesium oxide particles, including their morphology, are primarily governed by the precursor used for its synthesis. It has photocatalytic properties, and it has been tested as an adsorbent for toxic chemical agents [77]. Indeed, magnesium oxide is an effective and low-cost adsorbent for the removal of boron from aqueous solutions. Chemisorption on MgO can be primarily ascribed to the acid/base interaction of adsorbates with its surface. The reported pH_{PZC} for MgO lies between 9.8 and 12.7; consequently, most studies have been carried out at pH under 10 to have a positively charged surface. For example, Kameda et al. [55] attributed boron adsorption on the MgO surface to the electrostatic attractive force of positively charged MgO with $B(OH)_4^-$. The value of the maximum adsorption was 232.4 mg/g at 30 °C and a week of contact time following a pseudo-first-order reaction kinetics and a Langmuir-type adsorption isotherm. Boron desorption from MgO particles containing adsorbed $B(OH)_4^-$ species was possible using a 0.1 M NaOH solution as a rinse; however, the adsorption capacity of regenerated MgO was lower than that of the original MgO.

Li et al. [56] used MgO nanosheets for boron adsorption. The MgO nanosheets were fabricated by an ultrasonic method by magnesium salt solution precipitation, calcination, and further ultrasonic treatment. Adsorption capacity increased with the temperature reaching the maximum, 87 mg/g, at 90 °C. The results were better described by the Langmuir model than the Freundlich model. Characterization techniques have shown that both the hydration of MgO nanosheets and the polymerization of adsorbed boron species occur concurrently, leading to the coexistence of unbound surface H_2O molecules.

Masindi et al. [57] investigated boron adsorption using calcinated magnesite tailings. In contrast to other reported results, the removal yield remained nearly constant across the entire pH range of 1.0–8.0, not reaching a maximum at the highest pH. At pH values higher than 8, boron removal was reduced. The adsorption process took 30 min to reach equilibrium and obtained a capacity of 6 mg/g at pH 8. Kinetic and isotherm data were best modeled by the pseudo-second-order and Langmuir isotherm equations, respectively.

The research of De la Fuente et al. [58] showed that the removal rate improved as the pH increased, presenting a maximum adsorption capacity at a pH value between 9.5 and 10.5. For an initial concentration of 50 mg/L, the maximum adsorption capacity was obtained after 30 min of stirring and 6 h of repose at 70 °C. Under these optimum conditions, a boron removal rate of around 95% was obtained; however, this value decreased to 75% at

30 °C. In a subsequent article [78], they demonstrated that the adsorption process followed a pseudo-second-order kinetic model and the Langmuir isotherm model. Nevertheless, the reaction process between boron and magnesium compounds was found to be irreversible, and neither pH nor temperature adjustment could be used to regenerate the adsorbents. In their study, Fukuda et al. [59] developed a low crystalline magnesium oxide material (LC-MgO), exhibiting a remarkable boron adsorption capability of 27.3 mg/g with a 60 min contact time at pH 7 and with an initial boron concentration of 500 mg/L. This behavior was attributed to the formation of Mg-B layered compounds that precipitated on the MgO particle surface. One of the last works regarding boron adsorption onto MgO corresponds to Song et al. [60], where MgO was previously hydrated to form $Mg(OH)_2$, responsible for boron adsorption. This material exhibited the highest adsorption capacity, reaching 202.4 mg/g in 6 h at 25 °C and pH 9, and exhibited a 70% surface regeneration after treating it with NaOH.

2.4. Mixed Oxides

The use of metal oxide mixtures has attracted wider attention due to the possibility of achieving synergistic effects of the mixed metal oxides to develop better properties and achieve improved performance than individual oxides.

Mixed oxides have also been widely reported for boron adsorption [5]. As Guan et al. [19] highlighted, the boron adsorption capacity of layered double hydroxides (LDHs) can be enhanced through thermal activation via calcination at high temperatures. This process alters the LDHs' structure by removing interlayer anions and water, promoting the formation of a mixed metal oxide material. Various LDHs have been utilized for effective boron removal. The most important works in terms of adsorption capacity are the ones from Gao et al. [61], Liu et al. [62], Kentjono et al. [63], Kurashina et al. [64], and Kameda et al. [65].

In the study by Gao et al. [61] on ultrathin Mg-Al LDH, the maximum adsorption capacity was 77.8 mg/g at 25 °C and pH 7. However, it is worth noting that the initial boron concentration used was considerably lower than in other studies (see Table 2). Liu et al. [62] synthesized Mg-Al LDH with the highest adsorption capacity of 33 mg/g. Kentjono et al. [63] found that Mg-Al (NO_3) LDH exhibited a notable boron removal efficiency (37.9 mg/g) when optoelectronic wastewater was treated at pH 9.0–9.2. Kurashina et al. [64] employed Mg-Al LDH to treat a highly concentrated borate solution, achieving a 90% elimination rate at pH 10 with an adsorption capacity of 25 mg/g. Kameda et al. [65] conducted a comparative study on two Mg-Al LDHs intercalated Cl^- and NO_3^-, respectively, for boron uptake. They found that the maximum adsorption values were 38.9 mg/g for NO_3^-·Mg-Al LDH and 41.1 mg/g for Cl^-·Mg-Al LDH, respectively. All these works reached a similar adsorption capacity which can be due to the presence of hydroxyl groups on the surface that can provide sites for boron adsorption through surface complexation [79].

In other works, tertiary mixed oxides were studied. Heredia et al. [78] studied boron removal capacity from aqueous solutions utilizing MgAlFe mixed oxides. They achieved a maximum boron removal exceeding 85%, equivalent to 9.08 mg/g. They attributed boron adsorption to electrostatic attraction between the surface of mixed oxides and boron species. Other tertiary oxides focused on aluminum-based oxides were approached by Irawan et al. [66]. They studied three samples of aluminum-based residuals (Al-WTR1, Al-WTR2, and Al-WTR3) composed of Al_2O_3, Fe_2O_3, and SiO_2 in different ratios. The optimum pH was found at 8.2–8.5, and maximum adsorption capacities were 0.980, 0.700, and 0.190 mg/g, respectively, relatively low compared to Mg-Al LDH. The adsorption reaction was spontaneous and exothermic, and it was characterized by physical adsorption (electrostatic interactions and van der Waals forces).

Despite the extensive reporting of synthetic and naturally occurring mixed oxides as effective boron adsorbents, practical applications still face limitations. In the case of synthetic mixed oxides, a significant drawback is the inability to exercise precise control over the synthesis process that guarantees the same resulting product. This constraint arises from direct dependence on the proportions of constituent elements and the initial reaction conditions. Furthermore, the understanding of the surface chemistry of mixed oxides may be insufficient, leading to challenges in predicting and optimizing their adsorption behavior, especially with the existence of competing anions such as CO_3^{2-}, Cl^-, and NO_3^- [79]. Natural mixed oxides, on the other hand, face the primary challenge of heterogeneity and varying availability across different geographical locations. Consequently, despite their advantageous properties, obtaining uniform quantities of natural mixed oxides can prove to be a challenging task.

3. Modification of the Inorganic Boron Adsorbent

In recent studies, inorganic adsorbents have been extensively utilized due to their distinct advantages and outstanding performance. However, these adsorbents face a significant drawback in their relatively lower boron adsorption capacities compared to other metallic ions. The negative charge of boron species in natural waters requires the design of adsorbents with positive surfaces, enabling physical electrostatic interactions to enhance boron adsorption. More importantly, the primary driving force in chemical boron adsorption is the formation of surface complexations through ion/ligand exchange between adsorbents and boron ions [48]. Consequently, research efforts have been focused on exploring various surface modification strategies using different types of molecules to maximize boron adsorption.

3.1. Functional Groups Used for Boron Adsorbent Modification

According to the literature, molecules containing three or more hydroxyl groups located in the cis position, known as "vis-diols", tend to form stable complexes with boron species in both acidic and alkaline environments [19,80]. This suggests a promising functionalization strategy for the specific removal of boron species from water, involving the utilization of polyhydroxy groups as chelating agents. Previous studies have successfully employed sugar derivatives for boron removal [81,82]. These polyoxide compounds typically bind by forming boric acid esters or borate anion complexes with a proton as a counterion. Consistent with these results, tertiary amine groups are crucial for chelating boron because they can bind a proton that is released when borate is complexed with hydroxyl functionalities [83–85]. *N*-methyl-D-glucamine (NMDG), glucose, iminodipropylene glycol, polyalcohols, polyether sulfonates, hydrogels, and other similar substances are used to functionalize a variety of modified adsorbents based on this theory, as summarized in Table 3.

Table 3. Boron adsorption capacity reported for some modified adsorbents.

Adsorbents	Adsorption Capacity (mg/g)	Isotherm and Adsorption Kinetic Models	Ref.
Magnetic nanoparticles attached to NMDG	13.4 *	Langmuir, pseudo-second-order	[86]
Glycidol-modified magnetic chitosan beads (MCG)	128.5 *	Redlich–Peterson, pseudo-second-order	[87]
Magnesium ferrite magnetic with polyvinyl alcohol	45.5 at 100 mg/L	-	[88]
Magnesium ferrite magnetic with glycidol	68.9 at 100 mg/L	-	[88]
Alginate–alumina (CAAl)	56.2 *	Langmuir, pseudo-second-order	[50]
Silica matrix with gluconamide moieties	20.0 *	Langmuir–Freundlich model	[89]
Diol-functionalized silica particles	57.09 *	Sips, pseudo-second-order	[90]
NMDG-modified magnetic microparticles	0.02 at 0.9 mg/L	-	[91]
Silica-supported NMDG	16.7 *	Freundlich, pseudo-second-order	[92]
Poly(Si-NMDG)@MIL-101(Cr)	24.8 *	Langmuir	[93]
Silica–polyallylamine composites grafting NMDG	16.8 *	Freundlich, chemical reaction	[94]
N-methyl-D-glucamine-based hybrid	21.8 *	Langmuir	[95]
$Fe_3O_4@SiO_2$ functionalized with glycidol	25.6 *	Langmuir	[96]

* Maximum adsorption capacity obtained from fitting the adsorption isotherm.

Among all types of molecules, NMDG has been mostly studied. The presence of polyols and tertiary amine ends in NMDG can increase the possibility of complexation reactions with boron. The binding process of boron with NMDG-modified adsorbents is shown in Figure 2. Using the click coupling technique, Tural et al. [86] synthesized magnetic nanoparticles modified with NMDG. The maximum adsorption capacity was 13.44 mg/g, which was significantly higher than the adsorbent synthesized by direct coupling. The boron adsorption kinetics followed a pseudo-second-order model, and the adsorption was consistent with the Langmuir isotherm. Hong et al. [97]. reported an NMDG-modified cellulose acetate as an effective adsorbent for boron removal from aqueous solutions. The adsorbent capacity reached ~34 mg/g within 10 min, which overcame the disadvantage of the slow adsorption rate observed with the current commercial resin (Amberlite IRA743).

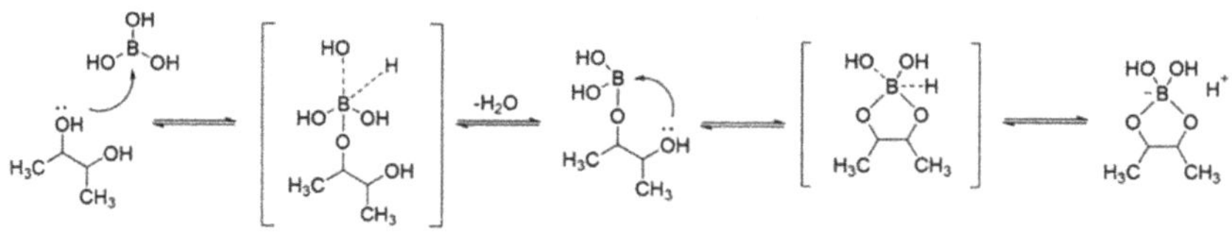

Figure 2. Example of the binding mechanism of boron with any glycol compound used as adsorbent. Adapted with permission from Ref. [83]. Copyright 2022, copyright Elsevier Ltd.

In addition to the extensive use of NMDG modifications, numerous other useful molecules have also been employed. Oladipo et al. [87] reported glycidol-modified magnetic chitosan beads (MCG) with cis-diol groups. After modification, the maximum adsorption capacity increased from 66.85 mg/g to 128.50 mg/g. MCG micro-spheres were easily separated from the solution and could be reused up to seven times. Even in the presence of competing ions such as Cu^{2+}, Fe^{3+}, Ni^{2+}, Mg^{2+}, Ca^{2+}, Na^{+}, and K^{+} ions, MCG still showed approximately 96% removal efficiency from the boric acid solution (125 mg/L). Their other work [88] reported functionalized magnesium ferrite magnetic nanopowder with polyvinyl alcohol and glycidol as boron-selective adsorbents. For polyvinyl alcohol-functionalized particles, the adsorption peaked at pH 7.0, while for glycidol-functionalized particles, the adsorption peaked at pH 10.0. At a low boron concentration of 5 mg/L, polyvinyl alcohol-functionalized polymers showed a higher adsorption capacity (45.5 mg/g). However, at a higher boron concentration (50 mg/L), glycidol-functionalized particles showed a higher adsorption capacity of 68.9 mg/g. Demey et al. [50] synthesized a new composite adsorbent consisting of alginate-alumina (CAAl) for boron removal. After modification, the alumina showed an increased maximum experimental adsorption capacity from 0.41 to 1.81 mg/g at pH 9.5. The experimental results of the equilibrium data were better fitted by the Langmuir equation, and the kinetic data followed a pseudo-second-order rate equation. The incorporation of alumina particles in the new alginate-based composite improves mechanical strength and slightly improves boron removal. Sanfeliu et al. [89] studied the removal of boron using nanosized mesoporous silica that had been crosslinked in successive steps with 3-aminopropyltriethoxysilane (APTES) and polyhydroxy compounds (gluconolactone), respectively. The maximum adsorption capacity obtained was 20.0 mg/g. Each anchored glucose group had multiple boron coordination, according to the results obtained. Under ambient conditions at pH 7, Tang et al. [90] investigated a group of diol-functionalized silica particles for boron removal. The adsorption capacity was less than 5 mg/g. The adsorption data were well fitted by the Sip's isotherm model and the pseudo-second-order kinetic model.

3.2. Methodologies Used for Boron Adsorbent Modification

The chemical properties of different functional groups require the use of diverse adsorbent modification techniques. This necessity arises from the varying levels of hydrophobicity and polarity exhibited by these functional groups, which, in turn, impact their interactions with adsorbent materials. The most frequently reported methods are suspension polymerization reactions, as well as polymerization or copolymerization reactions [98,99].

For instance, Adeyemi et al. [87] studied the modification of magnetic chitosan beads using glycidol. The process involved dispersing magnetic chitosan beads in ethanol, followed by the addition of a glycidol solution in *N*-methyl-2-pyrrolidone to the reaction flask. The suspension was then stirred for 10 h, and the final product was collected, washed with ethanol, and dried under vacuum at 70 °C. Similar modification approaches were reported by Liao et al. [100] and Wu et al. [101].

However, direct bonding to inorganic adsorbents such as alumina, iron oxide, and magnesium oxide may not always be practical, especially when using NMDG as the functional group. In such situations, organosiloxane groups can serve as a bridge between NMDG and inorganic adsorbents. Two primary strategies have been reported to achieve this objective.

One strategy involves modifying the surface of inorganic adsorbents with halogen or epoxy groups through silanization reactions and subsequently linking the NMDG functional groups via nucleophilic substitution reactions. However, this synthetic route imposes strict requirements on the pH of the solution because the functional groups contain a high number of amino groups. Excessive alkalinity ($pH > 11$) can result in the dissolution of the silica gel alkyl. The use of a single low-polarity solvent (e.g., dioxane) or a mixed solvent (e.g., dioxane and water) can reduce the solution's alkalinity ($pH < 10$) but may also hinder the effective progress of nucleophilic substitution reactions of halogen, epoxy groups, and functional groups. Tural et al. [91] used a two-step synthesis to functionalize γ-Fe_2O_3-silica particles with NMDG. In the first step, γ-Fe_2O_3-silica composites were functionalized with trimethoxysilylpropyl chloride. Subsequently, the obtained magnetic composite reacted with NMDG to introduce tertiary amine functionality into the material. A schematic illustration of these two steps for functionalizing γ-Fe_2O_3-silica particles with NMDG is shown in Figure 3.

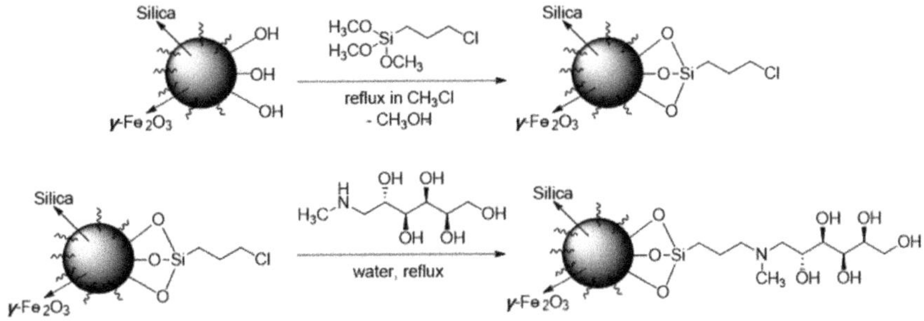

Figure 3. A schematic illustration of preparation steps for functionalizing γ-Fe_2O_3-silica particles with NMDG. Adapted with permission from Ref. [91]. Copyright 2010 copyright John Wiley and Sons.

Another strategy for connecting NMDG to inorganic adsorbents is to first form a modified silane coupling agent via the reaction of functional groups with halogen and epoxy groups, as shown in Figure 4. In this method, a silica gel coupling agent is generated to reduce the alkalinity of the reaction system and avoid the dissolution of silica gel, thereby improving the synthesis efficiency. However, this method results in a wide particle size distribution of the synthesized adsorbent. Xu et al. [92] successfully utilized this strategy to synthesize NMDG-modified silica-supported adsorbents. They synthesized NMDG-modified GPTMS (Si-NMDG) and then loaded it onto an activated silica surface via salinization.

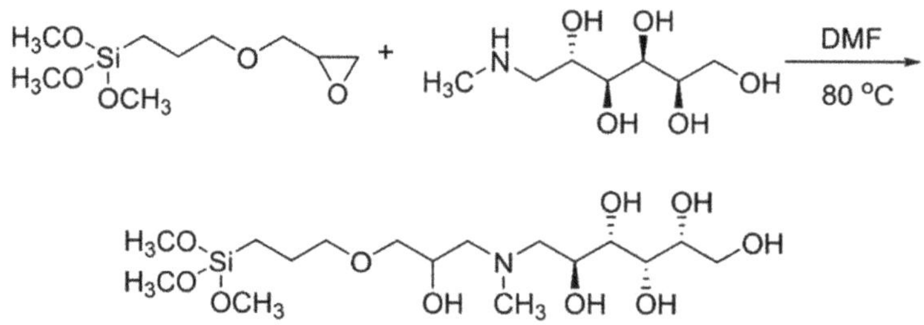

Figure 4. Synthesis pathway of organosiloxane Si-NMDG. Adapted with permission from Ref. [93]. Copyright 2016, copyright Wiley-VCH Verlag GmbH & Co. KGaA.

4. Challenges and Perspectives

Achieving efficient boron adsorption from water is significantly important for environmental protection, safeguarding human health, and resource recovery. Despite the encouraging advances made in recent years in the development of new adsorbents, there are still several issues that must be addressed and overcome before these adsorbents become a viable option for commercial use. In this section, the challenges and future perspectives of boron adsorption in water treatment systems are outlined.

Firstly, current adsorbents are limited by their low selectivity and capacity, requiring large quantities of adsorbents to achieve the desired removal efficiency. Moreover, their high cost, powdered form, and challenging recovery hamper their practical applications. Therefore, it is crucial to focus on the development of new adsorbents with high adsorption capacity, selectivity, recyclability, affordability, and scalable production in the future. Secondly, although ion/ligand exchange interactions between adsorbents and boron species are frequently reported, there are limited investigations of the underlying adsorption mechanism at the molecular level. The use of characterization techniques that provide a molecular picture of the absorbate-surface interaction (e.g., molecular spectroscopic techniques, density functional theory (DFT), and molecular dynamics simulations) is encouraged to improve the understanding of the adsorption and desorption processes. Finally, it is suggested that potential synergies between adsorption and other technologies, such as membrane processes, be explored to address the complexities of certain water systems and improve outcomes. Thus, there is a wide scope for future research to broaden our knowledge about the boron adsorption and desorption mechanisms and to promote their applications in water treatment.

Author Contributions: Writing—original draft preparation, X.-Y.L. and L.Y.; writing—review and editing, R.B., M.R.-G. and D.H.; supervision, R.B. and M.V.; project administration, M.V.; funding acquisition, M.V. All authors have read and agreed to the published version of the manuscript.

Funding: This research was financially supported by the project CTM2015-65414-C2-1-R from the MINECO Ministry of Spain. X-Y.L. and D.H. acknowledge the China Scholarship Council for fellowships No. 201708110179 and No. 201906450024, respectively. L.Y. acknowledges Agencia Nacional de Promoción Científica y Tecnológica for PICT funding (grant PICT 2019-00565) and the BEC.AR program (call 2019) for financial support. R.B. acknowledges funding support from the European Union's Horizon 2020 research and innovation program under the Marie Skłodowska-Curie grant agreement No. 665919.

Institutional Review Board Statement: Not applicable.

Informed Consent Statement: Not applicable.

Data Availability Statement: No new data were created or analyzed in this study. Data sharing is not applicable to this article.

Conflicts of Interest: The authors declare no conflicts of interest.

References

1. Wu, L.; Wei, Y.; Zhou, W.B.; Zhang, Y.S.; Chen, Q.H.; Liu, M.X.; Zhu, Z.P.; Zhou, J.; Yang, L.H.; Wang, H.M.; et al. Gene Expression Alterations of Human Liver Cancer Cells Following Borax Exposure. *Oncol. Rep.* **2019**, *42*, 115–130. [CrossRef] [PubMed]
2. Ulatowska, J.; Polowczyk, I.; Bastrzyk, A.; Koźlecki, T.; Sawiński, W. Fly Ash as a Sorbent for Boron Removal from Aqueous Solutions: Equilibrium and Thermodynamic Studies. *Sep. Sci. Technol.* **2020**, *55*, 2149–2157. [CrossRef]
3. Rodriguez-Espinosa, P.F.; Sabarathinam, C.; Ochoa-Guerrero, K.M.; Martínez-Tavera, E.; Panda, B. Geochemical Evolution and Boron Sources of the Groundwater Affected by Urban and Volcanic Activities of Puebla Valley, South Central Mexico. *J. Hydrol.* **2020**, *584*, 124613. [CrossRef]
4. Liu, M.; Guo, Q.; Luo, L.; He, T. Environmental Impacts of Geothermal Waters with Extremely High Boron Concentrations: Insight from a Case Study in Tibet, China. *J. Volcanol. Geotherm. Res.* **2020**, *397*, 106887. [CrossRef]
5. Bhagyaraj, S.; Al-Ghouti, M.A.; Kasak, P.; Krupa, I. An Updated Review on Boron Removal from Water through Adsorption Processes. *Emergent Mater.* **2021**, *4*, 1167–1186. [CrossRef]
6. Bolan, S.; Wijesekara, H.; Amarasiri, D.; Zhang, T.; Ragályi, P.; Brdar-Jokanović, M.; Rékási, M.; Lin, J.-Y.; Padhye, L.P.; Zhao, H.; et al. Boron Contamination and Its Risk Management in Terrestrial and Aquatic Environmental Settings. *Sci. Total Environ.* **2023**, *894*, 164744. [CrossRef] [PubMed]
7. Stein, S.; Sivan, O.; Yechieli, Y.; Kasher, R.; Nir, O. An Advantage for Desalination of Coastal Saline Groundwater over Seawater in View of Boron Removal Requirements. *Environ. Sci. Water Res. Technol.* **2021**, *7*, 2241–2254. [CrossRef]
8. Kayaci, S.; Tantekin-Ersolmaz, S.B.; Ahunbay, M.G.; Krantz, W.B. Technical and Economic Feasibility of the Concurrent Desalination and Boron Removal (CDBR) Process. *Desalination* **2020**, *486*, 114474. [CrossRef]
9. Zhang, X.; Luo, X.; Jiao, J.J.; Li, H.; Lian, E.; Yang, S.; Kong, F.; Kuang, X.; Zuo, J. Hydrogeochemistry and Fractionation of Boron Isotopes in the Inter-Dune Aquifer System of Badain Jaran Desert, China. *J. Hydrol.* **2021**, *595*, 125984. [CrossRef]
10. Kadam, A.; Wagh, V.; Umrikar, B.; Sankhua, R. An Implication of Boron and Fluoride Contamination and Its Exposure Risk in Groundwater Resources in Semi-Arid Region, Western India. *Environ. Dev. Sustain.* **2020**, *22*, 7033–7056. [CrossRef]
11. Mao, H.-R.; Liu, C.-Q.; Zhao, Z.-Q. Source and Evolution of Dissolved Boron in Rivers: Insights from Boron Isotope Signatures of End-Members and Model of Boron Isotopes during Weathering Processes. *Earth-Sci. Rev.* **2019**, *190*, 439–459. [CrossRef]
12. Lee, H.W.; Yoo, K.J.; Tran, M.T.; Moon, I.Y.; Oh, Y.S.; Kang, S.H.; Kim, D.K. Effect of Quenching Tempering-Postweld Heat Treatment on the Microstructure and Mechanical Properties of Laser-Arc Hybrid-Welded Boron Steel. *Materials* **2019**, *12*, 2862. [CrossRef] [PubMed]
13. Neochoritis, C.G.; Shaabani, S.; Ahmadianmoghaddam, M.; Zarganes-Tzitzikas, T.; Gao, L.; Novotná, M.; Mitríková, T.; Romero, A.R.; Irianti, M.I.; Xu, R.; et al. Rapid Approach to Complex Boronic Acids. *Sci. Adv.* **2019**, *5*, eaaw4607. [CrossRef] [PubMed]
14. Alkhudhiri, A.; Bin Darwish, N.; Hakami, M.W.; Abdullah, A.; Alsadun, A.; Abu Homod, H. Boron Removal by Membrane Distillation: A Comparison Study. *Membranes* **2020**, *10*, 263. [CrossRef] [PubMed]
15. Dai, L.; Wei, Y.; Xu, X.; Wu, P.; Zhang, M.; Wang, C.; Li, H.; Zhang, Q.; Li, H.; Zhu, W. Boron and Nitride Dual Vacancies on Metal-Free Oxygen Doping Boron Nitride as Initiating Sites for Deep Aerobic Oxidative Desulfurization. *ChemCatChem* **2020**, *12*, 1734–1742. [CrossRef]
16. Chen, D.; Chen, D.; Xue, R.; Long, J.; Lin, X.; Lin, Y.; Jia, L.; Zeng, R.; Song, Y. Effects of Boron, Silicon and Their Interactions on Cadmium Accumulation and Toxicity in Rice Plants. *J. Hazards Mater.* **2019**, *367*, 447–455. [CrossRef]
17. Brdar-Jokanović, M. Boron Toxicity and Deficiency in Agricultural Plants. *Int. J. Mol. Sci.* **2020**, *21*, 1424. [CrossRef]
18. Jarma, Y.A.; Karaoğlu, A.; Tekin, Ö.; Baba, A.; Ökten, H.E.; Tomaszewska, B.; Bostancı, K.; Arda, M.; Kabay, N. Assessment of Different Nanofiltration and Reverse Osmosis Membranes for Simultaneous Removal of Arsenic and Boron from Spent Geothermal Water. *J. Hazard. Mater.* **2021**, *405*, 124129. [CrossRef]
19. Guan, Z.; Lv, J.; Bai, P.; Guo, X. Boron Removal from Aqueous Solutions by Adsorption—A Review. *Desalination* **2016**, *383*, 29–37. [CrossRef]
20. Hussain, A.; Sharma, R.; Minier-Matar, J.; Hirani, Z.; Adham, S. Application of Emerging Ion Exchange Resin for Boron Removal from Saline Groundwater. *J. Water Process Eng.* **2019**, *32*, 100906. [CrossRef]
21. Ingri, N.; Kakolowicz, W.; Sillén, L.G.; Warnqvist, B. High-Speed Computers as a Supplement to Graphical Methods—V1Haltafall, a General Program for Calculating the Composition of Equilibrium Mixtures. *Talanta* **1967**, *14*, 1261–1286. [CrossRef] [PubMed]
22. Eriksson, G. An Algorithm for the Computation of Aqueous Multi-Component, Multiphase Equilibria. *Anal. Chim. Acta* **1979**, *112*, 375–383. [CrossRef]
23. Li, D.; Zhou, G.; Gu, S.; Zhang, T.; Meng, L.; Guo, Y.; Deng, T. Thermodynamic and Dynamic Modeling of the Boron Species in Aqueous Potassium Borate Solution. *ACS Omega* **2020**, *5*, 15835–15842. [CrossRef] [PubMed]
24. Mehanathan, S.; Jaafar, J.; Nasir, A.M.; Rahman, R.A.; Ismail, A.F.; Illias, R.M.; Othman, M.H.D.; A Rahman, M.; Bilad, M.R.; Naseer, M.N. Adsorptive Membrane for Boron Removal: Challenges and Future Prospects. *Membranes* **2022**, *12*, 798. [CrossRef] [PubMed]
25. Chen, T.; Lyu, J.; Wang, Q.; Bai, P.; Wu, Y.; Guo, X. Mechanistic Study on Boron Adsorption and Isotopic Separation with Magnetic Magnetite Nanoparticles. *J. Mater. Sci.* **2021**, *56*, 4624–4640. [CrossRef]
26. Wang, Z.; Jia, Y.; Song, W.; Li, X.; Xu, K.; Wang, Z. Optimization of Boron Adsorption from Desalinated Seawater onto UiO-66-NH2/GO Composite Adsorbent Using Response Surface Methodology. *J. Clean. Prod.* **2021**, *300*, 126974. [CrossRef]

27. Najid, N.; Fellaou, S.; Kouzbour, S.; Gourich, B.; Ruiz-García, A. Energy and Environmental Issues of Seawater Reverse Osmosis Desalination Considering Boron Rejection: A Comprehensive Review and a Case Study of Exergy Analysis. *Process Saf. Environ. Prot.* **2021**, *156*, 373–390. [CrossRef]
28. Yang-Zhou, C.-H.; Cao, J.-X.; Dong, S.-S.; Chen, S.-H.; Michael, R.N. Phosphorus Co-Existing in Water: A New Mechanism to Boost Boron Removal by Calcined Oyster Shell Powder. *Molecules* **2021**, *27*, 54. [CrossRef]
29. Almustafa, G.; Sulaiman, R.; Kumar, M.; Adeyemi, I.; Arafat, H.A.; AlNashef, I. Boron Extraction from Aqueous Medium Using Novel Hydrophobic Deep Eutectic Solvents. *Chem. Eng. J.* **2020**, *395*, 125173. [CrossRef]
30. Çermikli, E.; Şen, F.; Altıok, E.; Wolska, J.; Cyganowski, P.; Kabay, N.; Bryjak, M.; Arda, M.; Yüksel, M. Performances of Novel Chelating Ion Exchange Resins for Boron and Arsenic Removal from Saline Geothermal Water Using Adsorption-Membrane Filtration Hybrid Process. *Desalination* **2020**, *491*, 114504. [CrossRef]
31. Pinotti, C.N.; de Souza, L.M.; Marques, W.P.; Proveti, J.R.C.; Jesus, H.C.; Freitas, J.C.C.; Porto, P.S.S.; Muniz, E.P.; Passamani, E.C. A New Magnetic Composite with Potential Application in Boron Adsorption: Development, Characterization, and Removal Tests. *Mater. Chem. Phys.* **2022**, *277*, 125368. [CrossRef]
32. Seval, K.; Onac, C.; Kaya, A.; Akdogan, A. Separation of Boron from Geothermal Waters with Membrane System. *Membranes* **2021**, *11*, 291. [CrossRef] [PubMed]
33. Karthikeyan, P.; Elanchezhiyan, S.S.D.; Preethi, J.; Meenakshi, S.; Park, C.M. Mechanistic Performance of Polyaniline-Substituted Hexagonal Boron Nitride Composite as a Highly Efficient Adsorbent for the Removal of Phosphate, Nitrate, and Hexavalent Chromium Ions from an Aqueous Environment. *Appl. Surf. Sci.* **2020**, *511*, 145543. [CrossRef]
34. Baransi-Karkaby, K.; Bass, M.; Freger, V. In Situ Modification of Reverse Osmosis Membrane Elements for Enhanced Removal of Multiple Micropollutants. *Membranes* **2019**, *9*, 28. [CrossRef] [PubMed]
35. Figueira, M.; Reig, M.; Fernández de Labastida, M.; Cortina, J.L.; Valderrama, C. Boron Recovery from Desalination Seawater Brines by Selective Ion Exchange Resins. *J. Environ. Manag.* **2022**, *314*, 114984. [CrossRef] [PubMed]
36. Lyu, J.; Zeng, Z.; Zhang, N.; Liu, H.; Bai, P.; Guo, X. Pyrocatechol-Modified Resins for Boron Recovery from Water: Synthesis, Adsorption and Isotopic Separation Studies. *React. Funct. Polym.* **2017**, *112*, 1–8. [CrossRef]
37. Luo, Q.; Zeng, M.; Wang, X.; Huang, H.; Wang, X.; Liu, N.; Huang, X. Glycidol-Functionalized Macroporous Polymer for Boron Removal from Aqueous Solution. *React. Funct. Polym.* **2020**, *150*, 104543. [CrossRef]
38. Chen, Y.; Lyu, J.; Wang, Y.; Chen, T.; Tian, Y.; Bai, P.; Guo, X. Synthesis, Characterization, Adsorption, and Isotopic Separation Studies of Pyrocatechol-Modified MCM-41 for Efficient Boron Removal. *Ind. Eng. Chem. Res.* **2019**, *58*, 3282–3292. [CrossRef]
39. Bai, S.; Han, J.; Du, C.; Li, J.; Ding, W. Removal of Boron and Silicon by a Modified Resin and Their Competitive Adsorption Mechanisms. *Environ. Sci. Pollut. Res.* **2020**, *27*, 30275–30284. [CrossRef]
40. Jung, J.; Choi, H.; Hong, S.; Yoon, S.J.; Kim, T.H.; Lee, J.Y.; Hong, Y.T.; So, S. Surface-Initiated ATRP of Glycidyl Methacrylate in the Presence of Divinylbenzene on Porous Polystyrene-Based Resins for Boron Adsorption. *Desalination* **2020**, *473*, 114166. [CrossRef]
41. Kluczka, J.; Pudło, W.; Krukiewicz, K. Boron Adsorption Removal by Commercial and Modified Activated Carbons. *Chem. Eng. Res. Des.* **2019**, *147*, 30–42. [CrossRef]
42. Ismanto, A.E.; Liu, J.C. Enhanced Boron Adsorption Using PVA-Modified Carbonaceous Materials. *Compos. Interfaces* **2014**, *21*, 639–650. [CrossRef]
43. Cai, L.; Zhang, Y.; Peng, X.; Guo, F.; Ji, L.; Li, L. Preparation of Layered Double Hydroxide Intercalated by Gallic Acid for Boron Adsorption. *J. Water Process Eng.* **2021**, *44*, 102394. [CrossRef]
44. Song, T.; Luo, Q.; Gao, F.; Zhao, B.; Hao, X.; Liu, Z. Adsorption and Electro-Assisted Method Removal of Boron in Aqueous Solution by Nickel Hydroxide. *J. Ind. Eng. Chem.* **2023**, *118*, 372–382. [CrossRef]
45. Kumar, A.; Gangawane, K.M. Synthesis and Effect on the Surface Morphology & Magnetic Properties of Ferrimagnetic Nanoparticles by Different Wet Chemical Synthesis Methods. *Powder Technol.* **2022**, *410*, 117867. [CrossRef]
46. Konstantinou, M.; Kasseta, G.; Pashalidis, I. Boron Adsorption on Alumina (Al_2O_3) and Magnesia (MgO) in Aqueous Solutions: A Comparative Study. *Int. J. Environ. Technol. Manag.* **2006**, *6*, 466. [CrossRef]
47. Bouguerra, W.; Mnif, A.; Hamrouni, B.; Dhahbi, M. Boron Removal by Adsorption onto Activated Alumina and by Reverse Osmosis. *Desalination* **2008**, *223*, 31–37. [CrossRef]
48. Lou, X.-Y.; Boada, R.; Yohai, L.; Valiente, M. Outstanding Performance of Hierarchical Alumina Microspheres for Boron Removal in the Presence of Competing Ions. *J. Water Process Eng.* **2023**, *55*, 104218. [CrossRef]
49. Demetriou, A.; Pashalidis, I. Competitive Adsorption of Boric Acid and Chromate onto Alumina in Aqueous Solutions. *Water Sci. Technol.* **2014**, *69*, 378–384. [CrossRef]
50. Demey, H.; Barron-Zambrano, J.; Mhadhbi, T.; Miloudi, H.; Yang, Z.; Ruiz, M.; Sastre, A.M. Boron Removal from Aqueous Solutions by Using a Novel Alginate-Based Sorbent: Comparison with Al_2O_3 Particles. *Polymers* **2019**, *11*, 1509. [CrossRef]
51. Al-Ghouti, M.A.; Khan, M.; Malik, A.; Khraisheh, M.; Hijazi, D.; Mohamed, S.; Alsorour, S.; Eltayeb, R.; Al Mahmoud, F.; Alahmad, J. Development of Novel Nano-γ-Al_2O_3 Adsorbent from Waste Aluminum Foil for the Removal of Boron and Bromide from Aqueous Solution. *J. Water Process Eng.* **2022**, *50*, 103312. [CrossRef]
52. Demetriou, A.; Pashalidis, I. Adsorption of Boron on Iron-Oxide in Aqueous Solutions. *Desalination Water Treat.* **2012**, *37*, 315–320. [CrossRef]

53. Liu, H.; Qing, B.; Ye, X.; Li, Q.; Lee, K.; Wu, Z. Boron Adsorption by Composite Magnetic Particles. *Chem. Eng. J.* **2009**, *151*, 235–240. [CrossRef]
54. Chen, T.; Wang, Q.; Lyu, J.; Bai, P.; Guo, X. Boron Removal and Reclamation by Magnetic Magnetite (Fe_3O_4) Nanoparticle: An Adsorption and Isotopic Separation Study. *Sep. Purif. Technol.* **2020**, *231*, 115930. [CrossRef]
55. Kameda, T.; Yamamoto, Y.; Kumagai, S.; Yoshioka, T. Mechanism and Kinetics of Aqueous Boron Removal Using MgO. *J. Water Process Eng.* **2018**, *26*, 237–241. [CrossRef]
56. Li, P.; Liu, C.; Zhang, L.; Zheng, S.; Zhang, Y. Enhanced Boron Adsorption onto Synthesized MgO Nanosheets by Ultrasonic Method. *Ultrason. Sonochem.* **2017**, *34*, 938–946. [CrossRef] [PubMed]
57. Masindi, V.; Gitari, M.W. Removal of Boron from Aqueous Solution Using Cryptocrystalline Magnesite. *J. Water Reuse Desalination* **2017**, *7*, 205–213. [CrossRef]
58. de la Fuente García-Soto, M.M.; Muñoz Camacho, E. Boron Removal by Means of Adsorption Processes with Magnesium Oxide-Modelization and Mechanism. *Desalination* **2009**, *249*, 626–634. [CrossRef]
59. Fukuda, H.; Tsuchiya, K.; Toba, Y.; Eguchi, M.; Tokoro, C. Rapid Boron Removal from Wastewater Using Low-Crystalline Magnesium Oxide. *J. Environ. Chem. Eng.* **2020**, *8*, 104171. [CrossRef]
60. Song, T.; Gao, F.; Du, X.; Hao, X.; Liu, Z. Removal of Boron in Aqueous Solution by Magnesium Oxide with the Hydration Process. *Colloids Surfaces A Physicochem. Eng. Asp.* **2023**, *665*, 131211. [CrossRef]
61. Gao, Z.; Xie, S.; Zhang, B.; Qiu, X.; Chen, F. Ultrathin Mg-Al Layered Double Hydroxide Prepared by Ionothermal Synthesis in a Deep Eutectic Solvent for Highly Effective Boron Removal. *Chem. Eng. J.* **2017**, *319*, 108–118. [CrossRef]
62. Liu, J.; Guo, X.; Yuan, J. Synthesis of Mg/Al Double-Layered Hydroxides for Boron Removal. *Desalination Water Treat.* **2014**, *52*, 1919–1927. [CrossRef]
63. Kentjono, L.; Liu, J.C.; Chang, W.C.; Irawan, C. Removal of Boron and Iodine from Optoelectronic Wastewater Using Mg–Al (NO_3) Layered Double Hydroxide. *Desalination* **2010**, *262*, 280–283. [CrossRef]
64. Kurashina, M.; Inoue, T.; Tajima, C.; Kanezaki, E. Removal of Borate by Coprecipitation with Mg / Al Layered Double Hydroxide. *Mod. Phys. Lett. B* **2015**, *29*, 1540031. [CrossRef]
65. Kameda, T.; Oba, J.; Yoshioka, T. New Treatment Method for Boron in Aqueous Solutions Using Mg–Al Layered Double Hydroxide: Kinetics and Equilibrium Studies. *J. Hazard. Mater.* **2015**, *293*, 54–63. [CrossRef] [PubMed]
66. Irawan, C.; Liu, J.C.; Wu, C.-C. Removal of Boron Using Aluminum-Based Water Treatment Residuals (Al-WTRs). *Desalination* **2011**, *276*, 322–327. [CrossRef]
67. Said, S.; Mikhail, S.; Riad, M. Recent Progress in Preparations and Applications of Meso-Porous Alumina. *Mater. Sci. Energy Technol.* **2019**, *2*, 288–297. [CrossRef]
68. Demetriou, A.; Pashalidis, I.; Nicolaides, A.V.; Kumke, M.U. Surface Mechanism of the Boron Adsorption on Alumina in Aqueous Solutions. *Desalination Water Treat.* **2013**, *51*, 6130–6136. [CrossRef]
69. Zeebe, R.E.; Sanyal, A.; Ortiz, J.D.; Wolf-Gladrow, D.A. A Theoretical Study of the Kinetics of the Boric Acid-Borate Equilibrium in Seawater. *Mar. Chem.* **2001**, *73*, 113–124. [CrossRef]
70. Wang, B.; Guo, X.; Bai, P. Removal Technology of Boron Dissolved in Aqueous Solutions—A Review. *Colloids Surfaces A Physicochem. Eng. Asp.* **2014**, *444*, 338–344. [CrossRef]
71. Mohapatra, M.; Anand, S. Synthesis and Applications of Nano-Structured Iron Oxides/Hydroxides—A Review. *Int. J. Eng. Sci. Technol.* **2011**, *2*, 127–146. [CrossRef]
72. Lou, X.-Y.; Boada, R.; Verdugo, V.; Simonelli, L.; Pérez, G.; Valiente, M. Decoupling the Adsorption Mechanisms of Arsenate at Molecular Level on Modified Cube-Shaped Sponge Loaded Superparamagnetic Iron Oxide Nanoparticles. *J. Environ. Sci.* **2022**, *121*, 1–12. [CrossRef] [PubMed]
73. Kharissova, O.V.; Dias, H.V.R.; Kharisov, B.I. Magnetic Adsorbents Based on Micro- and Nano-Structured Materials. *RSC Adv.* **2015**, *5*, 6695–6719. [CrossRef]
74. Díez, A.G.; Rincón-Iglesias, M.; Lanceros-Méndez, S.; Reguera, J.; Lizundia, E. Multicomponent Magnetic Nanoparticle Engineering: The Role of Structure-Property Relationship in Advanced Applications. *Mater. Today Chem.* **2022**, *26*, 101220. [CrossRef]
75. Glasgow, W.; Fellows, B.; Qi, B.; Darroudi, T.; Kitchens, C.; Ye, L.; Crawford, T.M.; Mefford, O.T. Continuous Synthesis of Iron Oxide (Fe_3O_4) Nanoparticles via Thermal Decomposition. *Particuology* **2016**, *26*, 47–53. [CrossRef]
76. Yetgin, A.G.; Dündar, O.A.; Çakmakçı, E.; Arar, Ö. Removal of Boron from Aqueous Solution by Modified Cellulose. *Biomass Convers. Biorefinery* **2023**, *13*, 13081–13090. [CrossRef]
77. Falyouna, O.; Bensaida, K.; Maamoun, I.; Ashik, U.P.M.; Tahara, A.; Tanaka, K.; Aoyagi, N.; Sugihara, Y.; Eljamal, O. Synthesis of Hybrid Magnesium Hydroxide/Magnesium Oxide Nanorods [$Mg(OH)_2$/MgO] for Prompt and Efficient Adsorption of Ciprofloxacin from Aqueous Solutions. *J. Clean. Prod.* **2022**, *342*, 130949. [CrossRef]
78. Heredia, A.C.; de la Fuente García-Soto, M.M.; Narros Sierra, A.; Mendoza, S.M.; Gómez Avila, J.; Crivello, M.E. Boron Removal from Aqueous Solutions by Synthetic MgAlFe Mixed Oxides. *Ind. Eng. Chem. Res.* **2019**, *58*, 9931–9939. [CrossRef]
79. Akdağ, S.; Keyikoğlu, R.; Karagunduz, A.; Keskinler, B.; Khataee, A.; Yoon, Y. Recent Advances in Boron Species Removal and Recovery Using Layered Double Hydroxides. *Appl. Clay Sci.* **2023**, *233*, 106814. [CrossRef]
80. Korkmaz, M.; Özmetin, C.; Özmetin, E.; Süzen, Y. Modelling of Boron Removal from Solutions by Ion Exchange for Column Reactor Design in Boron Mine Wastewater Treatment. *Desalination Water Treat.* **2020**, *179*, 63–74. [CrossRef]

81. Torunoglu, T.G.; Burak, K.; Erdem, Y.; Bahire Filiz, S. A New Multi-Hydroxyl Methacrylate Based Adsorbent for Boron Removal from Aqueous Solutions. *Sep. Sci. Technol.* **2022**, *57*, 1702–1713. [CrossRef]
82. Duran, H.; Yavuz, E.; Sismanoglu, T.; Senkal, B.F. Functionalization of Gum Arabic Including Glycoprotein and Polysaccharides for the Removal of Boron. *Carbohydr. Polym.* **2019**, *225*, 115139. [CrossRef] [PubMed]
83. Zhang, Y.; Liang, X.; Jiang, T.; Chen, Z.; Ji, X. Removal of Boron from Aqueous Solutions by N-Methyl-D-Glucamine@γ-Glycidoxypropyl-Trimethoxysilane with Nonwoven Basalt Fibers Support. *Mater. Today Commun.* **2022**, *31*, 103611. [CrossRef]
84. Ke, Q.; Ulbricht, M. In Situ Reactive Coating of Porous Filtration Membranes with Functional Polymer Layers to Integrate Boron Adsorber Property. *J. Memb. Sci.* **2022**, *660*, 120851. [CrossRef]
85. Manjunatha Reddy, G.N.; Gerbec, J.A.; Shimizu, F.; Chmelka, B.F. Nanoscale Surface Compositions and Structures Influence Boron Adsorption Properties of Anion Exchange Resins. *Langmuir* **2019**, *35*, 15661–15673. [CrossRef]
86. Tural, S.; Ece, M.Ş.; Tural, B. Synthesis of Novel Magnetic Nano-Sorbent Functionalized with N-Methyl-D-Glucamine by Click Chemistry and Removal of Boron with Magnetic Separation Method. *Ecotoxicol. Environ. Saf.* **2018**, *162*, 245–252. [CrossRef]
87. Oladipo, A.A.; Gazi, M. Hydroxyl-Enhanced Magnetic Chitosan Microbeads for Boron Adsorption: Parameter Optimization and Selectivity in Saline Water. *React. Funct. Polym.* **2016**, *109*, 23–32. [CrossRef]
88. Oladipo, A.A.; Gazi, M. High Boron Removal by Functionalized Magnesium Ferrite Nanopowders. *Environ. Chem. Lett.* **2016**, *14*, 373–379. [CrossRef]
89. Sanfeliu, C.; Martínez-Máñez, R.; Sancenón, F.; Soto, J.; Amorós, P.; Azaïs, T.; Marcos, M.D. 11B-MAS NMR Approach to the Boron Adsorption Mechanism on a Glucose-Functionalised Mesoporous Silica Matrix. *Microporous Mesoporous Mater.* **2018**, *266*, 232–241. [CrossRef]
90. Tang, Y.P.; Chung, T.S.; Weber, M.; Maletzko, C. Development of Novel Diol-Functionalized Silica Particles toward Fast and Efficient Boron Removal. *Ind. Eng. Chem. Res.* **2017**, *56*, 11618–11627. [CrossRef]
91. Tural, B. Separation and Preconcentration of Boron with a Glucamine Modified Novel Magnetic Sorbent. *Clean-Soil Air Water* **2010**, *38*, 321–327. [CrossRef]
92. Xu, L.; Liu, Y.; Hu, H.; Wu, Z.; Chen, Q. Synthesis, Characterization and Application of a Novel Silica Based Adsorbent for Boron Removal. *Desalination* **2012**, *294*, 1–7. [CrossRef]
93. Wang, J.; Bao, Z.; Xing, H.; Su, B.; Zhang, Z.; Yang, Q.; Yang, Y.; Ren, Q.; Chen, B. Incorporation of N-Methyl-d-Glucamine Functionalized Oligomer into MIL-101(Cr) for Highly Efficient Removal of Boric Acid from Water. *Chem.-A Eur. J.* **2016**, *22*, 15290–15297. [CrossRef] [PubMed]
94. Li, X.; Liu, R.; Wu, S.; Liu, J.; Cai, S.; Chen, D. Efficient Removal of Boron Acid by N-Methyl-d-Glucamine Functionalized Silica-Polyallylamine Composites and Its Adsorption Mechanism. *J. Colloid Interface Sci.* **2011**, *361*, 232–237. [CrossRef]
95. Zhang, X.; Wang, J.; Chen, S.; Bao, Z.; Xing, H.; Zhang, Z.; Su, B.; Yang, Q.; Yang, Y.; Ren, Q. A Spherical N-Methyl-D-Glucamine-Based Hybrid Adsorbent for Highly Efficient Adsorption of Boric Acid from Water. *Sep. Purif. Technol.* **2017**, *172*, 43–50. [CrossRef]
96. Moorthy, M.S.; Seo, D.-J.; Song, H.-J.; Park, S.S.; Ha, C.-S. Magnetic Mesoporous Silica Hybrid Nanoparticles for Highly Selective Boron Adsorption. *J. Mater. Chem. A* **2013**, *1*, 12485. [CrossRef]
97. Hong, M.; Li, D.; Wang, B.; Zhang, J.; Peng, B.; Xu, X.; Wang, Y.; Bao, C.; Chen, J.; Zhang, Q. Cellulose-Derived Polyols as High-Capacity Adsorbents for Rapid Boron and Organic Pollutants Removal from Water. *J. Hazard. Mater.* **2021**, *419*, 126503. [CrossRef] [PubMed]
98. Luo, Q.; Wang, Y.; Li, L.; Huang, X.; Cheng, Z.; Wang, X.; He, L. Hydrothermal Synthesis of Hydroxyl Terminated Polymer Boron Adsorbents. *J. Solid State Chem.* **2021**, *296*, 121977. [CrossRef]
99. Cyganowski, P.; Şen, F.; Altıok, E.; Wolska, J.; Bryjak, M.; Kabay, N.; Arda, M.; Yüksel, M. Surface-Activated Chelating Resins Containing N-Methyl-D-Glucamine Functional Groups for Desalination of Geothermal Water Aimed for Removal of Boron and Arsenic. *Solvent Extr. Ion Exch.* **2021**, *39*, 584–603. [CrossRef]
100. Liao, X.; Zhang, Q. Mesoporous Polymer Nanosponges Immobilized with Functional Polyols for Rapid Removal of Boric Acid and Organic Micropollutants. *ACS Appl. Polym. Mater.* **2019**, *1*, 2089–2098. [CrossRef]
101. Wu, Q.; Liu, M.; Wang, X. A Novel Chitosan Based Adsorbent for Boron Separation. *Sep. Purif. Technol.* **2019**, *211*, 162–169. [CrossRef]

molecules

MDPI

Article

Significant Differences in the Effects of Nitrogen Doping on Pristine Biochar and Graphene-like Biochar for the Adsorption of Tetracycline

Lingling Rong [1,2,†], Ligui Wu [3,†], Tiao Zhang [2], Cui Hu [2], Haihui Tang [2], Hongcheng Pan [1,*] and Xiaoming Zou [2,*]

1 College of Environmental Science and Engineering, Guilin University of Technology, Guilin 541004, China; ronglingling1616@163.com
2 School of Life Science, Jinggangshan University, 28 Xueyuan Road, Ji'an 343009, China; zhangtiao2020@163.com (T.Z.); hucui912@126.com (C.H.); tanghaihui_1995@163.com (H.T.)
3 College of Environmental Science and Engineering, Tongji University, Shanghai 200092, China; wuligui1229@126.com
* Correspondence: hcpan@163.com (H.P.); zouxming_80@hotmail.com (X.Z.); Tel./Fax: +86-796-8100493 (X.Z.)
† These authors contributed equally to this work.

Abstract: To improve the adsorption efficiency of pollutants by biochar, preparing graphene-like biochar (GBC) or nitrogen-doped biochar are two commonly used methods. However, the difference in the nitrogen doping (N-doping) effects upon the adsorption of pollutants by pristine biochar (PBC) and GBC, as well as the underlying mechanisms, are still unclear. Take the tetracycline (TC) as an example, the present study analyzed the characteristics of the adsorption of TCs on biochars (PBC, GBC, N-PBC, N-GBC), and significant differences in the effects of N-doping on the adsorption of TCs by PBC and GBC were consistently observed at different solution properties. Specifically, N-doping had varied effects on the adsorption performance of PBC, whereas it uniformly improved the adsorption performance of GBC. To interpret the phenomenon, the N-doping upon the adsorption was revealed by the QSAR model, which indicated that the pore filling (V_M) and the interactions between TCs with biochars ($E_{ad\text{-}v}$) were found to be the most important two factors. Furthermore, the density functional theory (DFT) results demonstrated that N-doping slightly affects biochar's chemical reactivity. The van der Waals (vdWs) and electrostatic interactions are the main forces for TCs-biochars interactions. Moreover, N-doping mostly strengthened the electrostatic interactions of TCs-biochars, but the vdWs interactions of most samples remained largely unaffected. Overall, the revealed mechanism of N-doping on TCs adsorption by biochars will enhance our knowledge of antibiotic pollution remediation.

Keywords: nitrogen doping; biochar; tetracycline antibiotics; adsorption mechanism; density functional theory

1. Introduction

The water environment is threatened by various pollutants due to rapid industrialization, such as heavy metals and organic pollutants, which is considered the main problem affecting the sustainable development of human beings [1]. Especially, the global use of antibiotics in both livestock and humans has increased significantly since the outbreak of COVID-19. This has led to a concerning level of antibiotic pollution in water bodies, which poses a serious threat to public health and the environment [2]. Tetracycline antibiotics (TCs) are widely used antimicrobials that inhibit bacterial protein synthesis by hindering the binding process between aminoacyl-tRNA and ribosomes [3]. These antibiotics possess broad-spectrum antimicrobial properties and have found extensive application in human therapy, animal disease control, as well as feed additives in livestock and aquaculture [4–6]. However, the pervasive misuse and overuse of TCs in recent decades have led to their steady discharge into various aquatic environments. Regrettably, TCs have a stable chemical structure that hinders their biodegradation, resulting in the presence of approximately 75% of parent compounds in groundwater, wastewater, and drinking water at ng·L^{-1} to μg·L^{-1} concentrations [7]. This is a cause for significant concern, as residual TCs have been observed to exhibit toxic effects on aquatic organisms, including bacteria, algae, and crustaceans. Moreover, their presence stimulates the proliferation of antibiotic resistance genes (ARGs) and disrupts the ecosystem balance of the original microbial community [8]. In some cases, this disruption has even led to the emergence of superbugs [9]. Consequently, it is imperative to address the removal of TCs from aquatic environments to ensure the continued health of these ecosystems [10].

Numerous treatment techniques have been employed to remove TCs from the polluted aquatic environment [11], including biological wastewater treatment plants (WWTPs), electrochemistry, membrane separation, advanced oxidation processes (AOPs), and adsorption [12]. However, conventional WWTPs often have insufficient removal efficiency for TCs due to their stable chemical structure, which is toxic to bacteria and resistant to biodegradation [13]. On the other hand, AOPs can achieve high removal efficiency for TCs due to the presence of highly reactive radicals [14]. Nevertheless, the practical application of AOPs is limited by their expensive operation, the formation of intermediate products, and the low concentration of TCs in real environments [15]. Adsorption, on the other hand, provides a practical solution to water contaminated with TCs due to its safety, low cost, and ease of operation. Recently, a variety of absorbents, including activated carbon [16], zeolite [17], biochars [18], and other porous materials, have been used. Biochar, in particular, is a carbon-rich material with a large specific surface area (SSA), rich functional groups, and a high cation exchange capacity. It can be produced from waste feedstock [19]. However, concerns persist regarding the removal of TCs and other antibiotics by biochars due to their excellent adsorption performance and economic benefits [20].

In general, commonly used biochar exhibits natural deficiencies, such as low hydrophilicity, insufficient surface active sites, and the generation of numerous tar-like substances and amorphous carbon during the pyrolysis process. These factors have been observed to decrease the SSA of biochar, and, consequently, result in a reduction in its adsorption efficiency for pollutants [21]. To overcome these limitations, various strategies have been proposed to develop novel modified biochars, which can be classified into chemical modification (e.g., acid modification, alkalinity modification, oxidizing agent modification, metal salts) and physical modification (e.g., steam and gas purging) approaches [22]. Among these approaches, alkali treatment has been proven to effectively increase the surface area, pore volume, and number of functional groups on the surface of biochar [23]. Consequently, alkali-prepared graphene-like biochar (GBC), a novel type of material, has recently gained attention as an adsorbent and support material [24]. Furthermore, the surface properties of biochar can be enhanced by incorporating heteroatoms, such as nitrogen (N), sulfur (S), and phosphorus (P), which have been demonstrated to improve biochar's unique adsorption performance [25]. However, the specific effects of N-doping on the adsorption performance of GBC and pristine biochar (PBC) in the case of antibiotic adsorption have yet to be fully elucidated.

The adsorption process typically involves multiple stages, including solute transportation, diffusion across the liquid film, diffusion within the liquid, and interactions with active sites [18]. Various factors influence the adsorption performance, such as the structural characteristics of the adsorbate, the properties of the adsorbent surface, and the adsorption interface [26]. Specifically, previous studies showed that the π-π interaction and SSA of biochar are increased after N-doping, which improves the adsorption capacity for bisphenol A and phenol [27,28]. For the adsorption of TCs onto adsorbents, the π-π interaction is known to be crucial in the adsorption process of nitrogen-doped (N-doped) lignin-based carbon material [29]. However, the details of how N-doping affects the adsorption performance of GBC and PBC for TCs are yet to be fully understood. Understanding these mechanisms is crucial for the development of more effective adsorbents for TCs removal.

Density functional theory (DFT) is a potent method for quantitatively investigating the micro-mechanisms of adsorption between biochars and antibiotics [30]. Oxytetracycline (OTC), tetracycline (TC), and other tested TCs are frequently found in polluted aquatic environments and have been extensively studied to understand their adsorption behaviors [31,32]. Therefore, the objectives of this study are to (1) evaluate the influence of N-doping on PBC and GBC in terms of TC adsorption, and (2) elucidate the underlying mechanisms behind the differential impact of N-doping on TCs adsorption efficiency in PBC and GBC.

2. Results and Discussion

2.1. Characterization of the Tested Biochars

The basic physical and chemical properties of the four tested biochars were determined, including the surface morphology, atomic ratios, and point of zero charge (pH_{PZC}) (Figure 1 and Table S1). The elemental composition of the tested biochars is listed in Table S1. As shown, the order of the molar ratios for O/C, (O+N)/C, and H/C was N-GBC > GBC > N-PBC > PBC, N-GBC > N-PBC > GBC > PBC and GBC > N-GBC > PBC > N-PBC, respectively. It can be concluded that (1) N-doped biochars possessed a higher content of N element (>4.50%) than those of N-undoped biochars (<0.50%), which was consistent with the SEM results of Figure 1; (2) a higher total percentage of O combined N was obtained in the N-doped biochars (>34.40%) than that of N-undoped biochars (<19.60%). Those results suggested that higher hydrophobicity (O/C) and polarity ((N+O)/C) and lower aromaticity (H/C) were obtained in N-doped biochars (N-PBC, N-GBC) [33]. Consequently, it is reasonable to speculate that the N-doping process may cause different effects on the TCs adsorption performance of PBC and GBC.

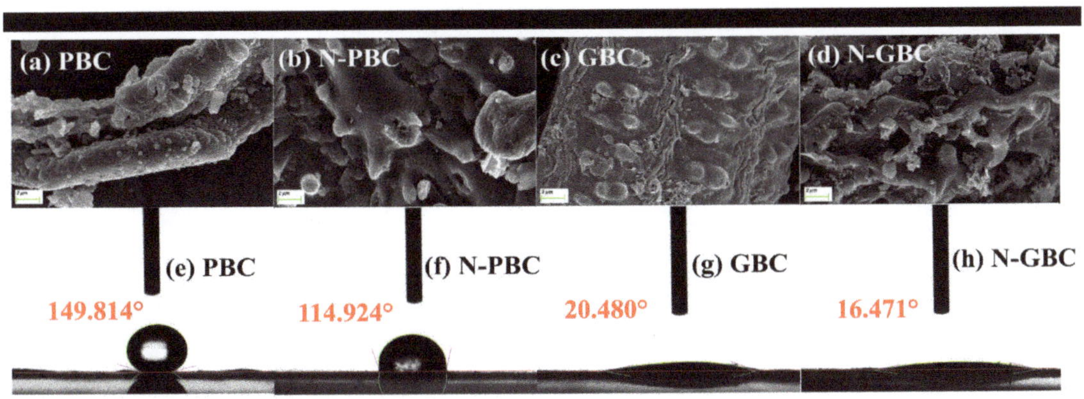

Figure 1. SEM images (5000-fold) (**a–d**) and contact angles (**e–h**) of tested biochars.

2.2. Adsorption Characteristics

2.2.1. Adsorption Kinetics

The kinetics of adsorption are frequently employed to investigate the time required to attain adsorption equilibrium and to discern potential adsorption mechanisms [34]. As shown in Figure 2 and Table S2, it was observed that the PSO model (R^2 > 0.90) provided a better fit for the adsorption data than the PFO model (R^2 > 0.80), indicating that chemisorption is likely the primary adsorption process involved in the tested biochars [35]. Notably, the fast adsorption process of five TCs (DMC, DOC, OTC, TTC, and DOX) onto the tested biochars (PBC, N-PBC, GBC, and N-GBC) was achieved with almost 90% adsorption quantity within 100 min and near equilibrium level at 200 min (Figure 2a–h). However, adsorption equilibrium was not reached until roughly 800 min for the other three TCs (CTC, MN, and TG). Furthermore, according to the fitted results of the PSO model, the adsorption efficiency of TCs (q_t) for PBC, N-PBC, GBC, and N-GBC ranged from 41.16 mg·g^{-1}

to 235.51 mg·g^{-1}, 17.80 mg·g^{-1} to 270.25 mg·g^{-1}, 83.58 mg·g^{-1} to 308.09 mg·g^{-1}, and 135.91 mg·g^{-1} to 327.20 mg·g^{-1}, respectively.

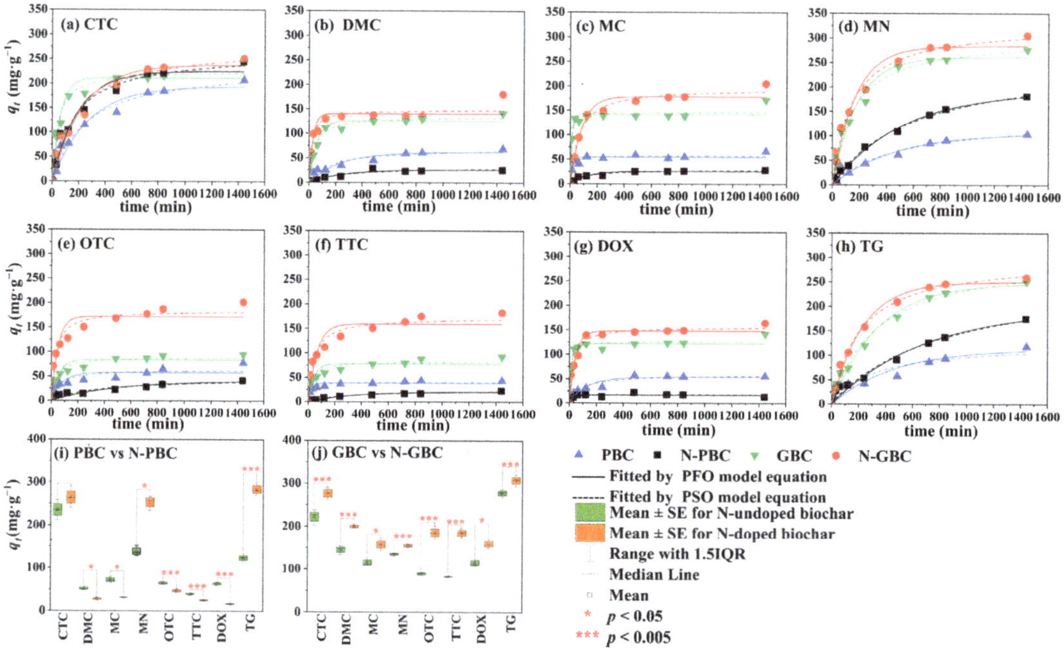

Figure 2. Fitted kinetic curves by the model of PFO and PSO of tested TCs adsorbed by N-doped and N-undoped biochars (**a–h**) and the statistical distribution of their TCs adsorption efficiency (q_t) (**i**,**j**).

To further analyze the difference among samples, paired t-test analysis was performed on the above data (Figure 2i,j). The results showed that except for the groups of GBC/N-GBC for MN, OTC, and TTC, there is a statistically significant difference ($p < 0.05$) or very significant difference ($p < 0.005$) between samples with the corresponding N-doped samples, indicating the significant results of N-doping upon the adsorption of TCs by PBC and GBC. Specifically, it was determined that N-doping had differential effects on adsorption efficiency, depending on whether the biochar was a pristine group (PBC, N-PBC) or a graphene-like group (GBC, N-GBC). In the case of the pristine group, N-doping improved or reduced adsorption performance, as evidenced by the fact that the q_t values of N-PBC increased from −69.988% to 116.096% compared with PBC. Conversely, N-doping consistently improved the adsorption performance for the graphene-like group, where the q_t values of N-GBC increased from 12.90% to 106.23% compared to GBC.

2.2.2. Adsorption Isotherm

Adsorption isotherm models are commonly utilized to characterize the interaction mechanisms between adsorbent and adsorbate, including the distribution of adsorbates in aqueous solutions [36]. In the present study, experimental data were fitted using both the Freundlich and Langmuir models, with the latter demonstrating superior accuracy in describing the adsorption of seven TCs (CTC, DOX, MC, TG, DMC, TTC, and MN) onto tested biochars (Figure 3). The fitting analysis revealed that the R^2 values for the Langmuir model ranged from 0.630 to 0.999, while those for the Freundlich model ranged from 0.930 to 0.998, indicating that the Freundlich model exhibited a better fit for describing the adsorption of OTC on biochar. These results suggested that the adsorption process of TCs onto the tested biochars was primarily monolayer rather than multilayer adsorption [18].

Furthermore, significant differences were observed between the effects of N-doping on the adsorption performance of the pristine group (PBC, N-PBC) and graphene-like group (GBC, N-GBC). Based on the fitted results of the Langmuir model (Table S3), it was observed that the maximum adsorption capacity (q_m) for PBC, N-PBC, GBC, and N-GBC varied from 60.260 mg·g^{-1} to 552.290 mg·g^{-1}, 32.800 mg·g^{-1} to 464.810 mg·g^{-1}, 77.860 mg·g^{-1} to 491.110 mg·g^{-1}, and 160.840 mg·g^{-1} to 942.710 mg·g^{-1}, respectively. When it comes to the pristine group (PBC, N-PBC), the effect of N-doping on the adsorption performance can be two-fold, as evidenced by the fact that the q_m values of N-PBC increased by −79.765% to 57.374% compared to PBC. On the other hand, for the graphene-like group (GBC, N-GBC), N-doping consistently improved the adsorption performance, as the q_m values of N-GBC increased by 7.108% to 141.934% compared to GBC. Additionally, it should be noted that the efficacy of N-doping for enhancing adsorption performance in graphene-like groups is influenced by the specific structural characteristics of the tested TCs. The outcome of the ANOVA revealed that there is a statistically significant distinction between the adsorption efficiency of TCs for GBC and N-GBC ($p = 0.0472$). Notably, N-GBC exhibits higher q_m values compared to GBC, with a pattern of increasing order as DMC < DOC < MN < CTC < MC < OTC < TTC < TG.

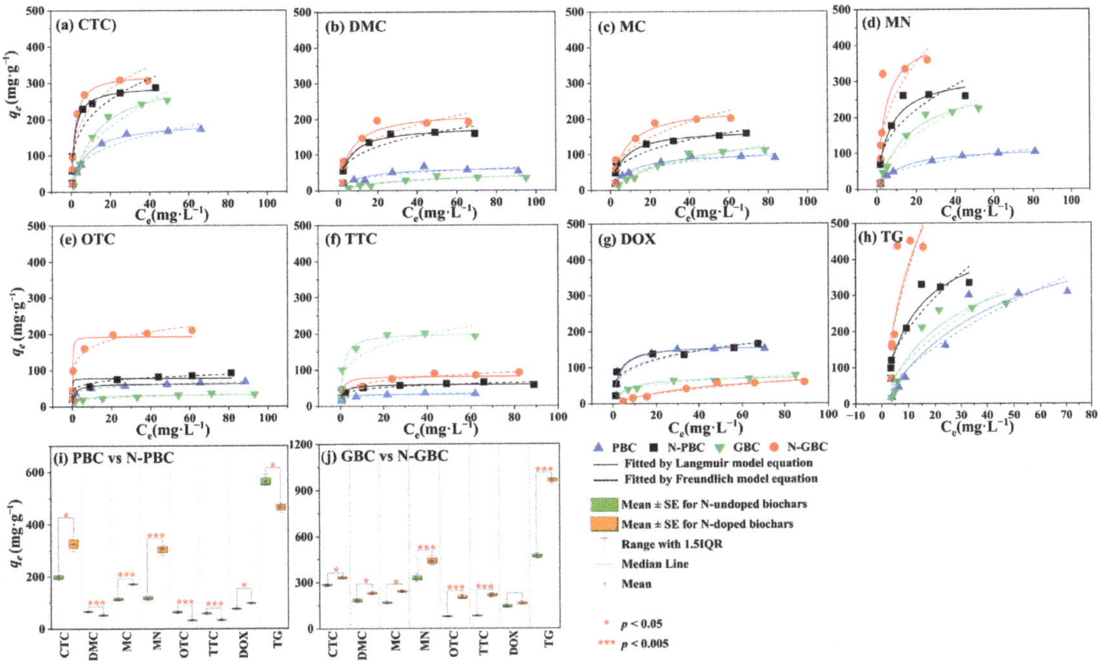

Figure 3. Fitted kinetic curves by the model of Langmuir and Freundlich of eight TCs.

Overall, N-doping exhibits a remarkable strengthening effect on the adsorption performance of the graphene-like group (N-GBC, GBC). This phenomenon is closely correlated with the structural characteristics of the tested TCs. However, in the pristine group (PBC, N-PBC), the impact of N-doping on the adsorption efficiency may vary.

2.3. *Influence of Solution Properties*

To further verify the aforementioned conclusion, the impact of solution properties on the adsorption of TCs by the tested biochars was assessed under conditions of varying pH values (3–11), humic acid (HA) concentrations (0–30 mg·L^{-1}), and salinity concentrations (0–10 mg·L^{-1}). Results presented in Figure S1a,b indicated that the adsorption of TTC and OTC onto the tested biochars was significantly influenced by solution pH ($p < 0.01$). Changes in salinity can significantly impact the electrostatic interactions between chemicals and biochars, subsequently altering their adsorption features [37]. As shown in Figure S1c,d, our study found that the addition of Ca^{2+} had no significant effects on the adsorption of OTC-GBC ($p > 0.05$), but did have significant effects on other adsorption processes ($p < 0.01$ or $p < 0.05$). HA is a typical dissolved organic matter with π electrons, aromatic rings, and fatty acids. As presented in Figure S1e,f, the HA has significant effects on the adsorption of tested TCs (OTC, TTC) ($p < 0.01$). Overall, results of solution properties indicate that (1) the significant strengthening and complex effects of N-doping on the TCs adsorption performance can be obtained for PBC and GBC, respectively, and (2) the electrostatic interaction and π-π interaction are important mechanisms of TCs interacting with tested biochars, which can be changed by the N-doping process. Detailed information about the influence of solution properties upon adsorption features was listed in Supplementary Materials (SM, Text S1).

2.4. *Mechanisms for the Effects of N-Doping upon the Adsorption Performance of TCs*

According to the previous results [38], N elements in the N-doped biochar are mainly classified as chemical N and structural N, which significantly modify the surface in the form of N-containing functional groups and thereby change the SSA with a developed pore structure. In the present study, the molar ratios of H/C, O/C, and (O+N)/C for the tested biochars were different (Table S4). Consequently, it could be speculated that the structural differentiation of the tested biochars is the original driver resulting in the different characterization of TCs adsorption.

2.4.1. Modification of Pore Structure

The porous characteristics of tested biochars were analyzed by N_2 adsorption and desorption isotherms, as shown in Figure 4a, which can be viewed as type I and IV adsorption isotherms. For all samples, the isotherms rapidly increased at $P/P_0 < 0.1$, strongly demonstrating the presence of micropore structure in the biochars. The isotherms of GBC and N-GBC also indicated multilayer adsorption followed by capillary condensation, as proved by the obviously increasing adsorbing capacity at $P/P_0 > 0.9$, indicating the presence of a mesopore structure in the GBC and N-GBC. It should be noted that the slightly increasing adsorbing capacity was obtained in N-PBC rather than in PBC. Furthermore, the pore size distribution of the tested biochars was proved to range from 2–5 nm (Figure 4b), confirming that the pores in the biochars were mainly mesopores. The positions of the most probable aperture size for PBC, N-PBC, GBC, and N-GBC were 3.122 nm, 3.611 nm, 4.931 nm, and 4.147 nm, respectively. The results indicate that the N-doping process slightly hinders the mesopore formation of N-PBC but aids the mesopore formation of N-GBC. A smaller pore diameter typically corresponds to a larger pore volume and SSA. As shown in Figure 4a, N-doping treatment significantly affected SSA and pore volume ($p < 0.01$). The SSA for PBC, N-PBC, GBC, and N-GBC was 374.67 $m^2 \cdot g^{-1}$, 203.11 $m^2 \cdot g^{-1}$, 807.41 $m^2 \cdot g^{-1}$, and 935.09 $m^2 \cdot g^{-1}$, respectively. The pore volumes for PBC, N-PBC, GBC, and N-GBC were 0.127 $m^2 \cdot g^{-1}$, 0.031 $m^2 \cdot g^{-1}$, 0.229 $m^2 \cdot g^{-1}$, and 0.294 $m^2 \cdot g^{-1}$, respectively. Obviously, those results consistently demonstrated that the N-doping process significantly decreased the adsorption sphere of PBC but significantly increased the adsorption sphere of N-GBC. Similar results were also obtained in the study of Diao et al. [39], which can be explained by the breaking down of carbon framework due to the gas produced by urea during pyrolysis [40].

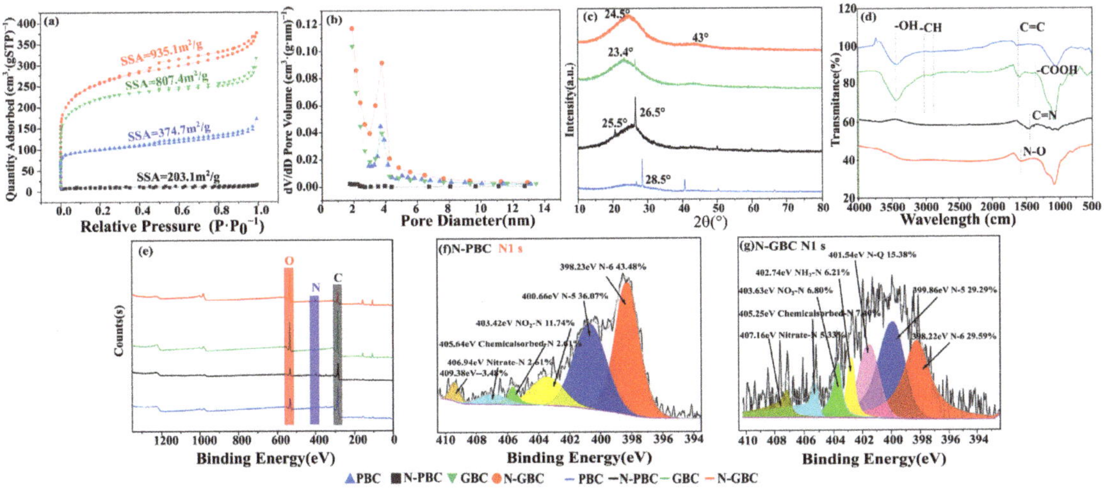

Figure 4. N_2 adsorption and desorption isotherms (a), pore size distribution (b), XRD patterns (c), FTIR images (d), XPS diagrams (e) of tested biochars and XPS regional spectra of nitrogen in N-PBC and N-GBC (f,g).

2.4.2. Modification of Surface Functional Groups

The short-range ordered structures of the tested biochars were revealed through XRD patterns, with the observed strong and sharp peaks indicating the presence of crystalline inorganic phases. Figure 4c shows that the broad peaks at 2θ = 23~29° were consistently observed and considered as amorphous structures of carbon [41,42], consistent with the SEM results in Figure 1. The peaks (2θ = 23~29°) of N-doped biochars (N-PBC, N-GBC) were flatter than those of PBC and GBC, indicating that N-doping improved the crystallinity of the carbon in these biochars [38]. Furthermore, FTIR spectra were determined to verify the main functional group of the tested biochars (Figure 4d). As shown, the N-undoped biochars exhibited similar FTIR spectra, with major bands occurring at peaks around 3436 cm^{-1} (–OH), 1514 cm^{-1} (–COOH), 1590 cm^{-1} (aromatic C=O and C=C), 1400 cm^{-1} and 2900 cm^{-1} (–CH), and 1050 cm^{-1} (–SiO_3^{2-}) [18]. Compared to the N-undoped group, the important peaks of N-PBC and N-GBC were shifted, strengthened, or weakened. Specifically, the vibration intensity of 3436 cm^{-1} (–OH) was weakened, and similar results were also obtained in the study of Li et al. [43], which can be mainly attributed to the replacement of O atoms by the N-doped structures (e.g., pyridinic-nitrogen, pyrrolic-nitrogen, and graphitic-nitrogen) [38]. The vibration intensity of around 1600 cm^{-1} (C=N) was strengthened, indicating the successful doping of N atoms in the carbon network structure [44]. The vibration intensity of around 1514 cm^{-1} (–COOH) was also modified, suggesting that the –COOH group was affected by the N-doping process [45].

To analyze the functional groups of the tested biochars on a quantitative level, the XPS spectra were further determined because XPS measurement is widely used to analyze the binding relationship between elements [46]. Figure 4e illustrates the dominance of electron energy spectra peaks associated with C1s and O1s primarily on the surfaces of PBC and GBC, delineating carbon (C) and oxygen (O) as the predominant constituents of rice husk. This observation aligns consistently with elemental analysis results. However, subsequent biochar modification employing $C_2H_4N_4$ induced notable alterations in elemental composition. Remarkably, N contents in N-PBC and N-GBC exhibited substantial escalation, with increments of 9.08% and 45.28%, respectively, in comparison to their precursor PBC. Zhou et al. utilized N dopants—such as amides, triethanolamine, and ethylenediamine—to augment biochar derived from corn stover, achieving heightened N content and bolstering CO_2 adsorption capabilities [47]. Additionally, scrutiny of the C1s,

O1s, and N1s spectra (Figure 4e,f and Figure S2) revealed the deconstruction of combined C1s and O1s relationships into four distinct C functional groups and three O functional groups. Notably, N1s spectroscopy predominantly identified three types of N functional groups: pyridinic-nitrogen (398.23 eV, 398.22 eV), pyrrolic-nitrogen (400.66 eV, 399.86 eV), and graphitic-nitrogen (401.54 eV). Previous studies elucidate that within N-doped biochar, pyridinic-nitrogen emerges as the principal active N species, pivotal in initiating the activation of peroxymonosulfate and facilitating the degradation of ciprofloxacin processes [48]. Consequently, this infers that within this study, the emergence of heterocyclic nitrogen configurations (e.g., pyrrolic, pyridinic, and graphitic nitrogen) subsequent to the further modification of GBC into N-GBC potentially underlies its augmented adsorption capabilities, given the higher reactivity exhibited by these heterocyclic nitrogen species [49].

2.4.3. Mechanistic Model Developments and Model Explanation

Previous studies demonstrated that biochar-driven adsorption of antibiotics is predominantly influenced by pore filling, ion exchange, hydrogen bonding, and π-π energy decomposition analyses (EDA) interactions [50]. As is well known, the adsorption process of TCs in solution involves several steps, including dissolution, migration or diffusion, and interaction with biochars [51]. To better understand the roles of the above steps, the $\log K_{ow}$ values of the tested TCs were calculated, as $\log K_{ow}$ has proven to be an important factor in determining the uptake, distribution, and elimination of antibiotics [52]. To describe the effects of pore filling, the parameters of SSA and molecular volume (V_M) were used because (1) the SSA values were derived from the curve of the pore-filling in the adsorption (Figure 4a), and (2) pore filling is tightly linked to the molecular volume of the tested chemicals [53]. To measure the interactions with the biochars, the valid adsorption energies ($E_{ad\text{-}v}$) were developed because, (1) the adsorption energy (E_{ad}) rationally describes the strength between such interactions as ion exchange, H-bond and π-π EDA interaction [18,54] and (2) $E_{ad\text{-}v}$ exactly describes the total adsorption interactions of TCs on the biochar [51], according to the proportion of different biochar models (i.e., pyridinic-nitrogen, pyrrolic-nitrogen and graphitic-nitrogen). Therefore, the adsorption efficiency of TCs (q_e) and three parameters relating to adsorption mechanisms ($\log K_{ow}$, $\log(SSA)$, V_M, $E_{ad\text{-}v}$; Tables S4 and S5) are used as independent and dependent, respectively. The mechanistic models were developed in Equations (1) and (2) for pristine biochars (PBC, N-PBC) and graphene-like biochars (GBC and N-GBC), respectively.

$$q_e = -1912.160 - 0.648 \times E_{ad-v} + 3.869 \times V_M - 12.571 \times \log K_{ow} \tag{1}$$

$n = 14$, $F = 10.715$, $R^2 = 0.763$, $Q^2_{(\text{cum})} = 0.618$, $SE = 130.159$, $p = 0.002$; $n_{(\text{Ext})} = 2$, $Q^2_{(\text{ext})} = 0.503$, $SE_{(\text{Ext})} = 171.178$

$$q_e = -2611.753 - 0.381 \times E_{ad-v} + 8.773 \times V_M + 79.3156 \times \log K_{ow} \tag{2}$$

$n = 14$, $F = 17.078$, $R^2 = 0.837$, $Q^2_{(\text{cum})} = 0.516$, $SE = 151.417$, $p = 0.000$; $n_{(\text{Ext})} = 2$, $Q^2_{(\text{ext})} = 0.518$, $SE_{(\text{Ext})} = 145.928$

Based on the analysis of various parameters and Williams plots, it can be concluded that the developed models are of good quality (Figure 5). This is attributed to the fact that the R^2 and Q^2 values exceeded the 0.50 threshold and there were no outliers observed for the response, as evidenced by the low standardized residuals (δ) of the tests (<3) (Figure 5d,e). Moreover, the h_{i} values of the tested data consistently exhibited lower values than the corresponding h^* values, indicating the absence of influential predicted values in the mode space [55]. The h^* is the warning value of h_i. Hence, it can be stated that the developed Equations (1) and (2) models can effectively elucidate the adsorption mechanism of TCs by tested biochars. Furthermore, the variable importance in the projection (VIP) values (Figure 5c) in Equations (1) and (2) is uniformly $V_M > E_{ad\text{-}v} > \log K_{ow}$, the VIP values of V_M and $E_{ad\text{-}v}$ are consistently above 1.0, suggesting the fact that the effects of pore filling (V_M) and the interactions between TCs with the tested biochars, tend to be important processes for the adsorption characteristics of TCs by the tested biochars [56].

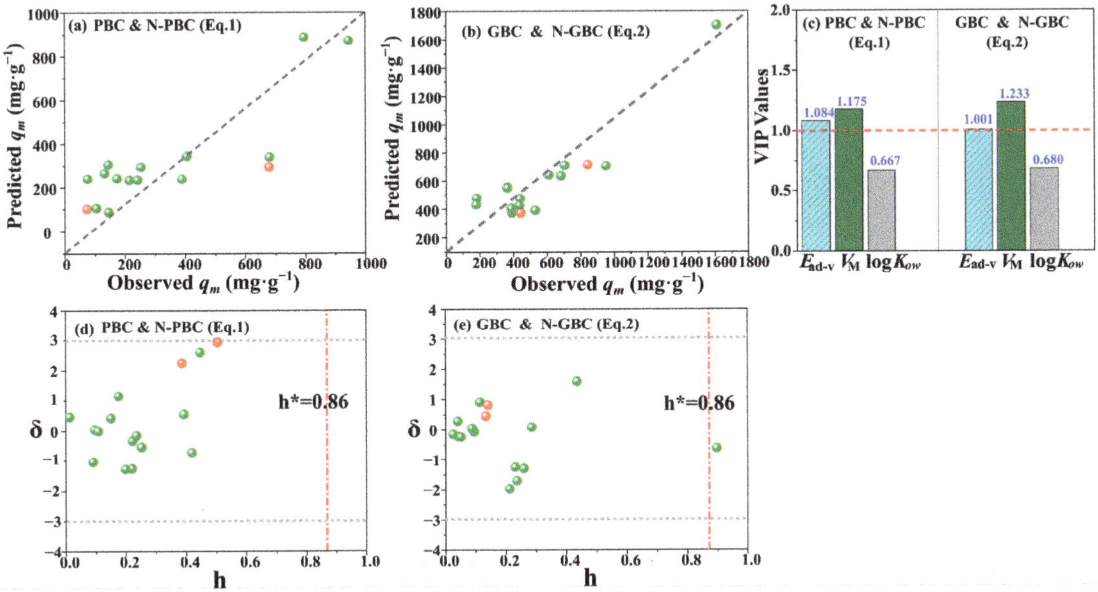

Figure 5. The correlation of predicted and observed adsorption capacity q_m (**a**,**b**), VIP values of variables (**c**), and Williams plots (**d**,**e**) for the developed QSAR model. The red and grey dotted lines present the threshold of h and δ.

2.5. The Micro-Mechanisms of N-Doping Effects upon PBC and GBC for the Adsorption of TCs

2.5.1. Effects of N-Doping upon the Chemical Reactivity of Tested Biochars

The significance of electrostatic interaction and π-π EDA interaction were highlighted as mechanisms for TCs interacting with the biochars in Figure 3. Therefore, The ESP of the studied absorbents was calculated to give an overview of the adsorption phenomenon at the point of charge distribution. As shown in Figure 6a–h, the tested biochars were colored blue, white, and red. According to the color transition description [57], the red, white and blue colors represent, respectively, a positive electrostatic potential, a vanished ESP value region, and a negative electrostatic potential. Thus, the red and blue colors indicate the region of an electron-deficient or an electron-rich, respectively. As shown in Figure 6, N-doping exhibits remarkable effects on the electron-deficient region and the electron-rich region. The values on the vdWs surface for PBC, py-PBC, pyr-PBC, and gra-PBC are ranging from −13.81 to 14.23 kcal·mol^{-1}, −17.12 to 47.66 kcal·mol^{-1}, 55.00 to 91.61 kcal·mol^{-1}, −36.99 to 17.27 kcal·mol^{-1}, respectively. Whereas the values on the vdWs surface for GBC, py-GBC, pyr-GBC, gra-GBC are ranging from −19.69 to 14.23 kcal·mol^{-1}, −35.89 to 19.61 kcal·mol^{-1}, 34.57 to 68.85 kcal·mol^{-1}, −17.40 to 51.15 kcal·mol^{-1}, respectively. Therefore, it can be concluded that N-doping strengthens the positive electrostatic potential of all tested biochars while improving the negative electrostatic potential of partial biochars (except for pyr-PBC and pyr-GBC). Furthermore, as shown in Figure 6, the maximum values are mostly located in the H atoms around the N containing groups, which should have a strong ability to electrostatically attract negatively charged atoms and act as a hydrogen bonding donor group (i.e., Figure 6b,c,g,h). Also, it can be observed that if the

N-containing groups have no H atoms, the minimum values were also detected around the N atoms and should be the most favorable nucleophilic sites when interacting with tested TCs (i.e., Figure 6d,f). It should be noted that the above results are mainly caused by the fact that, (1) the incorporation of the N atom in the lattice induces a change in the hybridization of carbon atoms (sp3 → sp2), which significantly changes the electronic distribution of tested biochars [58], and (2) the N atom poses strong electronegativity due to its lone pair electrons [59].

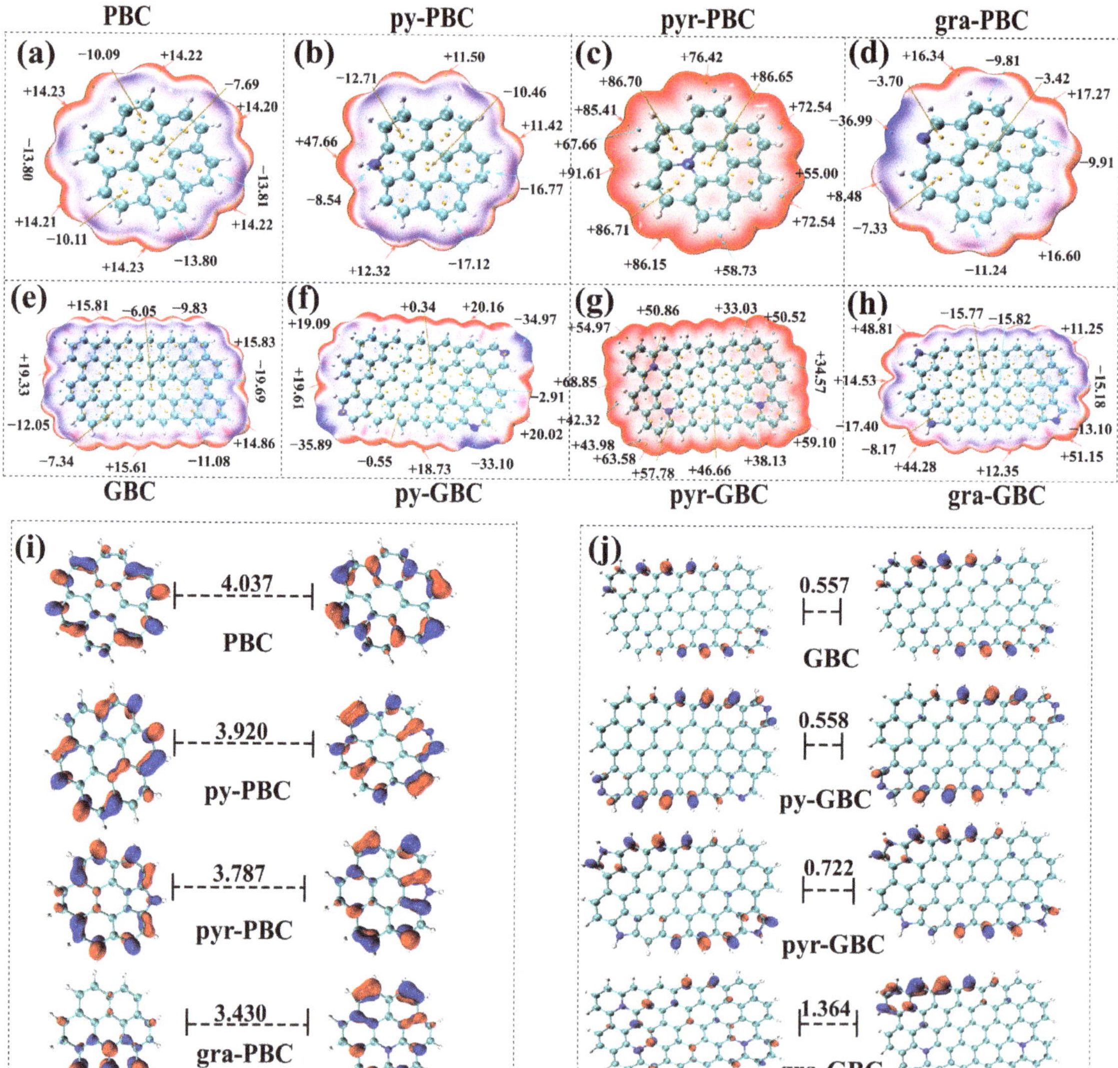

Figure 6. The electrostatic potential (**a**–**h**) and 3D isosurface plots of the HOMO and LUMO orbital (**i**,**j**) of tested biochars.

Furthermore, in the field of chemistry, the frontier electron density commonly uses molecular orbitals to describe the reactive characteristics of chemicals in π-electron systems. LUMO accepts electrons and HOMO is associated with the electron donor. A molecule with a small HOMO-LUMO gap ($E_{h\text{-}l}$) typically indicates its high global chemical reactivity [60]. As shown in Figure 6i,j, the $E_{h\text{-}l}$ values of pristine biochars and graphene-like biochars ranged from 3.430–4.037 eV and 0.557–1.364 eV, respectively. $E_{h\text{-}l}$ value of biochar is higher than that of other N-doped biochars (py-PBC, pyr-PBC, gra-PBC), whereas $E_{h\text{-}l}$ value of GBC is lower than that of other N-doped biochars (py-GBC, pyr-GBC, gra-GBC). It is obvious that the results suggested that (1) higher chemical reactivity can be obtained in graphene-like biochars compared to pristine biochars, and (2) the chemical reactivity of tested biochars was consistently affected by N-doping. In detail, the N-doping process slightly increased the chemical reactivity of pristine biochars whereas weakly decreased the chemical reactivity of graphene-like biochars.

2.5.2. Effects of N-Doping upon the Interactions of Biochars with TCs

Equilibrium Configurations and Adsorption Sites

To further investigate the impact of N-doping on the interactions between biochars and TCs, we employed graphene structures consisting of 7 and 24 aromatic rings as models for PBC and GBC, respectively [18]. To further illustrate this, we can refer to Figure 7a (the example of TTC), which demonstrates the following: (1) the lowest energy configuration uniformly exhibits a plane-shaped geometry when TTC is absorbed onto the biochars; (2) a high level of consistency is observed in the adsorption sites of TTC on the biochar surface. Specifically, the D ring, B ring, and the –C=O group in the C ring serve as common sites for TTC binding with biochars; (3) the impact of N-doping on the adsorption sites of TCs with biochars is also evident in certain samples. For instance, only the B ring and the –C=O group in the C ring showed interactions with biochar in the adsorption of pyr-PBC-TTC, while the –OH group of the C ring interacted with the N group of biochar during the adsorption of py-PBC-TTC. It is worth noting that a similar phenomenon can be observed in other TCs-biochar adsorptions (Figure S3) and other published works [61].

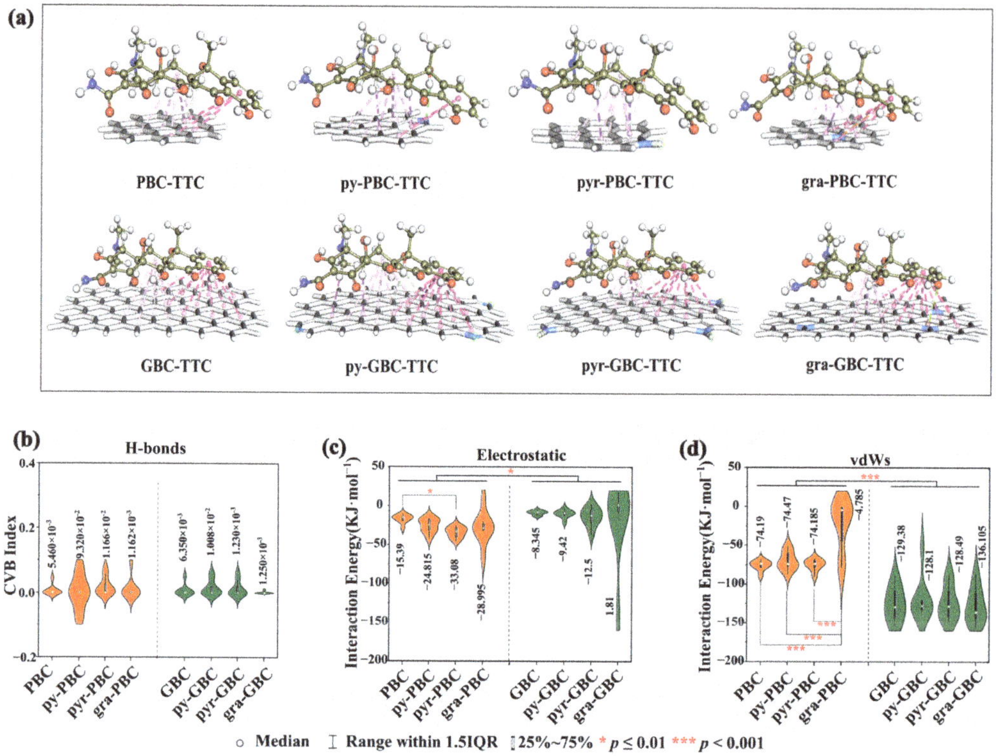

Figure 7. The lowest energy configurations of TTC-biochars (**a**), CVB index of H-bond (**b**), interaction energy of electrostatic (**c**) and vdWs (**d**) for TCs-biochars.

Weak Interaction Forces

Based on the acquired equilibrium configurations, we conducted DS and IGMH analyses to elucidate the intricate interactions between the analyzed TCs and biochars (Figure 7, Figure S3, Table S6). The DS results consistently revealed that the equilibrium configurations of TCs-biochars exhibited a small amount of hydrogen bonding (H-bond) and a large number of π-bonds. Specifically, the H-bonds were predominantly found in the interactions between the –C=O group in the C ring of TTC and biochar, while the π-bonds were primarily identified in the interactions between the D and B rings of TTC and biochar. For example, PBC-TTC and py-GBC-TTC formed one and three H-bonds, respectively (Figure 7a). Furthermore, the IGMH analysis results (Figure S4) also confirmed this conclusion. The spike with a blue color (H-bond) was hardly observed in the interactions of TCs-biochars, whereas the spikes with green (vdWs interaction) and red (electrostatic interaction) colors were present in all interactions of TCs-biochars. Therefore, the weak forces for the adsorption of TCs-biochars consisted mainly of vdWs interactions and electrostatic interactions.

To quantitatively analyze the characteristics of H-bond, vdWs, and electrostatic interactions in relation to adsorption, the CVB index was calculated based on topological characterization, and the vdWs interaction energies (E_{vdWs}) and the strength of electrostatic interactions (E_{es}) were obtained using EDA-FF [57,62]. As illustrated in Figure 7b, the average CVB values for the pristine group (PBC, N-PBC) and the graphene-like group (GBC, N-GBC) ranged from 5.460×10^{-3} to 1.166×10^{-2} and 1.250×10^{-3} to 1.230×10^{-2}, respectively. However, the CVB values in many samples were calculated to be zero. Generally, a smaller CVB index corresponds to a stronger H-bond interaction, and H-bond interactions with CVB values below zero are typically stronger [51]. Therefore, the CVB results indicated the weak H-bonding interactions in the adsorption of TCs-biochars. Regarding the electrostatic interaction (Figure 7c), the average E_{es} values for the pristine group (PBC, N-PBC) and the graphene-like group (GBC, N-GBC) ranged from -15.390 to -33.081 KJ·mol^{-1} and 1.810 to -12.510 KJ·mol^{-1}, respectively. This suggests that (1) the electrostatic interactions of the graphene-like group are consistently and significantly weaker than those of the pristine group ($p < 0.05$), and (2) the N-doping process typically strengthens the electrostatic interactions of TCs-biochars, with the exception of gra-GBC-TCs. For example, the average E_{es} value of PBC-TCs is -15.390 KJ·mol^{-1}, while the average E_{es} values of pyr-PBC-TCs and gra-PBC-TCs are -33.081 KJ·mol^{-1} and -28.995 KJ·mol^{-1}, respectively. As for the vdWs interaction (Figure 7d) in the pristine group (PBC, N-PBC), there is no significant difference between PBC with py-PBC and pyr-PBC, whereas the average E_{vdWs} value of gra-PBC-TCs is increased to -4.785 KJ·mol^{-1}. Interestingly, when it comes to the E_{vdWs} in the graphene-like group (GBC, N-GBC), there is no significant difference between GBC with py-GBC and pyr-GBC. However, the average E_{vdWs} value of gra-GBC-TCs decreases to -136.105 KJ·mol^{-1}, suggesting that in comparison to N-undoped samples, the vdWs interaction of gra-GBC is significantly strengthened, while that of gra-PBC is significantly weakened by the N-doping process ($p < 0.05$).

Key Quantum Chemical Descriptors of TCs for Adsorption Energies

Based on the results of the adsorption isotherm and kinetics, we have observed that N-doping has a significant impact on the adsorption performance, which is closely related to the structural characteristics of the tested TCs (Figures 2–4). In general, quantum chemical descriptors are commonly employed to quantitatively describe the structural features of TCs and unveil their mechanism of action [63,64]. Previous studies have also indicated that the quantum chemical descriptors of antibiotics indirectly influence their adsorption process [18]. Therefore, redundancy analysis (RDA) was conducted to unveil the relationships between quantum chemical descriptors and E_{CVB}, E_{vdWs}, and E_{es} (Figure S5). Specifically, in the pristine group (PBC, N-PBC), E_{vdWs} was principally affected by *ATSC5p* and *MATS5p*, q^{H+} and q^{-}, *G*, *MAXDP2* in the interactions of PBC-TCs, py-PBC-TCs, pyr-PBC-TCs and gra-PBC-TCs, respectively. The major influencing factors for E_{es} were q^{-}, E_{lumo} and *GATS6i*, *MATS4e*, *MinHBint* in the interactions of PBC-TCs, py-PBC-TCs, pyr-PBC-TCs and gra-PBC-TCs, respectively. On the other hand, in the graphene-like group (GBC, N-GBC), the principal influencing parameters for E_{vdWs} were *MATS5p*, q^{-} and q^{H+}, q^{-} and *MDEC-33*, q^{-} and q^{H+} in the interactions of GBC-TCs, py-GBC-TCs, pyr-GBC-TCs and gra-GBC-TCs, respectively. The E_{es} were primarily affected by the parameters of *E*, *SpMin1_B*, μ, *GATS5p* and μ in the interactions of GBC-TCs, py-GBC-TCs, pyr-GBC-TCs and gra-GBC-TCs, respectively. Therefore, we can conclude that: (1) there is a significant difference in the structural characteristics of the tested TCs between the N-doped biochars (N-PBC, N-GBC) and the N-undoped biochars (PBC, GBC) ($p < 0.01$), and (2) there is also a significant difference in the structural characteristics of the tested TCs that are affected by N-doping in the adsorption process between the pristine group biochars (PBC, N-PBC) and the graphene-like group biochars (GBC, N-GBC) ($p < 0.01$).

2.6. Application

Based on the N-undoped biochars (PBC, GBC) and N-doped biochars (py-PBC, pyr-PBC, gra-PBC, py-GBC, pyr-GBC, gra-GBC), the adsorption results (Figures 2 and 3) proved that N-doping exhibits a remarkable strengthening effect on the adsorption performance in graphene-like group (N-GBC, GBC), and N-doping improves or reduces adsorption performance in pristine groups. Previous similar results showed that N-doping biochars increased the adsorption efficiency of heavy metals (Cu^{2+} and Zn^{2+}) [65], dye [66], atrazine [59] and sulfonamide [24] on biochars. However, the previous works mainly focused on studying the adsorption kinetics and isotherm to reveal the mechanism. The current study not only examines the adsorption kinetics and isotherm but also incorporates QSAR models, molecular simulation results, and energy decomposition analysis. By doing so, the study provides a comprehensive understanding of the underlying mechanisms behind the substantial variations in the adsorption effects of N-doped pristine biochar and graphene-like biochar on TCs at a molecular level. This multi-faceted approach allows for a deeper interpretation of the observed differences and sheds light on the intricate molecular interactions involved in the adsorption process. As far as we know, this is the first report to comprehensively reveal the different N-doping effects upon the adsorption of antibiotic by pristine biochar and graphene-like biochar. Therefore, the present study undoubtedly provides more accurate theoretical guidance for further improving the adsorption efficiency of N-doped biochar.

3. Materials and Methods

3.1. Materials and Tested Biochars

In this study, eight TCs with a purity of over 99%, namely oxytetracycline (OTC, CAS: 79-57-2), tetracycline (TTC, CAS: 60-54-8), chlortetracycline (CTC, CAS: 57-62-5), minocycline (MN, CAS: 10118-90-8), tigecycline (TG, CAS: 220620-09-7), methacycline (MC, CAS: 914-00-1), doxycycline (DOX, CAS: 564-25-0), demeclocycline (DMC, CAS: 64-73-3) were purchased from Aladdin (https://www.aladdin-e.com/, accessed on 10 April 2021) and were used without any further purification. Table S4 lists the physicochemical properties and the quantum descriptors of the analyzed TCs.

The utilization of rice straw biomass material was favored due to its high adsorption effectiveness for TCs [67]. To prepare PBC, the rice straw was cleaned, dried, and pulverized. The pulverized powder was then placed into a tube furnace and heated in an N_2 environment. The N-doping PBC (N-PBC) was synthesized through a direct pyrolysis process. To synthesize GBC, a comprehensive process with K_2CO_3 (CAS: 584-08-7, a purity of over 99%, Aladdin) was implemented. The process for generating N-doped GBC (N-GBC) was carried out using GBC and following a similar preparation method as that of N-PBC. Detailed information about the preparation of tested biochars was listed in Supplementary Materials (Text S2).

3.2. Characterization of Biochar and TCs Measurements

To elucidate the structural characteristics of the prepared biochars, a variety of analytical techniques were employed to assess their surface morphology, elemental composition, and functional groups. Specifically, the surface morphology was observed using scanning electron microscopy (SEM, Zeiss, Oberkochen, Germany), while the degree of crystallinity was determined through X-ray diffraction (XRD, Bruker, Bremen, Germany). Moreover, the contact angle was measured using a dynamic contact angle meter (Dataphysics DCAT11, Filderstadt, Germany), and the SSA and pore volume were analyzed using a surface area and porosity analyzer (Brunauer-Emmett-Teller (BET), Micromeritics, Norcross, GA, USA). The concentrations of carbon, hydrogen, oxygen, nitrogen, and sulfur were determined using an Element Analyzer (FLASH 2000, Thermo Fischer Scientific, Waltham, MA, USA). Fourier transform infrared spectrometry (FT-IR, Bruker, Bremen, Germany) was employed to investigate the functional groups, while X-ray photoelectron spectroscopy (XPS, PHI 5000, Kanagawa, Japan) was used to analyze changes in carbon, oxygen, and nitrogen. The detailed information about determining of point of zero charge and contact angle was listed in Supplementary Materials (Texts S3 and S4).

The concentration of TCs was assessed using an HPLC system (Shimadzu, LC-20AD, Kyoto, Japan) coupled with a C_{18} reverse-phase column. Detailed information about the measurement of tested TCs and the standard curves were listed in Supplementary Materials (Text S5 and Table S6).

3.3. Batch Adsorption Experiments

The adsorption of TCs onto biochars was conducted using a batch-type adsorption system under dark conditions. Specifically, a mixture containing 50 mL of TC solution and 10 mg of biochar was added to a 100 mL conical flask and agitated at 180 rpm and 25 °C for 48 h. Afterward, the resulting samples were filtered through a 0.22 μm filter, and the concentrations of TCs in the supernatant were analyzed using HPLC. Control groups were also included in the study to eliminate any deviations caused by the hydrolysis of TCs [68]. The effective adsorption data for all samples were determined by calculating the difference between the samples with and without biochars using Equation (3).

$$C_i^r = C_i^C - C_i^S \tag{3}$$

where C_i^r, C_i^C and C_i^S ($mg \cdot L^{-1}$) are the real adsorption concentrations of TCs in the sample solution at i time, the determined concentrations of TCs in the control solution at i time, and the determined concentrations of TCs in the sample solution at i time, respectively.

Based on the obtained results of C_i^r, the adsorption capacity (q_t ($mg \cdot g^{-1}$)) and removal percentage (P_e) were calculated by Equations (4) and (5).

$$q_t = \frac{(C_0^r - C_e^r) \times V}{m} \tag{4}$$

$$P_e = \frac{(C_0^r - C_e^r)}{C_0^r} \times 100\% \tag{5}$$

where C_0^r ($mg \cdot L^{-1}$) and C_e^r ($mg \cdot L^{-1}$) are the initial and equilibrium concentrations of residual TCs in the solution, respectively. m (g) denotes the tested biochar weight, and V (L) denotes the volume of the TC solution.

To further characterize the adsorption features, the Langmuir model equation and the Freundlich model equation were employed to fit the adsorption isothermal data and evaluate the adsorption capacity of TCs on biochars. In addition, an investigation into the adsorption kinetics of biochars in relation to TCs was undertaken and the obtained adsorption kinetics data were fitted by the pseudo-first-order (PFO) model and the pseudo-second-order (PSO) model. Moreover, the influence of solution characteristics (pH values (3 to 11), HA concentrations (0 to 30 $mg \cdot L^{-1}$), and salinity concentrations (0 to 10 $mg \cdot L^{-1}$)) on the adsorption process was investigated. Detailed information about investigating adsorption features was listed in Supplementary Materials (Text S6). It is worth noting that the adsorption experiments in the present study were consistently conducted in triplicate, and control groups were established to evaluate the impact of the adsorbent's absence.

3.4. Adsorption Simulation and Micro-Mechanisms

3.4.1. Adsorption Simulation

To investigate the adsorption interfacial interactions between TCs and biochars, we employed graphene structures as models. The models comprised seven aromatic rings and 24 aromatic rings, representing the PBC model and the GBC model, respectively [18,69,70]. For N-doped biochars, we assigned specific names to distinguish them based on the modified groups of pyridinic-nitrogen, pyrrolic-nitrogen, and graphitic-nitrogen, namely py-PBC, pyr-PBC, and gra-PBC, respectively. Similarly, for the GBC models with pyridinic-nitrogen, pyrrolic-nitrogen, and graphitic-nitrogen, we named them py-GBC, pyr-GBC, and gra-GBC, respectively.

To determine the optimal low-energy configuration for TCs adsorption onto biochars, we employed the DMol3 program package within Materials Studio 2020 to conduct adsorption calculations using the models we have developed. As outlined in our previous investigation [18], the adsorption simulation specifics were as follows: The electronic basis set utilized was dual numerical polarization (DNP 4.4), the atomic basis function was an atom-centered grid, the self-consistent field convergence value was set at 1.0×10^{-6}, and the core treatment was DFT semi-core pseudopots. After optimizing the biochars and TCs models, we employed the Metropolis Monte Carlo method to calculate absorption within the framework, utilizing 10^5 Monte Carlo steps and 10 annealing cycles under the COMPASS II force field of the "adsorption locator calculation" program [68].

3.4.2. Adsorption Micro-Mechanisms

To investigate the micro-mechanisms of adsorption, various parameters were studied, such as the adsorption energy (E_{ad}), the valid adsorption energy ($E_{ad\text{-}v}$), the equilibrium configurations of adsorption and their corresponding interaction features (such as bond angle, bond distance, and interaction types), the strength of interactions (including the vdWs interaction energies (E_{vdWs}), the electrostatic interactions energies (E_{es}), the core valence bifurcation (CVB) index for hydrogen bonding (H-bond) energy [54], and the electrostatic potential (ESP) [71]. Detailed information about obtaining the parameters of adsorption micro-mechanisms was listed in Supplementary Materials (Text S7).

3.5. Mechanistic QSAR Model Development, Validations, and Statistical Analyses

The adsorption features were analyzed through the development of quantitative structure activity relationship (QSAR) models. These models were created using partial least squares regression and evaluated based on metrics such as the determination coefficient (R^2), the Fisher criterion (F), the standard error of estimate (SE), and the p-value test (p). In addition, cross-validation (Q^2) and a Williams plot (standardized residuals (δ) versus leverage values (h_i)) were used to verify the predictive ability and applicability of the model, respectively [55].

The statistical analysis was performed using SPSS 26.0 software (SPSS Inc., Chicago, IL, USA). A one-way analysis of variance and paired t-test were employed to compare the differences in samples, and a p-value lower than 0.05 was considered statistically significant.

4. Conclusions

In the adsorption of TCs onto tested biochars, N-doping may improve or reduce the adsorption performance of PBC, whereas consistently improves the adsorption performance of GBC. The above N-doping effects can obtain the same obtained in different solution properties. Furthermore, the N-doping process significantly modified the surface functional groups of biochars, decreased the adsorption sphere of PBC, and significantly increased the adsorption sphere of GBC. Moreover, the pore filling and the interactions between TCs and biochars were key factors in the effects of N-doping (VIP values > 1.0). Otherwise, electrostatic interactions and vdWs interactions were the primary weak forces, with the former being strengthened and the latter unaffected by the N-doping process. In conclusion, the proven mechanism for the different N-doping effects of TCs adsorption by biochars will enhance our understanding of the critical role of biochar in TCs pollution remediation.

Nevertheless, multiple contaminants often coexist in aquatic environments, necessitating a broader approach to water treatment. Although restricting research to a single pollutant provides valuable insights into its adsorption mechanisms, it would fail to address the complex challenges associated with water treatment comprehensively. Our study focused on the adsorption treatment of a specific pollutant (TCs), future research should prioritize the development of adsorbent materials capable of effectively removing multiple pollutants. To tackle this issue, future advancements should aim to synthesize or modify adsorption media to enhance their adsorption capacity and selectivity for a broader range of pollutants. Additionally, integrating advanced separation and treatment technologies, such as membrane filtration and adsorption resins, can facilitate the efficient removal and recovery of various pollutants. Interdisciplinary collaboration in research plays a crucial role in driving progress in the field of multi-pollutant adsorption treatment.

Supplementary Materials: The following supporting information can be downloaded at: https://www.mdpi.com/article/10.3390/molecules29010173/s1, Figure S1: The influence of environmental factors (pH, Ca^{2+}, HA) on TTC and OTC adsorption efficiency (q_e) by tested biochars (a–f) and the statistical distribution of their adsorption efficiency (g); Figure S2: C1s, O1s peaks of tested biochars by XPS regional spectra; Figure S3: Equilibrium configurations of tested TCs (except TTC) adsorbed by PBC and GBC (a), adsorbed by py-PBC and py-GBC (b), adsorbed by pyr-PBC and pyr-GBC (c), adsorbed by gra-PBC and gra-GBC (d); Figure S4: The scatterplots of reduced density gradients versus sign(λ2)ρ for the equilibrium configurations of TTC-biochars; Figure S5: The redundancy analysis (RDA) of quantum chemical descriptors and E_{CVB}, E_{vdWs}, and E_{es}; Table S1: Characteristic of the tested biochars; Table S2: The fitted parameters of PFO model and PSO model; Table S3: The fitted parameters for Langmuir model and Freundlich model; Table S4: The physicochemical properties and the quantum descriptors of tested TCs; Table S5: The valid adsorption energies of TCs of TCs with tested biochars; Table S6: The interaction types, distances and angels of TCs-biochars; Table S7: The standard curves and recovery rates of tested TCs; Text S1: Adsorption features; Text S2: Preparation of tested biochars; Text S3: Determination of point of zero charge; Text S4: Determination of contact angle; Text S5: TCs measurements; Text S6: Adsorption micro-mechanisms; Text S7: Effects of environmental conditions. References are cited in [18,37,57,72–74].

Author Contributions: L.R.: Writing, original draft, conceptualization, investigation, data curation. L.W.: investigation, data curation. T.Z.: investigation. C.H.: investigation. H.T.: investigation. H.P.: Writing, review and editing, conceptualization. X.Z.: Writing, review and editing, conceptualization. All authors have read and agreed to the published version of the manuscript.

Funding: This work was funded by the National Natural Science Foundation of China (Nos. 32160303 and 52370202), the Double Thousand Plan of Jiangxi Province (No. JXSQ2023201061), the Natural Science Foundation of Jiangxi Province (No. 20224ACB203017), the Educational Commission of Jiangxi Province (Nos. GJJ211001 and GJJ190540), and the Key Science and Technology program of Ji'an (No. 20211-025333). We are grateful for their support.

Institutional Review Board Statement: Not applicable.

Informed Consent Statement: Not applicable.

Data Availability Statement: Data will be made available on request.

Conflicts of Interest: The authors declare that they have no known competing financial interest or personal relationships that could have appeared to influence the work reported in this paper.

List of Abbreviations and Definitions

Graphene like biochar—GBC, pristine biochar—PBC, nitrogen doped GBC—N-GBC, nitrogen doped PBC—N-PBC, tetracycline antibiotics—TCs, valid adsorption energies—$E_{ad\text{-}v}$, quantitative structure activity relationship—QSAR, octanol-water partition coefficients—K_{ow}, specific surface area—SSA, molecular volume—V_M, the core valence bifurcation index—CVB, hydrogen bonding—H-bond, van der Waals—vdWs, density functional theory—DFT, humic acid—HA, pseudo-first-order—PFO, pseudo-second-order—PSO, redundancy analysis—RDA, electrostatic potential—ESP, energy decomposition analyses—EDA.

References

1. Sabando-Fraile, C.; Corral-Bobadilla, M.; Lostado-Lorza, R.; Somovilla-Gomez, F.J.S. Multiresponse Performance Evaluation and Life Cycle Assessment for the Optimal Elimination of Pb (II) from Industrial Wastewater by Adsorption Using Vine Shoot Activated Carbon. *Sustainability* **2023**, *15*, 11007. [CrossRef]
2. Zu, Z.Y.; Jiang, M.D.; Xu, P.P.; Chen, W.; Ni, Q.Q.; Lu, G.M.; Zhang, L.J. Coronavirus disease 2019 (COVID-19): A perspective from China. *Radiology* **2020**, *296*, E15–E25. [CrossRef] [PubMed]
3. Schnappinger, D.; Hillen, W.J. Tetracyclines: Antibiotic action, uptake, and resistance mechanisms. *Arch. Microbiol.* **1996**, *165*, 359–369. [CrossRef] [PubMed]
4. Pérez-Rodríguez, M.; Pellerano, R.G.; Pezza, L.; Pezza, H.R.J. An overview of the main foodstuff sample preparation technologies for tetracycline residue determination. *Talanta* **2018**, *182*, 1–21. [CrossRef] [PubMed]
5. Hutchinson, T.J.I. f.E.v.S. In *Proceedings of the International Conference on Heavy Metals in the Environment, Toronto, OR, Canada, 21–22 January 1975*; M5S1A4; University of Toronto: Toronto, OR, Canada, 1975; Volume 1, p. 3.
6. Wang, Y.; Han, Y.; Li, L.; Liu, J.; Yan, X.J. Distribution, sources, and potential risk of antibiotic resistance genes in wastewater treatment plant: A review. *Environ. Pollut.* **2022**, *31*, 119870. [CrossRef] [PubMed]
7. Xu, L.; Zhang, H.; Xiong, P.; Zhu, Q.; Liao, C.; Jiang, G.J. Occurrence, fate, and risk assessment of typical tetracycline antibiotics in the aquatic environment: A review. *Sci. Total Environ.* **2021**, *753*, 141975. [CrossRef]
8. Xu, L.; Wang, W.; Xu, W.J. Health, Effects of tetracycline antibiotics in chicken manure on soil microbes and antibiotic resistance genes (ARGs). *Environ. Geochem. Health* **2022**, *44*, 273–284. [CrossRef] .
9. Wright, D.G. Resisting resistance: New chemical strategies for battling superbugs. *Cell Chem. Biol.* **2000**, *7*, R127–R132. [CrossRef]

10. Gopal, G.; Alex, S.A.; Chandrasekaran, N.; Mukherjee, A. A review on tetracycline removal from aqueous systems by advanced treatment techniques. *RSC Adv.* **2020**, *10*, 27081–27095. [CrossRef]
11. Zhang, Y.; Pei, M.; Zhang, B.; He, Y.; Zhong, Y. Changes of antibiotic resistance genes and bacterial communities in the advanced biological wastewater treatment system under low selective pressure of tetracycline. *Water Res.* **2021**, *207*, 117834. [CrossRef]
12. Minale, M.; Gu, Z.; Guadie, A.; Kabtamu, D.M.; Li, Y.; Wang, X. Application of graphene-based materials for removal of tetracyclines using adsorption and photocatalytic-degradation: A review. *J. Environ. Manag.* **2020**, *276*, 111310. [CrossRef] [PubMed]
13. Scaria, J.; Anupama, K.; Nidheesh, P. Tetracyclines in the environment: An overview on the occurrence, fate, toxicity, detection, removal methods, and sludge management. *Sci. Total Environ.* **2021**, *771*, 145291. [CrossRef] [PubMed]
14. Yi, X.H.; Wang, T.Y.; Chu, H.Y.; Gao, Y.; Wang, C.C.; Li, Y.J.; Chen, L.; Wang, P.; Fu, H.; Zhao, C. Effective elimination of tetracycline antibiotics via photoactivated SR-AOP over vivianite: A new application approach of phosphorus recovery product from WWTP. *Chem. Eng. J.* **2022**, *449*, 137784. [CrossRef]
15. Dutta, J.; Mala, A.A. Removal of antibiotic from the water environment by the adsorption technologies: A review. *Water Sci. Technol.* **2020**, *82*, 401–426. [CrossRef] [PubMed]
16. Fu, H.; Li, X.; Wang, J.; Lin, P.; Chen, C.; Zhang, X.; Suffet, I.M. Activated carbon adsorption of quinolone antibiotics in water: Performance, mechanism, and modeling. *J. Environ. Sci.* **2017**, *56*, 145–152. [CrossRef] [PubMed]
17. Ötker, H.M.; Akmehmet-Balcıoğlu, I. Adsorption and degradation of enrofloxacin, a veterinary antibiotic on natural zeolite. *J. Hazard. Mater.* **2005**, *122*, 251–258. [CrossRef]
18. Fan, Y.; Huang, L.; Wu, L.; Zhang, C.; Zhu, S.; Xiao, X.; Li, M.; Zou, X. Adsorption of sulfonamides on biochars derived from waste residues and its mechanism. *J. Hazard. Mater.* **2021**, *406*, 124291. [CrossRef]
19. Xiang, W.; Zhang, X.; Chen, J.; Zou, W.; He, F.; Hu, X.; Tsang, D.C.; Ok, Y.S.; Gao, B. Biochar technology in wastewater treatment: A critical review. *Chemosphere* **2020**, *252*, 126539. [CrossRef]
20. Jang, H.M.; Kan, E. A novel hay-derived biochar for removal of tetracyclines in water. *Bioresour. Technol.* **2019**, *274*, 162–172. [CrossRef]
21. Cheng, N.; Wang, B.; Wu, P.; Lee, X.; Xing, Y.; Chen, M.; Gao, B. Adsorption of emerging contaminants from water and wastewater by modified biochar: A review. *Environ. Pollut.* **2021**, *273*, 116448. [CrossRef]
22. Wang, J.; Wang, S. Preparation, modification and environmental application of biochar: A review. *J. Clean. Prod.* **2019**, *227*, 1002–1022. [CrossRef]
23. Godwin, P.M.; Pan, Y.; Xiao, H.; Afzal, M.T. Progress in preparation and application of modified biochar for improving heavy metal ion removal from wastewater. *J. Bioresour. Bioprod.* **2019**, *4*, 31–42. [CrossRef]
24. Qi, Y.; Ge, B.; Zhang, Y.; Jiang, B.; Wang, C.; Akram, M.; Xu, X. Three-dimensional porous graphene-like biochar derived from Enteromorpha as a persulfate activator for sulfamethoxazole degradation: Role of graphitic N and radicals transformation. *J. Hazard. Mater.* **2020**, *399*, 123039. [CrossRef] [PubMed]
25. Yuan, X.; Xiao, J.; Yılmaz, M.; Zhang, T.C.; Yuan, S. N, P Co-doped porous biochar derived from cornstalk for high performance CO_2 adsorption and electrochemical energy storage. *Sep. Purif. Technol.* **2022**, *299*, 121719. [CrossRef]
26. Chen, Q.; Zheng, J.; Xu, J.; Dang, Z.; Zhang, L. Insights into sulfamethazine adsorption interfacial interaction mechanism on mesoporous cellulose biochar: Coupling DFT/FOT simulations with experiments. *Chem. Eng. J.* **2019**, *356*, 341–349. [CrossRef]
27. Yi, L.; Zuo, L.; Wei, C.; Fu, H.; Qu, X.; Zheng, S.; Xu, Z.; Guo, Y.; Li, H.; Zhu, D. Enhanced adsorption of bisphenol A, tylosin, and tetracycline from aqueous solution to nitrogen-doped multiwall carbon nanotubes via cation-π and π-π electron-donor-acceptor (EDA) interactions. *Sci. Total Environ.* **2020**, *719*, 137389. [CrossRef]
28. Li, Y.; Xing, B.; Wang, X.; Wang, K.; Zhu, L.; Wang, S. Nitrogen-doped hierarchical porous biochar derived from corn stalks for phenol-enhanced adsorption. *Energy Fuel* **2019**, *33*, 12459–12468. [CrossRef]
29. Zhou, H.; Li, X.; Jin, H.; She, D. Mechanism of a double-channel nitrogen-doped lignin-based carbon on the highly selective removal of tetracycline from water. *Bioresour. Technol.* **2022**, *346*, 126652. [CrossRef]
30. Wang, Q.; Zhang, Z.; Xu, G.; Li, G. Pyrolysis of penicillin fermentation residue and sludge to produce biochar: Antibiotic resistance genes destruction and biochar application in the adsorption of penicillin in water. *J. Hazard. Mater.* **2021**, *413*, 125385. [CrossRef]
31. Zhang, X.; Lin, X.; He, Y.; Chen, Y.; Luo, X.; Shang, R. Study on adsorption of tetracycline by Cu-immobilized alginate adsorbent from water environment. *Int. J. Biol. Macromol.* **2019**, *124*, 418–428. [CrossRef]
32. Parolo, M.E.; Savini, M.C.; Valles, J.M.; Baschini, M.T.; Avena, M.J. Tetracycline adsorption on montmorillonite: pH and ionic strength effects. *Appl. Clay. Sci.* **2008**, *40*, 179–186. [CrossRef]
33. Leng, L.; Li, J.; Yuan, X.; Li, J.; Han, P.; Hong, Y.; Wei, F.; Zhou, W. Beneficial synergistic effect on bio-oil production from co-liquefaction of sewage sludge and lignocellulosic biomass. *Bioresour. Technol.* **2018**, *251*, 49–56. [CrossRef] [PubMed]
34. Revellame, E.D.; Fortela, D.L.; Sharp, W.; Hernandez, R.; Zappi, M.E. Adsorption kinetic modeling using pseudo-first order and pseudo-second order rate laws: A review. *Clean. Eng. Technol.* **2020**, *1*, 100032. [CrossRef]
35. Kajjumba, G.W.; Emik, S.; Öngen, A.; Özcan, H.K.; Aydın, S. Modelling of adsorption kinetic processes—Errors, theory and application. In *Advanced Sorption Process Applications;* IntechOpen: London, UK, 2018; pp. 1–19.
36. Al-Ghouti, M.A.; Da'ana, D.A. Guidelines for the use and interpretation of adsorption isotherm models: A review. *J. Hazard. Mater.* **2020**, *393*, 122383. [CrossRef]

37. Lin, L.; Jiang, W.B.; Xu, P. Comparative study on pharmaceuticals adsorption in reclaimed water desalination concentrate using biochar: Impact of salts and organic matter. *Sci. Total Environ.* **2017**, *601*, 857–864. [CrossRef] [PubMed]
38. Huang, P.; Zhang, P.; Wang, C.; Tang, J.; Sun, H. Enhancement of persulfate activation by Fe-biochar composites: Synergism of Fe and N-doped biochar. *Appl. Catal. B Environ.* **2022**, *303*, 120926. [CrossRef]
39. Diao, Y.; Shan, R.; Li, M.; Gu, J.; Yuan, H.; Chen, Y. Efficient Adsorption of a Sulfonamide Antibiotic in Aqueous Solutions with N-doped Magnetic Biochar: Performance, Mechanism, and Reusability. *ACS Omega* **2022**, *8*, 879–892. [CrossRef]
40. Zhong, Q.; Lin, Q.; He, W.; Fu, H.; Huang, Z.; Wang, Y.; Wu, L. Study on the nonradical pathways of nitrogen-doped biochar activating persulfate for tetracycline degradation. *Sep. Purif. Technol.* **2021**, *276*, 119354. [CrossRef]
41. Singh, B.; Raven, M.D. 21 X-ray diffraction analysis of biochar. In *Biochar: A Guide to Analytical Methods*; CRC Press: Boca Raton, FL, USA, 2017; p. 245.
42. Kang, J.; Duan, X.; Wang, C.; Sun, H.; Tan, X.; Tade, M.O.; Wang, S. Nitrogen-doped bamboo-like carbon nanotubes with Ni encapsulation for persulfate activation to remove emerging contaminants with excellent catalytic stability. *Chem. Eng. J.* **2018**, *332*, 398–408. [CrossRef]
43. Li, R.; Hu, D.; Hu, K.; Deng, H.; Zhang, M.; Wang, A.; Qiu, R.; Yan, K.J. Coupling adsorption-photocatalytic reduction of Cr (VI) by metal-free N-doped carbon. *Sci. Total Environ.* **2020**, *704*, 135284. [CrossRef] [PubMed]
44. Leng, L.; Xu, S.; Liu, R.; Yu, T.; Zhuo, X.; Leng, S.; Xiong, Q.; Huang, H. Nitrogen containing functional groups of biochar: An overview. *Bioresour. Technol.* **2020**, *298*, 122286. [CrossRef] [PubMed]
45. Kasera, N.; Kolar, P.; Hall, S.G. Nitrogen-doped biochars as adsorbents for mitigation of heavy metals and organics from water: A review. *Biochar* **2022**, *4*, 17. [CrossRef]
46. Qian, L.; Chen, B. Interactions of aluminum with biochars and oxidized biochars: Implications for the biochar aging process. *J. Agr. Food Chem.* **2014**, *62*, 373–380. [CrossRef] [PubMed]
47. Zhou, Y.; Wang, J.; Sun, M.; Li, W.; Hu, X. Adsorption of CO_2 by nitrogen doped corn straw based biochar. *Arab. J. Geosci.* **2021**, *14*, 1875. [CrossRef]
48. Qu, S.; Yuan, Y.; Yang, X.; Xu, H.; Mohamed, A.K.; Zhang, J.; Zhao, C.; Liu, L.; Wang, B.; Wang, X.J.C.E.J. Carbon defects in biochar facilitated nitrogen doping: The significant role of pyridinic nitrogen in peroxymonosulfate activation and ciprofloxacin degradation. *Chem. Eng. J.* **2022**, *441*, 135864. [CrossRef]
49. Ortiz-Medina, J.; Wang, Z.; Cruz-Silva, R.; Morelos-Gomez, A.; Wang, F.; Yao, X.; Terrones, M.; Endo, M.J. Defect engineering and surface functionalization of nanocarbons for metal-free catalysis. *Adv. Mater.* **2019**, *31*, 1805717. [CrossRef]
50. Du, L.Q.; Ahmad, S.; Liu, L.A.; Wang, L.; Tang, J.C. A review of antibiotics and antibiotic resistance genes (ARGs) adsorption by biochar and modified biochar in water. *Sci. Total Environ.* **2023**, *858*, 14. [CrossRef]
51. Jiang, Z.; Huang, X.; Wu, Q.; Li, M.; Xie, Q.; Liu, Z.; Zou, X. Adsorption of sulfonamides on polyamide microplastics in an aqueous solution: Behavior, structural effects, and its mechanism. *Chem. Eng. J.* **2023**, *454*, 140452. [CrossRef]
52. Yang, S.S.; Lu, W.C.; Gu, T.H.; Yan, L.M.; Li, G.Z. QSPR Study of n-Octanol/Water Partition Coefficient of Some Aromatic Compounds Using Support Vector Regression. *QSAR Comb. Sci.* **2009**, *28*, 175–182. [CrossRef]
53. Chen, Q.; Zheng, J.; Yang, Q.; Dang, Z.; Zhang, L. Effect of carbon chain structure on the phthalic acid esters (PAEs) adsorption mechanism by mesoporous cellulose biochar. *Chem. Eng. J.* **2019**, *362*, 383–391. [CrossRef]
54. Jiang, Z.; Huang, L.; Fan, Y.; Zhou, S.; Zou, X. Contrasting effects of microplastic aging upon the adsorption of sulfonamides and its mechanism. *Chem. Eng. J.* **2022**, *430*, 132939. [CrossRef]
55. Eriksson, L.; Jaworska, J.; Worth, A.P.; Cronin, M.T.; McDowell, R.M.; Gramatica, P. Methods for reliability and uncertainty assessment and for applicability evaluations of classification-and regression-based QSARs. *Environ. Health Persp.* **2003**, *111*, 1361–1375. [CrossRef] [PubMed]
56. Zou, X.; Xiao, X.; Zhou, H.; Chen, F.; Zeng, J.; Wang, W.; Feng, G.; Huang, X. Effects of soil acidification on the toxicity of organophosphorus pesticide on Eisenia fetida and its mechanism. *J. Hazard. Mater.* **2018**, *359*, 365–372. [CrossRef] [PubMed]
57. Lu, T.; Chen, F. Multiwfn: A multifunctional wavefunction analyzer. *J. Comput. Chem.* **2012**, *33*, 580–592. [CrossRef] [PubMed]
58. Ahmad, S.; Liu, L.; Zhang, S.; Tang, J. Nitrogen-doped biochar (N-doped BC) and iron/nitrogen co-doped biochar (Fe/N co-doped BC) for removal of refractory organic pollutants. *J. Hazard. Mater.* **2023**, *446*, 130727. [CrossRef] [PubMed]
59. Cheng, Y.; Wang, B.; Shen, J.; Yan, P.; Kang, J.; Wang, W.; Bi, L.; Zhu, X.; Li, Y.; Wang, S. Preparation of novel N-doped biochar and its high adsorption capacity for atrazine based on π–π electron donor-acceptor interaction. *J. Hazard. Mater.* **2022**, *432*, 128757. [CrossRef]
60. Miar, M.; Shiroudi, A.; Pourshamsian, K.; Oliaey, A.R.; Hatamjafari, F. Theoretical investigations on the HOMO–LUMO gap and global reactivity descriptor studies, natural bond orbital, and nucleus-independent chemical shifts analyses of 3-phenylbenzo [d] thiazole-2 (3 H)-imine and its para-substituted derivatives: Solvent and substituent effects. *J. Chem. Res.* **2021**, *45*, 147–158.
61. Mei, Y.; Xu, J.; Zhang, Y.; Li, B.; Fan, S.; Xu, H. Effect of Fe–N modification on the properties of biochars and their adsorption behavior on tetracycline removal from aqueous solution. *Bioresour. Technol.* **2021**, *325*, 124732. [CrossRef]
62. Fuster, F.; Silvi, B. Does the topological approach characterize the hydrogen bond? *Theor. Chem. Acc.* **2000**, *104*, 13–21. [CrossRef]
63. Han, C.-H.; Park, H.-D.; Kim, S.-B.; Yargeau, V.; Choi, J.-W.; Lee, S.-H.; Park, J.-A. Oxidation of tetracycline and oxytetracycline for the photo-Fenton process: Their transformation products and toxicity assessment. *Water Res.* **2020**, *172*, 115514. [CrossRef] [PubMed]

64. Leichtweis, J.; Vieira, Y.; Welter, N.; Silvestri, S.; Dotto, G.L.; Carissimi, E. A review of the occurrence, disposal, determination, toxicity and remediation technologies of the tetracycline antibiotic. *Process Saf. Environ.* **2022**, *160*, 25–40. [CrossRef]
65. Yu, W.; Lian, F.; Cui, G.; Liu, Z. N-doping effectively enhances the adsorption capacity of biochar for heavy metal ions from aqueous solution. *Chemosphere* **2018**, *193*, 8–16. [CrossRef] [PubMed]
66. Lian, F.; Cui, G.; Liu, Z.; Duo, L.; Zhang, G.; Xing, B. One-step synthesis of a novel N-doped microporous biochar derived from crop straws with high dye adsorption capacity. *J. Environ. Manag.* **2016**, *176*, 61–68. [CrossRef] [PubMed]
67. Fan, S.S.; Wang, Y.; Li, Y.; Wang, Z.; Xie, Z.X.; Tang, J. Removal of tetracycline from aqueous solution by biochar derived from rice straw. *Environ. Sci. Pollut. R* **2018**, *25*, 29529–29540. [CrossRef] [PubMed]
68. Sun, H. COMPASS: An ab initio force-field optimized for condensed-phase applications overview with details on alkane and benzene compounds. *J. Phys. Chem. B* **1998**, *102*, 7338–7364. [CrossRef]
69. Cheng, L.; Ji, Y.H.; Liu, X.M. Insights into interfacial interaction mechanism of dyes sorption on a novel hydrochar: Experimental and DFT study. *Chem. Eng. Sci.* **2021**, *233*, 116432. [CrossRef]
70. Anderson, P.D.; Weber, L.J. In Proceedings of the International Conference on Heavy Metals in the Environment, Toronto, OR, Canada, 21–22 January 1975; p. 933.
71. Pal, R.; Chattaraj, P.K. Chemical reactivity from a conceptual density functional theory perspective. *J. Indian Chem. Soc.* **2021**, *98*, 100008. [CrossRef]
72. Mahmood, T.; Saddique, M.T.; Naeem, A.; Westerhoff, P.; Mustafa, S.; Alum, A. Comparison of different methods for the point of zero charge determination of NiO. *Ind. Eng. Chem. Res.* **2011**, *50*, 10017–10023. [CrossRef]
73. Liu, Z.; Feng, F.; Li, Y.; Sun, Y.; Tagawa, K. A corncob biochar-based superhydrophobic photothermal coating with micro-nano-porous rough-structure for ice-phobic properties. *Surf. Coat. Technol.* **2023**, *457*, 129299. [CrossRef]
74. Sergio, M.; Lu, T. The geometry and electronic structure of Aristolochic acid: Possible implications for a frozen resonance. *J. Phys. Org. Chem.* **2013**, *26*, 473–483.

molecules

MDPI

Article

Effective Detoxification of Olive Mill Wastewater Using Multi-Step Surfactant-Based Treatment: Assessment of Environmental and Health Impact

Yazan Akkam [1,*], Mohammad Zaitoun [1], Islam Aljarrah [1], Aiman Jaradat [2], Ali Hmedat [3], Hassan Alhmoud [3], Taha Rababah [4], Ali Almajwal [5] and Numan Al-Rayyan [6]

1 Department of Medicinal Chemistry and Pharmacognosy, Faculty of Pharmacy, Yarmouk University, Irbid 21163, Jordan; m.zaitoun@yu.edu.jo (M.Z.); islamsubhi12@gmail.com (I.A.)
2 Department of Civil Engineering, Hijjawi Faculty for Engineering Technology, Yarmouk University, Irbid 21163, Jordan; ayman.j@yu.edu.jo
3 Department of Pharmaceutical Technology and Pharmaceutics, Faculty of Pharmacy, Yarmouk University, Irbid 21163, Jordan; ali.hmedat@yu.edu.jo (A.H.); hassan.alhmoud@yu.edu.jo (H.A.)
4 Department of Nutrition and Food Technology, Jordan University of Science and Technology, Irbid 22110, Jordan; trababah@just.edu.jo
5 Department of Community Health Sciences, College of Applied Medical Sciences, King Saud University, P.O. Box 10219, Riyadh 11433, Saudi Arabia; aalmajwal@ksu.edu.sa
6 Department of Pediatrics, School of Medicine and Public Health, University of Wisconsin, Madison, WI 53706, USA; alrayyan@wisc.edu
* Correspondence: yazan.a@yu.edu.jo

Abstract: Olive mill wastewater (OMW) poses a significant environmental challenge and health concern in olive-producing countries, including Jordan. Surfactant micelles are frequently employed as solubilizing agents to enhance the water solubility of chemical compounds. This study aims to leverage the sodium dodecyl sulfate (SDS) micelles in a multi-step process to detoxify OMW for agricultural and industrial uses and reduce its impact. The OMW was treated in multiple steps: screening, coagulation with different chemicals, and distillation with different surfactants. The treatment steps were monitored using LC–MS, GC–MS, ICP–MS, chemical oxygen demand contents, and total phenolic compounds. The detoxification of OMW was evaluated using standard germination assays, MTT assays using tissue culture, and toxicity assays using fluorescence bacteria. Following the treatment, the seed growth rate improved significantly from 0% to 100%. The GC–MS revealed a substantial decrease in pollutants. The concentration of polyphenols was reduced to 2.5%, while the COD level decreased to 35%. The toxicity in bacteria was significantly reduced in a time-dependent manner, and the toxicity in human cells decreased by 95%. Additionally, between 50% and 95% of metals in OMW were removed. The multi-step SDS-based approach successfully detoxified the OMW and enhanced water quality, which would pave the road for its direct application in industry and agriculture.

Keywords: olive mill wastewater; detoxification; sodium dodecyl sulfate; micelles; tissue culture; antibacterial; characterization

1. Introduction

The Mediterranean countries are known as the primary global producers of olive oil, as they account for 98% of global olive oil production [1]. For instance, Jordan produced more than 20 thousand tons of olive oils in 2022 [2]. Unfortunately, the olive oil extraction process produces large amounts of polluted water as a liquid by-product called olive mill wastewater (OMW) [3], the estimated volume of which is about 30 million m^3 per year [4]. Subsequently, thousands of cubic meters of OMW and thousands of tons of solid waste are produced in Jordan [5].

Olive oil mill wastewater (OMW) is a significant environmental challenge due to its high chemical oxygen demand (COD), organic load, and phenolic content, all of which can cause phytotoxicity and impede microbial growth [6–8]. Moreover, as OMW is discharged into surface water or spread on the land, it can pollute surface and groundwater bodies, harm agricultural fields and the environment, and harm public health [2,4,9–12]. Furthermore, researchers have found that OMW is toxic to plants even after 100-fold dilution and affects seed germination [13].

Various methods have been explored to address this problem, such as electrocoagulation [14], electro-Fenton reaction combined with anaerobic digestion [15], and the integration of Fenton's reaction with anaerobic treatment [16]. Additionally, technologies like reverse osmosis [17], electrochemical oxidation [18], catalytic oxidation [19], and distillation [20] have been investigated for their efficacy in treating OMW. More about the current methods of treatment were discussed in detail previously [2]. However, despite advancements, each treatment approach encounters its own set of challenges and limitations, hindering their widespread adoption and effectiveness [2], such as high operational costs [21], incomplete treatment [22], seasonal variability [23], environmental concerns and safety [24], and limited applicability [25].

The efficiency of conventional treatment methods is limited due to the presence of a large amount of organic matter [22]. Moreover, the presence of substances like tyrosol, hydroxytyrosol, and polyphenols in OMW further complicates the treatment process [26,27]. Therefore, the treatment of OMW is complex and expensive [28,29]. Moreover, the current treatment methods may not fully comply with discharge limits and regulatory standards [30,31]. Thus, a unique, safe, and cost-effective treatment of OMW that produces harmless by-products without harming plants, humans, or microorganisms is required [28].

Micelles have been extensively studied for their ability to enhance the solubility of poorly water-soluble compounds, including drugs. The formation of micelles using surfactants has been shown to significantly improve the water solubility of drugs, leading to enhanced bioavailability, reduced toxicity, improved drug stability, and altered drug distribution [32,33].

Micelles are self-assembling colloidal systems composed of a hydrophobic core surrounded by a hydrophilic corona, which can enhance the solubility of poorly water-soluble compounds [4]. Hence, micelles can effectively capture hydrophobic molecules from water, assisting in the elimination of pollutants [34]. Furthermore, micelles have been investigated for their capacity to eliminate both anionic and neutral pollutants from water [35].

Sodium dodecyl sulfate (SDS) is a strong anionic surfactant with the formula $NaC_{12}H_{25}SO_4$. SDS is often selected for the effective removal of heavy metals in water [36]. Also, SDS was used to extract the phenolic compounds from OMW for industrial uses by a mixture of nonionic/anionic surfactants [36].

Previous applications of surfactants were observed in micellar-enhanced ultrafiltration (MEUF). The proposed approach involves the introduction of a surfactant into contaminated water at a concentration exceeding the critical micelle concentration (CMC). The micelles should be larger in size to be retained in a membrane with a pore size larger than required for pollutant retention [37]. However, this method has many limitations, such as frequent pore obstruction. This may lead to permeate flux reduction and permeation of surfactant monomers and, subsequently, potential toxicity due to the leakage of the surfactant monomer [38].

In this work, a multi-step method (*coagulation–flocculation–sedimentation*) along with SDS micelle-distillation was used to reduce the toxicity of OMW, prevent potential secondary toxicity from the surfactant, and collect the polyphenols from OMW. The detoxification steps were monitored using liquid chromatography–mass spectrometry (LC–MS), gas chromatography–mass spectrometry (GC–MS), inductively coupled plasma–mass spectrometry (ICP–MS), high chemical oxygen demand contents, and total phenolic compounds.

The detoxification of OMW was evaluated using standard germination assay, MTT assay using tissue culture, and toxicity assay using fluorescence bacteria.

2. Results and Discussion

2.1. Coagulation–Flocculation–Sedimentation

This step is to manipulate the electrical charges to promote the aggregation of suspended solid materials into larger flocs, which can then settle more readily. This floccule can capture various hazardous compounds and remove them from the liquid [39]. Several chemicals were tested: alum, ferrous sulfate, eggshell, and lime ($CaOH_2$). The best result was achieved using lime ($CaOH_2$) at a concentration of 30 g/L; the volume of sediment was 60% of the total volume, whereas it was less in other agents. This result suits well with the previously published works [39].

Typically, the pH of untreated OMW falls within the acidic range of 4–6 [40]. In this study, the pH of the OMW was measured at 4.5. Upon coagulation, the pH increased significantly, reaching a pH = 12. The significant increase in pH is due to the use of a high concentration of lime [41]. This high alkalinity may facilitate further degradation of compounds that are hard to remove. It is important to mention that the treated water after this step did not neutralize. The neutralization would add an extra step, and it has been reported that electrolytes that may be generated from the neutralization step affect the micelle formation [42].

2.2. Selection of Surfactant and Determination of Total Phenolic Compounds

According to the results of total phenolic compounds (Table 1), the coagulation step alone with lime ($CaOH_2$) successfully removed 56% of the total polyphenols. Subjecting the coagulated OMW to distillation (without the presence of any surfactant) reduced the polyphenols by 88%. To further remove the polyphenol contaminants, distillation with three different types of surfactants (anionic, cationic, and neutral) was evaluated. The neutral surfactants were used to keep the reaction as environmentally safe as possible. Neutral surfactants, in general, are biodegradable and based on natural fatty acids and sugar alcohol sorbitol [43].

Table 1. The effect of different surfactants on the residual polyphenol contents. Experiments were conducted in triplicate, and the error represents the means ± standard deviation.

Treatment Step	Residual Phenols %
Coagulation	43 ± 7.1
Distillation of coagulated OMW	11.9 ± 2.2
Distillation of coagulated OMW + SDS	2.5 ± 0.6
Distillation of coagulated OMW + span 80	14 ± 2.1
Distillation of coagulated OMW + span 20	11 ± 1.6
Distillation of coagulated OMW + CTAB	11.2 ± 2.4

Distillation with sodium lauryl sulfate was the best in removing polyphenolic compounds, as the residual percentage decreased to approximately 2.5%. The results of other surfactants (span 80, span 20, and CTAB) were 14%, 11%, and 11.2%, respectively. Therefore, SDS was solely used in further experiments.

Furthermore, to determine the optimum concentration of SDS, five different concentrations, CMC, CMC*2, CMC*4, CMC/2, and CMC/4, were evaluated. The highest clearance percentage was observed at all concentrations that were either equal to or more than the CMC, or 2.31 g/L or 0.008 mol/L. Micelles form at these concentrations.

Micelles have a hydrophobic core where many types of lipids and lipophilic compounds are entrapped, and the hydrophilic surface can form many interactions with polar compounds [3]. Micelles have been extensively studied for their ability to enhance the solubility of poorly water-soluble compounds, including drugs. The formation of micelles using surfactants has been shown to significantly improve water solubility [32,33]. For instance, the use of micelles has been reported to increase the solubility of drugs like imiquimod, fluocinolone acetonide, docetaxel, paclitaxel, curcumin, and thymoquinone [44–48]. Moreover, the solubility enhancement achieved through micelles has been linked to their ability to form nano-sized structures [49]. The formation of micelles is influenced by factors such as surfactant properties, critical micelle concentration (CMC), and the structure of the micelles themselves [50,51]. These micelles, typically ranging around 80 nm in size, provide a favorable environment for solubilizing poorly soluble compounds [52,53]. The micelle formation process involves interactions between hydrophobic and hydrophilic segments of surfactants, leading to the creation of stable structures that can encapsulate the chemicals or other hydrophobic molecules [54,55].

Sodium dodecyl sulfate (SDS) micelles have been extensively studied for their ability to enhance the solubility of various chemicals [56]. These micelles can interact with different compounds to form micelle–chemical aggregates, improving solubility. For instance, SDS has been used to enhance the solubility and dissolution rate of poorly soluble active pharmaceutical ingredients [53,57], such as cosmetics [58].

Furthermore, the use of SDS micelles has been explored in the removal of lead and mercury from aqueous solutions, showcasing the diverse applications of SDS micelles beyond solubility enhancement [59,60].

OMW contains a high load of organic compounds, including the following organic groups: benzoic acids and derivatives, aromatic conpounds, fatty acids, cinnamic acids, phenyl ethyl alcohols, other phenolic acids, flavonols, flavones, lignans, isochromans, and secoiridoids [61]. It is expected that distillation may remove 90% of the polyphenols and organic compounds. However, OMW contains a high concentration of volatile organic compounds, including organic acids. It has been reported that more than 60 volatile chemical compounds are present in OMW [61], including short-chain fatty acids (six or fewer carbons) [62]. Hence, by distillation, all volatile organic compounds will be concentrated in the distilled product. SDS-micelle will form extra interactions with such compounds; therefore, a large portion of the volatile compounds will be removed.

Different types of bonds are established during the interaction between micelles and the contaminants. For example, fatty acids possessing an extended alkyl chain, such as stearic acid, may establish a hydrophobic interaction with the micelle's core. Additionally, aromatic compounds such as styrene are also bound to the core. The toxic compound, alcohol tyrosol, is more likely to interact with the head and tail of the micelle in a hydrophilic and hydrophobic manner, respectively (Figure 1).

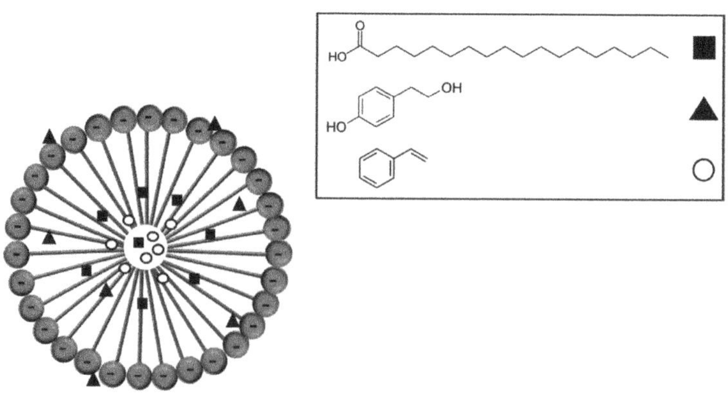

Figure 1. Possible site of interaction between micelle and some OMW, where the site of interactions differs according to hydrophobicity and the tendency of hydrogen bond. The rectangle is stearic acid, the triangle is tyrosol, and the circle is styrene.

Several studies have shown the efficacy of micelles in eliminating volatile organic contaminants from water [34,35,63]. Knowing that, along with the results, the addition of SDS micelles to OWM effectively slowed down the release of toxic compounds during the distillation. Hence, micelles can potentially augment the effectiveness of remediation procedures by augmenting the solubility of pollutants, thereby facilitating their removal or degradation.

2.3. Germination Assay

The germination assay was conducted for each treatment, and several parameters were measured (stem length and fungal contamination). Tap water was used as a positive control. Both filtered raw OMW and coagulated did not exhibit any signs of growth, and the mold was observed after a few days (Figure 2; Table 2). On the other hand, there was a significant increase in stem growth and germination percentage when distillation was applied immediately to raw and coagulated OMW; however, this was still 50% less than in the control group. Additionally, after one week, the presence of hazy molds became apparent. Moreover, as illustrated in Figure 2, the lime coagulation has demonstrated its effectiveness in promoting growth in both shoot and root length and preventing mold formation compared to the distilled raw OMW sample.

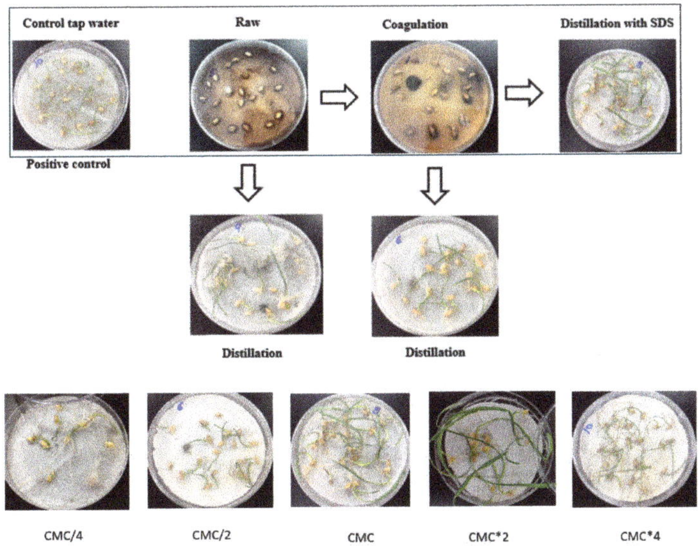

Figure 2. Germination assay for treatment steps.

Table 2. Germination assay parameters for all treatment steps.

Sample Name	Germination %	Stem Length %	Fungi Growth %
Control tap water	100	100	−
Raw OMW	-	-	+
Distilled Raw OMW	40	50	+++
Coagulated OMW	-	-	+++
Distilled Coagulated OMW	50	80	+
CMC Distillation	100	100	−

CMC: Critical micelle concentration. Experiments were conducted in triplicate. The (+) sign represents the presence and severity of fungal growth, the (−) sign represents the absence of the growth.

When SDS was applied at CMC, the outcomes closely resembled those of the control sample (tap water) in terms of the length of roots and shoots and the absence of any fungal development.

The effect on germination has not shown a significant difference in the growth between CMC and concentrations above it (Table 3). Therefore, SDS at CMC concentration was used for further steps.

Table 3. Germination assay parameters for different concentrations of SDS.

Sample Name	Germination %	Stem Length %	Fungi Growth %
CMC/4	60	40	+++
CMC/2	60	40	+
CMC	100	100	–
CMC*2	100	100	–
CMC*4	100	100	–

CMC: Critical micelle concentration. Experiments were conducted in triplicate.

It has been reported that the seed can germinate in a wide range of pH (5–8) [64]; in this experiment, the pH of treated water after all steps was 9.0. Hence, the effect of basic pH on growth was tested, and the results showed no differences. Subsequently, there was no need to neutralize the pH for any further experiments.

A hypothesis was formulated suggesting that the pollutants could disrupt the earliest stage of germination and cause the death of the embryo. Therefore, the seeds were soaked in tap water for 8 h to initiate germination, followed by moistening with OMW over a week. Based on the results, although the seeds effectively developed stems and roots, their growth was slower, and the growth rate was smaller than that of the positive control. In addition, clouds of fungal contamination were observed.

OMW contains a high concentration of phytotoxic and phenolic compounds that have antioxidant and antimicrobial properties, such as hydroxytyrosol, tyrosol, hydroquinone, 4-aminophenol, phenol, gallic acid, caffeic acid, 3,5-di-tert-butylcatechol, quercetin, oleuropein, and catechol [65,66]. It has been reported that OMW can inhibit the growth of several fungi, such as Trichoderma spp., Fusarium spp., and Aspergillus spp. [13].

According to the fungal growth, it appears that only using SDS at CMC and above can stop the fungal growth. So fungal contamination may already be in the OWW, and the contaminants (polyphenols) kill all microorganisms except the fungi, or the Fungai is already on the seeds, and the OMW pollutants facilitate the growth. In both cases, the pollutants that were removed by SDS are responsible for both actions.

2.4. Liquid Chromatography–Mass Spectrometry Analysis

LC–MS analysis was conducted to determine the compounds present in OMW and to evaluate the efficiency of the detoxification treatment. The list of highest-scoring compounds identified in the OMW at the different treatment steps is shown in Table 4 and the total ion chromatogram (positive and negative ionization modes) is shown in Figure 3.

It has been reported that simple biophenols (tyrosol, hydroxytyrosol, and homovanillic alcohol) and complex biophenols (decarbomethoxy ligstroside aglycone and decarbomethoxy oleuropein aglycone) are the most abundant analytes in OMW [61].

The most suggested compound to have the highest toxicity in OMW is hydroxyl tyrosol, which is thought to be produced from the hydrolysis reaction of oleuropein during the milling step [4]. As shown in Table 4, hydroxy tyrosol was successfully removed via the CMC step while existing in all other steps.

Table 4. The highest-score compounds were identified in the OMW using LC–MS at the different treatment steps.

Compound Name	Raw-Distilled	Raw-Coagulated-Distilled	Raw-Coagulated-Distilled-CMC	Source	Aromatic	Toxicity
Scopoletin	Yes	Yes	No	Natural	Yes	No
Capsaicin	Yes	Yes	Yes	Natural	Yes	No
Humulone	Yes	No	No	Natural	No	No
Caffeic Acid	Yes	No	No	Natural	Yes	No
Umbelliferone	Yes	Yes	Yes	Natural	Yes	No
4-Tert-butyl 2-methylphenol	Yes	Yes	No	Synthetic	Yes	No
Tetramethrin	Yes	Yes	Yes	Synthetic—potent insecticide	No	Yes
Eusolex 6007	Yes	Yes	No	Synthetic—Essential component in sunscreens	Yes	Yes
Isopentyl-4-methoxycinnamate	Yes	Yes	No	Synthetic—sunscreening agent.	Yes	Yes
Dodecylphenol	Yes	No	No	Synthetic—used in the manufacturing of epoxy resins	Yes	Yes
4-Hydroxybenzoic acid n-butyl ester	Yes	Yes	No	Synthetic—bactericidal/fungicidal additives in cosmetics.	Yes	Yes
4-n-Propylphenol	Yes	No	No	Synthetic—Food additive	Yes	No
Dibutylphthalate	Yes	Yes	Yes	Synthetic—plasticizer	Yes	Yes
Octocrylene	Yes	Yes	No	Synthetic—sunscreens and cosmetics	Yes	Yes
Hydrocortisone	Yes	Yes	Yes	Synthetic—Anti-inflammatory agent	No	No
(4 or 7) Hydroxy-Coumarin Plus Hydrate	Yes	Yes	No	Synthetic—Insecticides	Yes	Yes
Trans-nonachlor	Yes	Yes	No	Synthetic—Insecticides	No	No
3-(2,2-Dichlorovinyl)-2,2-dimethylcyclopropene carboxylic acid	Yes	Yes	No	Synthetic—Pesticide	No	Yes
4-hydroxybenzoic acid propyl ester	Yes	Yes	Yes	Natural	Yes	No
4-Nonylphenol	Yes	Yes	Yes	Synthetic	Yes	Yes
Cyprodinil	Yes	Yes	No	Synthetic Fungicide	Yes	Yes
2-ethylhexyl 3-(methoxyphenyl)-2-propenoate	Yes	Yes	Yes	Synthetwunscreen	Yes	Yes
Ethyl-4-aminobenzoate	Yes	Yes	Yes	Synthetic Local anesthetic	Yes	Yes
3-Hydroxy-4-methoxycinnamic acid (isoferulic acid)	Yes	Yes	Yes	Natural	Yes	No
Hydroxyl tyrosol	Yes	Yes	No	Natural	yes	yes

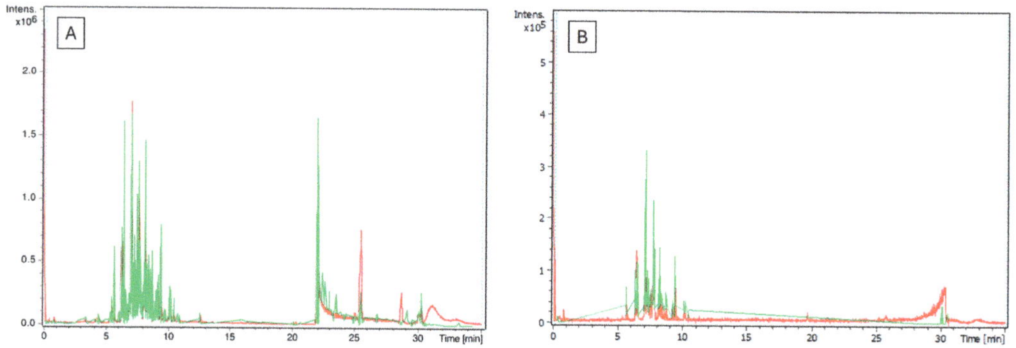

Figure 3. The LC–MS chromatogram SDS-treated OMW. (**A**) The positive ion mode. (**B**) The negative ion mode. The red color represents the total ion chromatogram and the green color represents the elective ion chromatogram.

According to the results, there were unexpected compounds such as sunscreens, pesticides, food additives, plasticizers, and pharmaceutical drugs. These chemicals may come from water contaminants during the washing step, preparation steps, or the actual compounds that were not included in the library. To avoid the possibility that the compound is not included, the main olive compounds were used as a standard for identification (Supplement Data S2).

In conclusion, the LC–MS results were inconclusive, so we conducted further instrumental analysis procedures.

2.5. Gas Chromatography–Mass Spectrometry Analysis

The OMW samples were assessed after the last stage of treatment (SDS) and in their untreated state (raw). Significant variations can be seen in the chromatograms before and after treatment. The CMC-treated OMW showed a considerable reduction in both peak count and intensity, as shown in Figure 4.

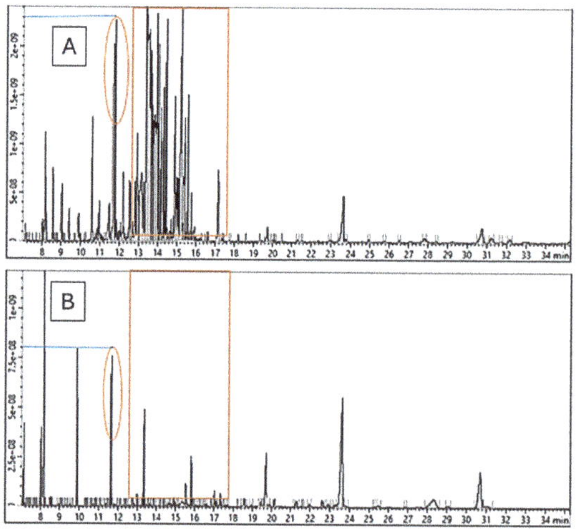

Figure 4. The GC–MS total ion chromatogram. (**A**) Untreated OMW. (**B**) OMW after SDS treatment.

For example, the tyrosol peak (Figures 4 and 5A), which was initially observed at 11.6–11.8 min, fell to one-third of its original intensity, and other neighboring peaks vanished. Furthermore, the peaks that were present between 12.5 and 17.5 min have completely disappeared.

The compounds in that area include organic acids like Octanoic acid, Aconitic acid, 4-Coumaric acid, Citric acid, Tranexamic acid, Quininic acid, Labdanolic acid, Fumaric acid, Octadecenoic acid, and Stearic acid. Additionally, there are other Glucosides like Syringaresinol diglucoside, hexopyranoside, Matairesinoside, and a wide variety of compounds like Brefeldin, Triphenylene, Homovanillyl alcohol, Lapachol, Styrene, Rhodamine B cation, Isobenzofuran, Gibban, Trimethoxyflavone, Prostaglandin A3, dimethoxyflavanone, Oxytetracycline. The complete list of identified compounds is in the Supplement Data S3.

It is important to clarify that not all peaks in the chromatogram are chemicals from OWM, as some of them are due to the derivatization process in Figure 5B,C. Both compounds were identified in the area between 15.88 and 15.91 min.

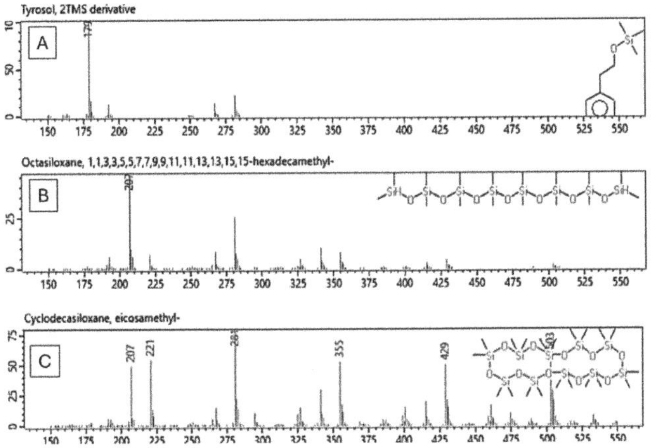

Figure 5. Mass spectrum of (**A**) the TMS derivative of tyrosol. (**B**,**C**) are examples of by-products of the derivatization process.

2.6. Inductively Coupled Plasma Mass Spectrometry Analysis

It has been reported that OMW contains a significant concentration of heavy metals [67,68]. Heavy metal contamination can take place during the handling and processing of olive fruits as well as from the soil [68]. Hence, all possible types of metal were measured, including heavy metals. The mechanism of adsorption of SDS towards heavy metals involves several complex interactions primarily driven by electrostatic forces and hydrophobic interactions. The primary mechanism is through electrostatic attraction. The negatively charged sulfate groups of SDS can bind to positively charged metal ions, such as Pb^{2+}, Cd^{2+}, and Cu^{2+}, leading to the formation of metal–surfactant complexes on the surface of adsorbents [69–71]. Moreover, hydrophobic interaction also plays a significant role in the adsorption process. The hydrophobic tail of SDS can interact with organic contaminants, which might be chelated or interact with heavy metals, promoting the aggregation of metal ions within the micellar structure of SDS [72,73].

The initial concentrations of metals in raw OMW are shown in Table 5. There were many variations in metal percentages at the different stages of treatment; however, all of the following heavy metals (Hg, Pb, Cd, As, and Mo) were under the detection limit in all samples (Figure 6). The highest concentration of metals was in the filtered raw sample. After the coagulation step, Ca ions have disappeared completely; it might be due to the precipitation by lime ($CaOH_2$), and there was a sudden increase in Sr concentration. After distillation, K ions approach zero, and metals like Cu and Fe have increased greatly compared to raw samples. This increase in the concentration of these ions can be explained by their transfer to the product container of distillation, accompanied by the decrease in the water volume [74]. At the final SDS-CMC step, the total amount of ions has decreased significantly; Ni, Fe, and Cu approach half of their original amounts in the raw sample due to their interaction with SDS micelles [75]. SDS is already known for its capability for the removal of heavy metal ions [76].

Table 5. The concentrations of metals in raw OMW. Experiments were conducted in triplicate, and the error represents the means ± standard deviation.

Treatment Step	mg/L	Typical OMW Values [70]
Ni	0.29 ± 0.05	-
Cu	0.25 ± 0.03	0.0021 (%)
Zn	0.575 ± 0.09	0.0057 (%)
Mg	376.036 ± 11.5	100–400 mg/L
K	1343.48 ± 23.54	2700–7200 mg/L
Ca	690.40 ± 12.4	120–750 mg/L
Mn	0.88 ± 0.20	0.0015 (%)
Fe	0.636 ± 0.08	-
Sr	6.789 ± 1.4	-
Ba	0.739 ± 0.17	-
Li	0.013 ± 0.006	-
Na	40.09 ± 2.4	40–900 mg/L
Hg	BDL	
Pb	BDL	-
Cd	BDL	-
As	BDL	-
Co	BDL	-
Mo	BDL	-

BDL: below detection limits.

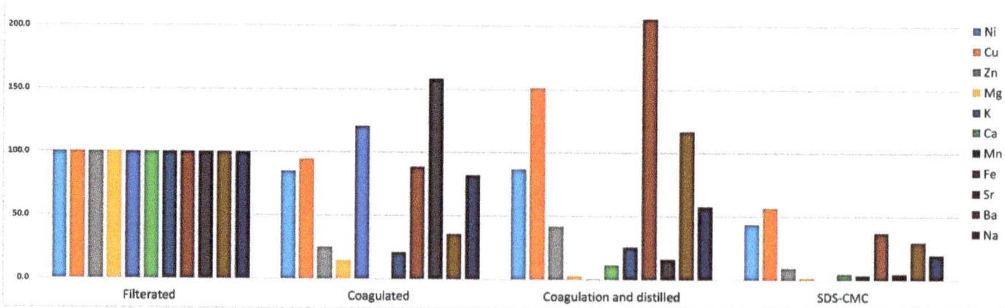

Figure 6. Percentage of metal in OMW samples at different stages.

2.7. *High Chemical Oxygen Demand (COD)*

The Chemical Oxygen Demand (COD) test is utilized to predict the amount of oxygen needed by the effluent. It is employed for the purpose of monitoring and regulating discharges, as well as evaluating the effectiveness of treatment plants [77]. OMW contains high concentrations of chemical oxygen demand (COD), reaching 220 g/L, mainly consisting of phenol, lipids, acids, and sugars [18]. Hence, it has a negative impact on aquatic and terrestrial ecosystems upon disposal [78]. Since the chemical composition of olive mill wastewater is highly variable both qualitatively and quantitatively according to the country and method of extraction [2,79], the COD was measured in all treatment stages. According to the data presented in Table 6, the concentration of COD decreases gradually during the treatment process, ultimately achieving a 65% reduction following the SDS-CMC phase.

Table 6. The concentration of COD and the percentage of removal for each treatment step. Experiments were conducted in triplicate, and the bars represent the means ± standard deviation.

Treatment Step	COD (g/L)	% Removal
Filtered OMW	124.6 ± 15.3	0
Coagulated OMW	108.0 ± 9.7	13.3
Distilled OMW	86.58 ± 7.5	30.5
CMC step	44.14 ± 4.3	65.6

2.8. Bacteria Toxicity Assay

The presence of OMW in the environment can disrupt microbial consortia and bacterial populations [78]. Moreover, the antimicrobial properties of OMW, particularly in its liquid form, can inhibit the growth of plants and microorganisms, affecting the overall biodiversity of ecosystems [78,80]. Two samples of OMW were analyzed: the first sample was filtered, and the final sample was treated with SDS-CMC. The survival percentage was determined by measuring the luminescence intensity of Vibrio fischeri at various time intervals over a period of 45 min. As shown in Figure 7, extended duration has significantly amplified the toxicity and mortality rates in filtered samples as compared to CMC samples. For instance, the survival rate in the filter sample decreased substantially, reaching 50% after the incubation period, while it remained constant in the SDS-CMC-treated sample. The toxicity develops in a time-dependent manner. As per the kit instructions, the assay can last for a maximum of 45 min before the fluorescence signal fades. The toxicity results of filtered OMW on photobacterium Vibrio fischeri suit well with previously published [81,82].

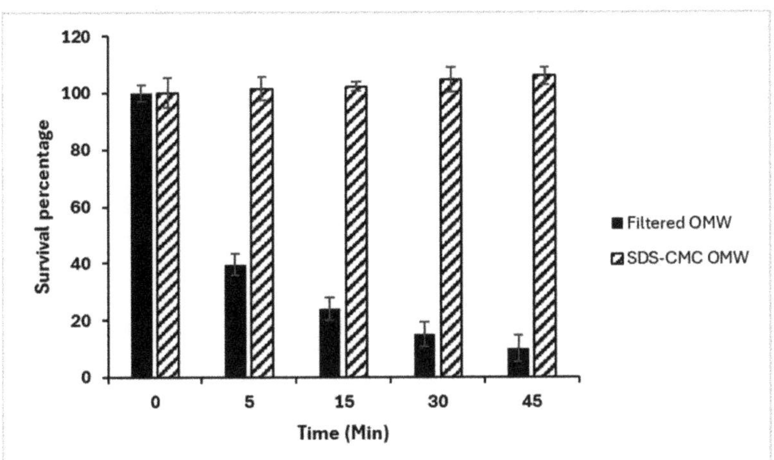

Figure 7. Toxicity assay using fluorescence bacteria over a 45 min period. Experiments were conducted in triplicates, and the error bars represent the means ± standard deviation.

According to reports, the main cause of toxicity in V. fischeri is the polar portion of OMW. Therefore, it has been recommended that diluting OMW should not be considered a viable remedy for disposing of OMW, as the toxicity remains even at low concentrations [82]. The incubation period with tested bacteria was extended to 45 min to confirm that there was no toxicity, as previous studies have only examined toxicity for a duration of 10 to 15 min.

2.9. MTT Assay

In addition to its phytotoxic and bactericidal effects, OMW also exhibits high toxicity towards human cells, with the polyphenols being the main contributors to this activity [83]. However, multiple studies reported that the polyphenol extracted from OMW (esp. hydroxy tyrosol) has anti-cancer and antioxidant activity and improves the chemotherapy and cytoprotective [84–86]. Furthermore, OMW extract also showed a promising ingredient for dermal applications to improve skin health and skin protection [87].

The purification steps were evaluated using a human fibroblast (non-cancer cells), one of the most abundant cell types in the stroma. It has a variety of functions and composes the basic framework for tissues and organs [88]. The survival rate relative to the control sample was quantified following each treatment stage. Figure 8 illustrates that the survival percentage, relative to the control sample of distilled water with equivalent tonicity, was just 3% for filtered OMW. However, it gradually climbed and reached approximately 95% with SDS-CMC.

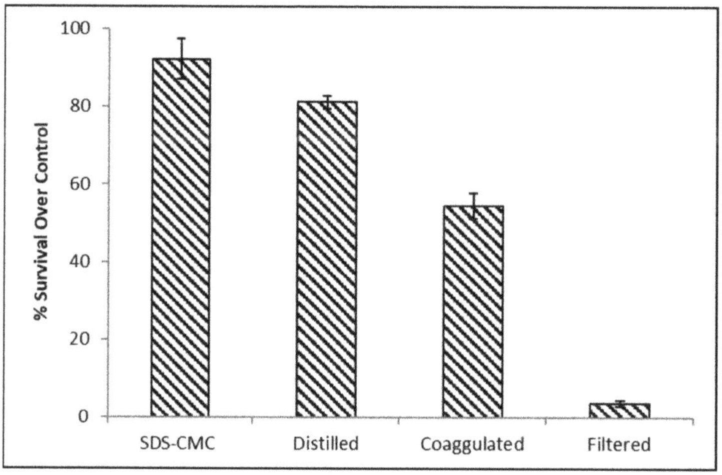

Figure 8. Toxicity assay of OMW treatment steps using human fibroblast and MTT assay. Experiments were conducted in triplicate, and the error bars represent the means ± standard deviation.

2.10. Possible Use for Mass Productions

Several factors can affect the feasibility of mass production, including energy consumption, capacity, production rate, and SDS (cost and toxicity). Figure 9 illustrates the process of detoxification, including the by-products and possible solutions. The debris, sludge, and oily waste can serve as fuel for distillation [89]. To decrease the consumption of fossil fuels and reduce the emission of carbon dioxide, it is recommended to utilize vacuum distillation as a means to lower the boiling point and, thus, minimize the requirement for heating during the distillation process. Organic compound distillation has been conducted using vacuum distillation [89]. Moreover, solar energy can be harnessed to produce the electrical energy required for vacuum distillation. In addition, OMW can be preheated before distillation by utilizing an indirect solar water heating system. This technology can raise the temperature of the water to 90 °C. Additionally, the impact of OMW on the efficiency of the system is negligible [90]. Moreover, it is essential to economically recover and recycle the SDS since the surfactant comprises a significant proportion of expenses and to avoid additional contamination of the environment. There are multiple techniques for recovering the surfactant. These include the addition of an excess of multivalent cations (such as Ca+2) beyond the stoichiometric value, which can lead to a significant amount (45–55%) of surfactant precipitation [91]. Another method is to lower the temperature of the SDS solution to 4 °C, which is below the reported Kriff point of SDS (16 °C) [92]. Other techniques include foam fractionation [93], electrochemical treatment [94], and acidification at Ph = 1 [36]. The capacity and rate of detoxification have to be further evaluated.

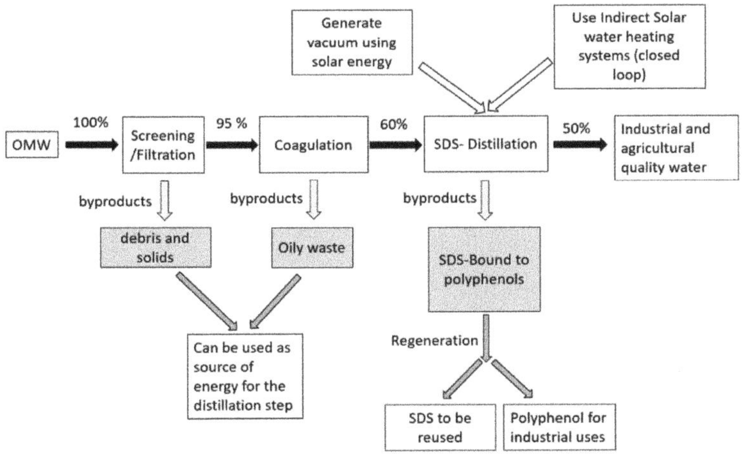

Figure 9. Diagram showing the detoxification steps of OMW, by-products, and possible solutions. The percent represents the percentage of treated water after each step compared to the initial volume.

3. Materials and Methods

3.1. Chemicals and Materials

Analytical-grade reagents, chemicals, and HPLC-grade solvents were obtained from Sigma-Aldrich and the local market. OMW was freshly collected in 20 L polyethylene containers from the decenter outlet of three-phase olive oil mill processing systems located in Irbid/Jordan during the olive harvesting season and then stored at 4 °C until use in laboratory experiments.

The natural eggshell waste materials (ES) were obtained from local homes and restaurants to serve as coagulant material. The ES was washed with water to remove impurities, dried for 3 h at 60 °C, and crushed mechanically in a mixer. The dried eggshell product was calcined (CES) at 800 °C for 3 h using a muffle furnace (Carbolite, Sheffield-England), sieved in a mesh number (40–60) to obtain grain size diameters ranging from 425–600 μm, and stored in a dry box until use.

3.2. Methods

3.2.1. Purification Steps

Pretreatment

Samples were completely stationary for 24 h and divided into two net layers at the macro level, and then the water layer was filtered using locally made fabric mesh to remove large debris and particles.

Coagulation–Flocculation–Sedimentation

The process was conducted as described elsewhere [36] using the optimum concentration of different agents: alum (600 mg/L) [37], ferrous sulfate (300 mg/L) [37], eggshell (6 g/L) [38], and lime (CaOH2) (15 g/L) [36].

The OMW samples were added first into a graduated Imhoff Cone for one hour to allow for any readily settleable solids to settle down by gravity. The supernatant was then taken and treated by coagulation and flocculation using a conventional 6-station Jar test apparatus (Stuart-SW6) supplied by six beakers (1.0 L each). Alum is used as a coagulant material, and the optimum dose for TSS removal was predetermined first. A total of 500 mL of OMW was transferred to each jar-test beaker, and then six different doses of alum (0.5–3.0 at a 0.5 g/L interval) were added into all beakers. All beakers were subjected to rapid mixing initially at 120 rpm for 3 min and slow mixing thereafter at 20 rpm for 30 min. Then, the contents of the beakers were allowed to settle by gravity for 1.0 h. The alum dose corresponded to the least turbidity, and the TSS value was taken as an optimum dose. The predetermined optimum dose of alum was then added to all OMW samples and treated in the same manner using a jar-test apparatus. The supernatant solution was then withdrawn, collected, and used in the next experiments. Part of this supernatant solution was filtered through filter paper (Whattmann 40) and tested for the same parameters tested for raw OMW to assess the effect of coagulation on OMW quality. All the experiments and sample analysis were performed in triplicate at room temperature of 25 ± 1 °C without pH adjustment.

Micellar Distillation Treatment

Different types of surfactants at the critical micelle concentration (CMC) were evaluated separately: Span 80 at 0.014 g/L, Span 20 at 0.07 mg/L, hexadecyltrimethylammonium (CTAB) at 0.334 g/L, and sodium dodecyl sulfate at 2.3 g/L (Figure 10). Moreover, SDS was tested at different concentrations (above or below CMC: half CMC concentration, quarter CMC concentration, double CMC concentration, and fourfold CMC concentrations). The surfactant was dissolved in 100 mL of OMW and then distilled at 90 °C using standard distillation apparatus and method [39].

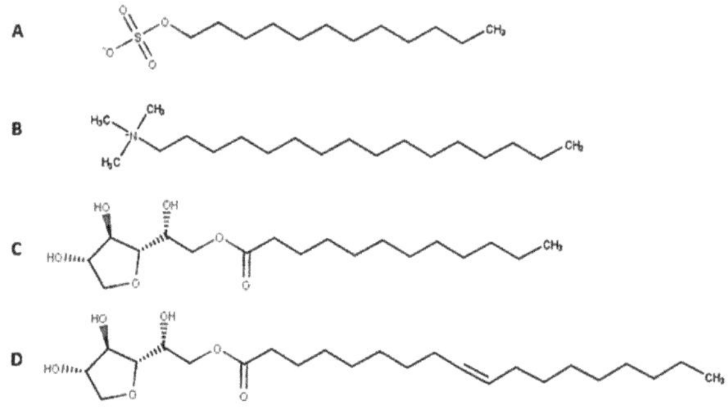

Figure 10. Surfactants (**A**) SDS, (**B**) CTAB, (**C**) Span 20, (**D**) Span 80.

3.2.2. Total Phenolic Compounds

The test was conducted as described previously elsewhere [40,41] using Spectrophotometer-UV 1800, Biotech Engineering Management Co., Ltd., Milton Keynes, UK. Briefly, 125 µL of 100-diluted extract was mixed with 500 µL of distilled water and 125 µL of Folin–Ciocalteu reagent. Following 3 min of stirring, 1250 µL of 7% sodium carbonate solution was added to the mixture. The mixture was adjusted to 3 mL with ultrapure water and left at room temperature for 90 min in the dark. The results were represented as mg gallic acid per mL of extract.

3.2.3. Germination Assay

The experiment was carried out following the previously described protocol, with some modifications [36]. Ten locally obtained wheat seeds were presoaked in the tested OMWW for 8 h before germinating under the light of a fume hood at room temperature. To prevent drying during germination, 5 mL of treated water was added periodically to the seeds. The entire process of germination lasted around two weeks.

3.2.4. High Chemical Oxygen Demand (COD)

The COD experiment was performed using the Lovibond COD high-range kit (200–15,000 mg/L) following the manufacturer's instructions [42]. The samples underwent a 10-fold dilution, and the experiment was conducted using the Lovibond thermal reactor for 120 min at a temperature of 150 °C. The Lovibond Spectro Direct was used at a wavelength of 602 nm (lambda, λ).

3.2.5. Liquid Chromatography–Mass Spectrometry Analysis (LC–MS)

The unknown sample (0.1 mL) was diluted with 0.9 mL of distilled water. Then, 1.0 mL was transferred to the autosampler, and inject 3.0 µL. The samples were analyzed on a Bruker Daltonik (Bremen, Germany) Impact II ESI-Q-TOF System equipped with Bruker Dalotonik (Bremen, Germany) using direct injection. The instrument was operated using the Ion Source Apollo II ion Funnel electrospray source. The capillary voltage was 2500 V, the nebulizer gas was 2.0 bar, the dry gas (nitrogen) flow was 8 L/min, and the dry temperature was 200 °C. The mass accuracy was <1 ppm; the mass resolution was 50,000 FSR (Full Sensitivity Resolution), and the TOF repetition rate was up to 20 kHz. Standards were used for the identification of m/z with high resolution. To evaluate the methods, standards were used (Supplement Data S1).

3.2.6. Gas Chromatography–Mass Spectrometry Analysis (GC–MS)

The OMWW samples were extracted using ethyl acetate and silylated using BSTFA [43]. The GC–MS analysis was conducted following the procedure described previously [44,45]. Briefly, one µL of the sample was subjected to the GC–MS (Chromatec Crystal GC-MSD, Yoshkar-Ola, Russia) equipped with a CR-5 MS column (5% diphenyl, 95% dimethyl polysiloxane, 30 m × 0.25 mm, 0.25-µm film thicknesses). In the MS detector, 70 eV electron ionization was used. MS source temperature was 300 °C, and the transfer line temperature was 230 °C. The temperature column was controlled from 40 °C for 1 min (isothermal) to 280 °C at 3 °C/min, maintaining constant lower and upper temperatures for 3 min. The carrier gas was helium (1.0 mL/min). Estimated compound percentage concentrations were based on relative peak areas. The same chromatographic conditions were used to evaluate a C8–C30 n-alkane standard solution. The chemical constituents were identified by comparing their computed Kovats retention index (KI), matching their mass spectra with the built-in library spectra.

3.2.7. Inductively Coupled Plasma Mass Spectrometry Analysis (ICP–MS)

The experiment was conducted at the Jordanian Atomic Energy Commission as previously published [46,47] on the Thermo Scientific™ iCAP™ TQ spectrometer (Bremen, Germany).

3.2.8. Anti-Bacterial Activity

BioTox™ WaterTox™ EVO Kit was used to determine the toxic effect of OMW on the living bacteria Vibrio fischeri. Samples were prepared according to the manufacturer's protocol and using the kit reagents. The luminescence was measured using a multi-well plate reader (Synergy HTX Multimode Reader, BioTek, Santa Clara, CA, USA).

3.2.9. Cell Culture Assay

The Human Fibroblast (PDL) cell line was obtained as a kind gift from Prof. Khaled Al-Qaoud (Yarmouk University, Irbid, Jordan). PDL cells were cultured at 37 °C under a 5% CO2 humidified atmosphere in DMEM (with high glucose and sodium pyruvate) supplemented with 10% fetal bovine serum, 1% penicillin/streptomycin, and 1% NEAA.

Cell viability was evaluated using 3-(4,5-dimethylthiazol-2-yl)-2,5-diphenyltetrazolium bromide (MTT) assays to assess cell metabolic activities [95]. Exponentially growing PDL cells were washed with PBS, shortly trypsinated, and counted. Then, cells were seeded at a density of 5000 cells/100 µL/well in 96-well plates and cultured overnight. Afterward, the cells were treated for 72 h with 100 μL of the water samples diluted in DMEM (1:2.5 dilutions).

The absorbance of the samples was recorded at 570 nm in a multiwall plate reader (Synergy HTX Multimode Reader, BioTek, USA). Results were plotted as the mean values of duplicates from a representative experiment that was repeated three independent times. Survival is expressed as a percentage of control.

3.2.10. Statistical Analysis

Data were presented as mean ± SD and analyzed using Microsoft Excel with one-way analysis of variance (ANOVA) ($p < 0.05$).

4. Conclusions

The multi-step SDS-based approach successfully detoxified Olive Mill Wastewater (OMW) and significantly improved water quality, making it suitable for direct use in industry and agriculture.

- The treatment process, involving screening, coagulation with various chemicals, and distillation with different surfactants, effectively reduced the environmental and health impacts of OMW.
- Monitoring via LC–MS, GC–MS, ICP–MS, chemical oxygen demand (COD), and total phenolic compounds confirmed the efficacy of the process.
- GC–MS and LC–MS show that the total number of contaminants was reduced significantly.
- The chemical oxygen demand in treated OMW was reduced by 66%.
- The polyphenol contents were reduced by 98%.
- The treatment process interfered with the metals, resulting in the following reductions: Cu (45%), Ni (57%), Fe and Ba (70%), Na (80%), Zn, Mg, K, Ca, Mn, and Sr (more than 95%). All other heavy metals (Hg, Pb, Cd, As, and Mo) were under the detection limit in the raw samples.
- The toxicity of treated OMW was successfully improved: 100% in germination assay, 100% elimination of mold growth, 95% on human cells, and 100% on fluorescence bacteria.
- The potential use on a large scale was also discussed, and solutions were offered.
- The overall reduction in pollutants and toxicity highlights the potential of using SDS micelles for OMW detoxification, offering a viable solution for environmental management in olive-producing regions.
- Future work may focus on mass production.

Supplementary Materials: The following supporting information can be downloaded at: https://www.mdpi.com/article/10.3390/molecules29184284/s1, S1: The list of standards used to validate the LC-MS method, S2: The olive oil compounds used for the identification in the LC-MS method, S3: The complete list of identified compounds using the GC-MS method.

Author Contributions: Conceptualization, Y.A., M.Z. and A.J.; methodology, I.A., A.J., M.Z., T.R., A.H. and Y.A.; software, Y.A. and I.A.; validation, Y.A., A.H, M.Z. and A.J.; formal analysis, Y.A., M.Z., I.A. and A.J.; investigation, Y.A., M.Z., A.H, I.A. and A.J.; resources, M.Z. and A.A.; data curation, Y.A., M.Z., T.R., A.H. and I.A.; writing—original draft preparation, I.A., A.J., A.H. and Y.A.; writing—review and editing, H.A., A.A. and N.A.-R.; visualization, Y.A.; supervision, Y.A.; project administration, Y.A., M.Z. and A.J.; funding acquisition, M.Z. and A.A. All authors have read and agreed to the published version of the manuscript.

Funding: This research was funded by the Deanship of Scientific Research at Yarmouk University, Irbid, Jordan.

Institutional Review Board Statement: Not applicable.

Informed Consent Statement: Not applicable.

Data Availability Statement: The data presented in this study are available upon request from the corresponding author.

Acknowledgments: The authors extend their appreciation to the Researchers Supporting Project number (RSP2024R502), King Saud University, Riyadh, Saudi Arabia, for further funding this project. Moreover, the authors would like to thank all reviewers for their constructive advice.

Conflicts of Interest: The authors declare no conflicts of interest.

References

1. Yangui, A.; Abderrabba, M.; Sayari, A. Amine-modified mesoporous silica for quantitative adsorption and release of hydroxytyrosol and other phenolic compounds from olive mill wastewater. *J. Taiwan Inst. Chem. Eng.* **2017**, *70*, 111–118. [CrossRef]
2. Jamrah, A.; Al-Zghoul, T.M.; Darwish, M.M. A comprehensive review of combined processes for olive mill wastewater treatments. *Case Stud. Chem. Environ. Eng.* **2023**, *8*, 100493. [CrossRef]
3. Raiti, J.; Hafidi, A. Mixed micelles-mediated dephenolisation of table olive processing's wastewaters. *Water Sci. Technol.* **2015**, *72*, 2132–2138. [CrossRef]
4. Bazoti, F.N.; Gikas, E.; Skaltsounis, A.L.; Tsarbopoulos, A. Development of a liquid chromatography–electrospray ionization tandem mass spectrometry (LC–ESI MS/MS) method for the quantification of bioactive substances present in olive oil mill wastewaters. *Anal. Chim. Acta* **2006**, *573*, 258–266. [CrossRef]
5. Al-Qodah, Z.; Al-Shannag, M.; Bani-Melhem, K.; Assirey, E.; Alananbeh, K.; Bouqellah, N. Biodegradation of olive mills wastewater using thermophilic bacteria. *Desalination Water Treat.* **2015**, *56*, 1908–1917. [CrossRef]
6. Sarris, D.; Rapti, A.; Papafotis, N.; Koutinas, A.A.; Papanikolaou, S. Production of added-value chemical compounds through bioconversions of olive-mill wastewaters blended with crude glycerol by a Yarrowia lipolytica strain. *Molecules* **2019**, *24*, 222. [CrossRef] [PubMed]
7. Neffa, M.; Hanine, H.; Lekhlif, B.; Taourirt, M.; Habbari, K. Treatment of wastewaters olive mill by electrocoagulation and biological process. *Linnaeus Eco-Tech* **2010**, 295–304. [CrossRef]
8. Roig, A.; Cayuela, M.L.; Sánchez-Monedero, M.A. An overview on olive mill wastes and their valorisation methods. *Waste Manag.* **2006**, *26*, 960–969. [CrossRef]
9. Khdair, I.A.; Abu-Rumman, G. Evaluation of the environmental pollution from olive mills wastewater. *Fresenius Environ. Bull* **2017**, *26*, 2537–2540.
10. da Silva, A.N. *Treatment of Wastewater from the Olive Pomace Oil Extraction Industry by Fenton*; Federal University of Technology: Paraná, Brazil, 2023.
11. Jakobek, L. Interactions of polyphenols with carbohydrates, lipids and proteins. *Food Chem.* **2015**, *175*, 556–567. [CrossRef]
12. Souilem, S.; El-Abbassi, A.; Kiai, H.; Hafidi, A.; Sayadi, S.; Galanakis, C.M. Olive oil production sector: Environmental effects and sustainability challenges. In *Olive Mill Waste*; Elsevier: Amsterdam, The Netherlands, 2017; pp. 1–28.
13. Cibelli, F.; Bevilacqua, A.; Raimondo, M.L.; Campaniello, D.; Carlucci, A.; Ciccarone, C.; Sinigaglia, M.; Corbo, M.R. Evaluation of fungal growth on olive-mill wastewaters treated at high temperature and by high-pressure homogenization. *Front. Microbiol.* **2017**, *8*, 2515. [CrossRef] [PubMed]
14. Inan, H.; Dimoglo, A.; Şimşek, H.; Karpuzcu, M. Olive oil mill wastewater treatment by means of electro-coagulation. *Sep. Purif. Technol.* **2004**, *36*, 23–31. [CrossRef]
15. Khoufi, S.; Aloui, F.; Sayadi, S. Treatment of olive oil mill wastewater by combined process electro-Fenton reaction and anaerobic digestion. *Water Res.* **2006**, *40*, 2007–2016. [CrossRef] [PubMed]
16. El-Gohary, F.A.; Badawy, M.I.; El-Khateeb, M.A.; El-Kalliny, A.S. Integrated treatment of olive mill wastewater (OMW) by the combination of Fenton's reaction and anaerobic treatment. *J. Hazard. Mater.* **2009**, *162*, 1536–1541. [CrossRef]
17. Ochando-Pulido, J.M.; Hodaifa, G.; Victor-Ortega, M.D.; Martinez-Ferez, A. Performance modeling and cost analysis of a pilot-scale reverse osmosis process for the final purification of olive mill wastewater. *Membranes* **2013**, *3*, 285–297. [CrossRef]
18. Gotsi, M.; Kalogerakis, N.; Psillakis, E.; Samaras, P.; Mantzavinos, D. Electrochemical oxidation of olive oil mill wastewaters. *Water Res.* **2005**, *39*, 4177–4187. [CrossRef]
19. Rocha, K.O.; Brandão, F.; átia Mendes, C.; Carvalho, M.G.V.S.; Mazierski, P.; Zaleska-Medynska, A.; Gomes, J.; Martins, R.C.; Domingues, E. Olive mill waste bio-based catalyst application in advanced oxidation processes for wastewater treatment. *Catal. Today* **2024**, *432*, 114618. [CrossRef]
20. Jaradat, A.Q.; Gharaibeh, S.; Abu Irjei, M. The application of solar distillation technique as a mean for olive mill wastewater management. *Water Environ. J.* **2018**, *32*, 134–140. [CrossRef]
21. Ioannou-Ttofa, L.; Michael-Kordatou, I.; Fattas, S.C.; Eusebio, A.; Ribeiro, B.; Rusan, M.; Amer, A.R.B.; Zuraiqi, S.; Waismand, M.; Linder, C.; et al. Treatment efficiency and economic feasibility of biological oxidation, membrane filtration and separation processes, and advanced oxidation for the purification and valorization of olive mill wastewater. *Water Res.* **2017**, *114*, 1–13. [CrossRef] [PubMed]
22. Amor, C.; Lucas, M.S.; García, J.; Dominguez, J.R.; De Heredia, J.B.; Peres, J.A. Combined treatment of olive mill wastewater by Fenton's reagent and anaerobic biological process. *J. Environ. Sci. Health Part A* **2015**, *50*, 161–168. [CrossRef]

23. Zorpas, A.A.; Inglezakis, V.J. Intergraded applied methodology for the treatment of heavy polluted waste waters from olive oil industries. *Appl. Environ. Soil Sci.* **2011**, *2011*, 537814. [CrossRef]
24. Aboutaleb, E.; Kamel, G.; Hellal, M. Investigation of effective treatment techniques for olive mill wastewater. *Egypt. J. Chem.* **2018**, *61*, 415–422.
25. Yassine, W.; Zyade, S.; Akazdam, S.; Essadki, A.; Gourich, B.; Left, D.B. A study of olive mill waste water removal by a biosorbent prepared by olive stones. *Mediterr. J. Chem.* **2019**, *8*, 420–434. [CrossRef]
26. Sygouni, V.; Pantziaros, A.G.; Iakovides, I.C.; Sfetsa, E.; Bogdou, P.I.; Christoforou, E.A.; Paraskeva, C.A. Treatment of two-phase olive mill wastewater and recovery of phenolic compounds using membrane technology. *Membranes* **2019**, *9*, 27. [CrossRef]
27. Agalias, A.; Magiatis, P.; Skaltsounis, A.-L.; Mikros, E.; Tsarbopoulos, A.; Gikas, E.; Spanos, I.; Manios, T. A new process for the management of olive oil mill waste water and recovery of natural antioxidants. *J. Agric. Food Chem.* **2007**, *55*, 2671–2676. [CrossRef]
28. Rusan, M.J.; Albalasmeh, A.A.; Zuraiqi, S.; Bashabsheh, M. Evaluation of phytotoxicity effect of olive mill wastewater treated by different technologies on seed germination of barley (*Hordeum vulgare* L.). *Environ. Sci. Pollut. Res.* **2015**, *22*, 9127–9135. [CrossRef]
29. Duarte, K.R.; Freitas, A.C.; Pereira, R.; Pinheiro, J.C.; Gonçalves, F.; Azaari, H.; El Azzouzi, M.; Zrineh, A.; Zaydoun, S.; Duarte, A.C.; et al. Treatment of olive oil mill wastewater by silica–Alginate–Fungi biocomposites. *Water Air Soil Pollut.* **2012**, *223*, 4307–4318. [CrossRef]
30. Arvaniti, E.C.; Zagklis, D.P.; Papadakis, V.G.; Paraskeva, C.A. High-added value materials production from OMW: A technical and economical optimization. *Int. J. Chem. Eng.* **2012**, *2012*, 607219. [CrossRef]
31. Paraskeva, C.A.; Arvaniti, E.C.; Papadakis, V.G. Fundamentals and Technoeconomical Analysis of Exploitation of Olive Mill Wastewater to Highadded Vale By-Products. *Linnaeus Eco-Tech* **2010**, 216–223. [CrossRef]
32. Maddiboyina, B.; Ramaiah; Nakkala, R.K.; Roy, H. Perspectives on cutting-edge nanoparticulate drug delivery technologies based on lipids and their applications. *Chem. Biol. Drug Des.* **2023**, *102*, 377–394. [CrossRef]
33. Ramyasree, J.; Hindustan, A.A.; Chinthaguinjala, H.; Reshma, T.C.; Venkata, H.V.Y.; Bharath, K. Solubility enhancement of drugs with aid of surfactants: Research done since last two decades. *Int. J. Pharma Bio. Sci.* **2020**, *10*, 11–16. [CrossRef]
34. Gokhale, D.; Chen, I.; Doyle, P.S. Micelle-Laden Hydrogel Microparticles for the Removal of Hydrophobic Micropollutants From Water. *ACS Appl. Polym. Mater.* **2022**, *4*, 746–754. [CrossRef]
35. Polubesova, T.; Nir, S.; Zadaka, D.; Rabinovitz, O.; Serban, C.; Groisman, L.; Rubin, B. Water Purification From Organic Pollutants by Optimized Micelle–Clay Systems. *Environ. Sci. Technol.* **2005**, *39*, 2343–2348. [CrossRef]
36. Huang, J.; Qi, F.; Zeng, G.; Shi, L.; Li, X.; Gu, Y.; Shi, Y. Repeating recovery and reuse of SDS micelles from MEUF retentate containing Cd2+ by acidification UF. *Colloids Surf. A Physicochem. Eng. Asp.* **2017**, *520*, 361–368. [CrossRef]
37. Moreno, M.; Mazur, L.P.; Weschenfelder, S.E.; Regis, R.J.; de Souza, R.A.F.; Marinho, B.A.; da Silva, A.; de Souza, S.M.A.G.U.; de Souza, A.A.U. Water and wastewater treatment by micellar enhanced ultrafiltration—A critical review. *J. Water Process Eng.* **2022**, *46*, 102574. [CrossRef]
38. Welch, J.; Wallace, J.; Lansley, A.B.; Roper, C. Evaluation of the toxicity of sodium dodecyl sulphate (SDS) in the MucilAir™ human airway model in vitro. *Regul. Toxicol. Pharmacol.* **2021**, *125*, 105022. [CrossRef] [PubMed]
39. Elayadi, F.; El Adlouni, C.; El Herradi, M.A.E.; El Krati, M.; Tahiri, S.; Naman, M.N.F. Effects of raw and treated olive mill wastewater (OMW) by coagulation-flocculation, on the germination and the growth of three plant species (wheat, white beans, lettuce). *Moroc. J. Chem.* **2019**, *7*, 111–122.
40. Issa, A.; El Riachy, M.; Bou-Mitri, C.; Doumit, J.; Skaff, W.; Karam, L. Influence of geographical origin, harvesting time and processing system on the characteristics of olive-mill wastewater: A step toward reducing the environmental impact of the olive oil sector. *Environ. Technol. Innov.* **2023**, *32*, 103365. [CrossRef]
41. Muhammad, G.; Marri, A.; Sami, H.A. Chemical Analysis of Reinforced Stabilized Soil. *Indian J. Sci. Technol.* **2018**, *11*, 1–8.
42. Bahri, M.A.; Hoebeke, M.; Grammenos, A.; Delanaye, L.; Vandewalle, N.; Seret, A. Investigation of SDS, DTAB and CTAB micelle microviscosities by electron spin resonance. *Colloids Surf. A Physicochem. Eng. Asp.* **2006**, *290*, 206–212. [CrossRef]
43. Nagtode, V.S.; Cardoza, C.; Yasin, H.K.A.; Mali, S.N.; Tambe, S.M.; Roy, P.; Singh, K.; Goel, A.; Amin, P.D.; Thorat, B.R.; et al. Green surfactants (biosurfactants): A petroleum-free substitute for Sustainability– Comparison, applications, market, and future prospects. *ACS Omega* **2023**, *8*, 11674–11699. [CrossRef] [PubMed]
44. Ghezzi, M.; Pescina, S.; Delledonne, A.; Ferraboschi, I.; Sissa, C.; Terenziani, F.; Remiro, P.D.F.R.; Santi, P.; Nicoli, S. Improvement of imiquimod solubilization and skin retention via TPGS micelles: Exploiting the co-solubilizing effect of oleic acid. *Pharmaceutics* **2021**, *13*, 1476. [CrossRef] [PubMed]
45. Limsitthichaikoon, S.; Soontaranon, S.; Hanpramukkun, N.; Thumanu, K.; Priprem, A. Polymeric Micelles Enhance Mucosal Contact Time and Deposition of Fluocinolone Acetonide. *Polymers* **2022**, *14*, 2247. [CrossRef] [PubMed]
46. Yang, Y.-N.; Cheng, J.-J.; He, J.; Lu, W.-G. Novel Docetaxel-Loaded Micelles Based on all-trans-Retinoic Acid: Preparation and Pharmacokinetic Study in Rats. *Pharm. Front.* **2022**, *4*, e188–e196. [CrossRef]
47. Dantas Lopes dos Santos, D.; Besegato, J.F.; de Melo, P.B.G.; Oshiro Junior, J.A.; Chorilli, M.; Deng, D.; Bagnato, V.S.; de Souza Rastelli, A.N. Curcumin-loaded Pluronic® F-127 micelles as a drug delivery system for curcumin-mediated photodynamic therapy for oral application. *Photochem. Photobiol.* **2021**, *97*, 1072–1088. [CrossRef]

48. Bergonzi, M.C.; Vasarri, M.; Marroncini, G.; Barletta, E.; Degl'Innocenti, D. Thymoquinone-loaded Soluplus®-Solutol® HS15 mixed micelles: Preparation, in vitro characterization, and effect on the SH-SY5Y cell migration. *Molecules* **2020**, *25*, 4707. [CrossRef]
49. Fang, J.; Chen, Z.; Song, J.; Li, J.; Han, Y.; Hou, W.; Wang, W.; Ruan, B.H. Biodegradable self-assembly micelles significantly enhanced the solubility, biological stability and in vivo antitumor efficacy of Hexylselen. *RSC Chem. Biol.* **2021**, *2*, 1669–1681. [CrossRef]
50. Juhász, Á.; Seres, L.; Varga, N.; Ungor, D.; Wojnicki, M.; Csapó, E. Detailed calorimetric analysis of mixed micelle formation from aqueous binary surfactants for design of nanoscale drug carriers. *Nanomaterials* **2021**, *11*, 3288. [CrossRef]
51. Bhowmik, S.; Pham, T.T.; Takahashi, R.; Kim, D.; Matsuoka, H.; Ishihara, K.; Yusa, S.-i. Preparation of Water-Soluble Polyion Complex (PIC) Micelles with Random Copolymers Containing Pendant Quaternary Ammonium and Sulfonate Groups. *Langmuir* **2023**, *39*, 8120–8129. [CrossRef]
52. Katona, G.; Sipos, B.; Ambrus, R.; Csóka, I.; Szabó-Révész, P. Characterizing the drug-release enhancement effect of surfactants on Megestrol-acetate-loaded granules. *Pharmaceuticals* **2022**, *15*, 113. [CrossRef]
53. Aswal, V.K.; Chatterjee, S.; Bhattacharyya, J. In vitro solubilization of antibiotic drug sulfamethazine: An investigation on drug–micelle aggregate formation by spectroscopic and scattering techniques. *J. Surfactants Deterg.* **2021**, *25*, 331–339.
54. Zhan, X.; Wu, Z.; Chen, Z.; Cui, X. Mechanism of the micellar solubilization of curcumin by mixed surfactants of SDS and Brij35 via NMR spectroscopy. *Molecules* **2022**, *27*, 5032. [CrossRef] [PubMed]
55. Aliabadi, A.; Hasannia, M.; Vakili-Azghandi, M.; Araste, F.; Abnous, K.; Taghdisi, S.M.; Ramezani, M.; Alibolandi, M. Synthesis approaches of amphiphilic copolymers for spherical micelles preparation: Application in drug delivery. *J. Mater. Chem. B* **2023**, *11*, 9325–9368. [CrossRef]
56. Correia, E.L.; Brown, N.; Ervin, A.; Papavassiliou, D.V.; Razavi, S. Contamination in sodium dodecyl sulfate solutions: Insights from the measurements of surface tension and surface rheology. *Langmuir* **2022**, *38*, 7179–7189. [CrossRef]
57. Mai, N.N.S.; Otsuka, Y.; Kawano, Y.; Hanawa, T. Preparation and characterization of solid dispersions composed of curcumin, hydroxypropyl cellulose and/or sodium dodecyl sulfate by grinding with vibrational ball milling. *Pharmaceuticals* **2020**, *13*, 383. [CrossRef]
58. Shoaib, T.; Ha, J.-M.; Han, Y.; Chen, W.-R.; Do, C. SANS characterization of time dependent, slow molecular exchange in an SDS micellar system. *Phys. Chem. Chem. Phys.* **2022**, *24*, 16988–16996. [CrossRef]
59. Na, M.; Kim, K.; Oh, K.; Choi, H.J.; Ha, C.; Chang, S. Sodium Cholate-Based Active Delipidation for Rapid and Efficient Clearing and Immunostaining of Deep Biological Samples. *Small Methods* **2022**, *6*, 2100943. [CrossRef] [PubMed]
60. Cedillo-Cruz, E.; Garcia-Ramos, D.; Dominguez, H. Lead and mercury removal from aqueous solution using Sodium Dodecyl Sulfate micelles: A molecular dynamics study. *Chem. Phys. Lett.* **2021**, *767*, 138340. [CrossRef]
61. Poerschmann, J.; Weiner, B.; Baskyr, I. Organic compounds in olive mill wastewater and in solutions resulting from hydrothermal carbonization of the wastewater. *Chemosphere* **2013**, *92*, 1472–1482. [CrossRef]
62. Yarimtepe, C.C.; Oz, N.A.; Ince, O. Volatile fatty acid production dynamics during the acidification of pretreated olive mill wastewater. *Bioresour. Technol.* **2017**, *241*, 936–944. [CrossRef]
63. Deriszadeh, A.; Husein, M.M.; Harding, T.G. Produced Water Treatment by Micellar-Enhanced Ultrafiltration. *Environ. Sci. Technol.* **2010**, *44*, 1767–1772. [CrossRef] [PubMed]
64. Mandić, V.; Simić, A.; Petričević, M.; Gogić, M.; Brankov, M.; Stanojković, A. Effect of pH on germination and seedling growth of maize. *Biotechnol. Anim. Husb.* **2023**, *39*, 195–203. [CrossRef]
65. Vougogiannopoulou, K.; Angelopoulou, M.T.; Pratsinis, H.; Grougnet, R.; Halabalaki, M.; Kletsas, D.; Deguin, B.; Skaltsounis, L.A. Chemical and biological investigation of Olive Mill Waste Water–OMWW secoiridoid lactones. *Planta Med.* **2015**, *81*, 1205–1212. [CrossRef]
66. Delisi, R.; Saiano, F.; Pagliaro, M.; Ciriminna, R. Quick assessment of the economic value of olive mill waste water. *Chem. Cent. J.* **2016**, *10*, 1–5. [CrossRef]
67. Mekersi, N.; Kadi, K.; Hackenberger, D.K.; Casini, S.; Addad, D.; Thangaraj, R.; Chen, Z.; Zhang, C.; Yuvaraj, A. Accumulation of heavy metals from single and combined olive mill wastewater and pomace in soil and bioaccumulation in tissues of two earthworm species: Endogeic (Aporrectodea trapezoides) and Epigeic (Eisenia fetida). *Environ. Sci. Pollut. Res.* **2024**, *31*, 1–15. [CrossRef] [PubMed]
68. Khalil, J.; Jaafar, A.A.K.; Habib, H.; Bouguerra, S.; Nogueira, V.; Rodríguez-Seijo, A. The impact of olive mill wastewater on soil properties, nutrient and heavy metal availability–A study case from Syrian vertisols. *J. Environ. Manag.* **2024**, *351*, 119861. [CrossRef]
69. Chavez-Martinez, E.H.; Cedillo-Cruz, E.; Dominguez, H. Adsorption of metallic ions from aqueous solution on surfactant aggregates: A molecular dynamics study. *arXiv* **2021**, arXiv:2106.13141. [CrossRef]
70. Chen, S.B.; Ma, Y.B.; Chen, L.; Xian, K. Adsorption of aqueous Cd2+, Pb2+, Cu2+ ions by nano-hydroxyapatite: Single-and multi-metal competitive adsorption study. *Geochem. J.* **2010**, *44*, 233–239. [CrossRef]
71. Furukawa, M.; Tateishi, I.; Katsumata, H.; Kaneco, S. Application of sodium dodecyl sulfate/activated carbon onto the preconcentration of cadmium ions in solid-phase extraction flow system. *ChemEngineering* **2019**, *3*, 67. [CrossRef]
72. Ohfuchi, M. Ab Initio Study on Electronic Sorting of Single-Wall Carbon Nanotubes Using Sodium Dodecyl Sulfate. *J. Phys. Chem. C* **2018**, *122*, 4691–4697. [CrossRef]

73. Wang, X.; Qi, F.; Xiong, J.; Zhao, J.; Zhang, G.; Afzal, S.; Gu, X.; Li, Q.; Luo, S.; Mo, H. Synthesis of a Novel Dithiocarbamate Surfactant Derivative Adsorbent for Efficient Removal of Heavy Metal Ions. *ACS Omega* **2023**, *8*, 41512–41522. [CrossRef] [PubMed]
74. Wang, Z.; Wang, Y.; Zhang, J.; Qu, D.; Liu, X. Grey correlation analysis of corrosion on oil atmospheric distillation equipment. In Proceedings of the 2008 Fifth International Conference on Fuzzy Systems and Knowledge Discovery (FSKD), Shandong, China, 18–20 October 2008; Volume 5, pp. 13–17.
75. Kim, H.; Baek, K.; Lee, J.; Iqbal, J.; Yang, J.-W. Comparison of separation methods of heavy metal from surfactant micellar solutions for the recovery of surfactant. *Desalination* **2006**, *191*, 186–192. [CrossRef]
76. Wołowicz, A.; Staszak, K. Study of surface properties of aqueous solutions of sodium dodecyl sulfate in the presence of hydrochloric acid and heavy metal ions. *J. Mol. Liq.* **2020**, *299*, 112170. [CrossRef]
77. Mekki, A.; Dhouib, A.; Aloui, F.; Sayadi, S. Olive wastewater as an ecological fertiliser. *Agron. Sustain. Dev.* **2006**, *26*, 61–67. [CrossRef]
78. Ntougias, S.; Bourtzis, K.; Tsiamis, G. The microbiology of olive mill wastes. *BioMed Res. Int.* **2013**, *2013*, 784591. [CrossRef]
79. Khdair, A.I.; Abu-Rumman, G.; Khdair, S.I. Pollution estimation from olive mills wastewater in Jordan. *Heliyon* **2019**, *5*, e02386. [CrossRef]
80. Hachicha, S.; Chtourou, M.; Medhioub, K.; Ammar, E. Compost of poultry manure and olive mill wastes as an alternative fertilizer. *Agron. Sustain. Dev.* **2006**, *26*, 135–142. [CrossRef]
81. Charistou, A.; Kyriakopoulou, K.; Anastasiadou, P.; Fokialakis, N.; Skaltsounis, A.L.; Machera, K. Acute toxicity of Olive Mill Wastewater on rats, Vibrio fischeri and Artemia fransiscana. *Hell. Plant Prot. J.* **2010**, *3*, 17–24.
82. Babić, S.; Malev, O.; Pflieger, M.; Lebedev, A.T.; Mazur, D.M.; Kužić, A.; Čož-Rakovac, R.; Trebše, P. Toxicity evaluation of olive oil mill wastewater and its polar fraction using multiple whole-organism bioassays. *Sci. Total Environ.* **2019**, *686*, 903–914. [CrossRef]
83. Gallazzi, M.; Festa, M.; Corradino, P.; Sansone, C.; Albini, A.; Noonan, D.M. An Extract of Olive Mill Wastewater Downregulates Growth, Adhesion and Invasion Pathways in Lung Cancer Cells: Involvement of CXCR4. *Nutrients* **2020**, *12*, 903. [CrossRef]
84. Schaffer, S.; Müller, W.E.; Eckert, G.P. Cytoprotective effects of olive mill wastewater extract and its main constituent hydroxytyrosol in PC12 cells. *Pharmacol. Res.* **2010**, *62*, 322–327. [CrossRef] [PubMed]
85. Benedetto, N.; Calabrone, L.; Gutmańska, K.; Macrì, N.; Cerrito, M.G.; Ricotta, R.; Pelosi, G.; Bruno, A.; Noonan, D.M.; Albini, A. An Olive Oil Mill Wastewater Extract Improves Chemotherapeutic Activity Against Breast Cancer Cells While Protecting from Cardiotoxicity. *Front. Cardiovasc. Med.* **2022**, *9*, 867867. [CrossRef] [PubMed]
86. Posadino, A.M.; Cossu, A.; Giordo, R.; Piscopo, A.; Abdel-Rahman, W.M.; Piga, A.; Pintus, G. Antioxidant properties of olive mill wastewater polyphenolic extracts on human endothelial and vascular smooth muscle cells. *Foods* **2021**, *10*, 800. [CrossRef] [PubMed]
87. Schlupp, P.; Schmidts, T.M.; Pössl, A.; Wildenhain, S.; Lo Franco, G.; Lo Franco, A.; Lo Franco, B. Effects of a phenol-enriched purified extract from olive mill wastewater on skin cells. *Cosmetics* **2019**, *6*, 30. [CrossRef]
88. Dick, M.K.; Miao, J.H.; Limaiem, F. *Histology, Fibroblast*; StatPearls Publishing: Treasure Island, FL, USA, 2023.
89. Nkuna, S.G.; Olwal, T.O.; Chowdhury, S.P.D.; Ndambuki, J.M. A review of wastewater sludge-to-energy generation focused on thermochemical technologies: An improved technological, economical and socio-environmental aspect. *Clean. Waste Syst.* **2024**, *7*, 100130. [CrossRef]
90. Kumar, L.; Hasanuzzaman, M.; Rahim, N.A. Global advancement of solar thermal energy technologies for industrial process heat and its future prospects: A review. *Energy Convers. Manag.* **2019**, *195*, 885–908. [CrossRef]
91. Juang, R.-S.; Xu, Y.-Y.; Chen, C.-L. Separation and removal of metal ions from dilute solutions using micellar-enhanced ultrafiltration. *J. Membr. Sci.* **2003**, *218*, 257–267. [CrossRef]
92. Geanta, R.M.; Olga Ruiz, M.; Escudero, I. Micellar-enhanced ultrafiltration for the recovery of lactic acid and citric acid from beet molasses with sodium dodecyl sulphate. *J. Membr. Sci.* **2013**, *430*, 11–23. [CrossRef]
93. Qu, Y.-H.; Zeng, G.-M.; Huang, J.-H.; Xu, K.; Fang, Y.-Y.; Li, X.; Liu, H.-L. Recovery of surfactant SDS and Cd^{2+} from permeate in MEUF using a continuous foam fractionator. *J. Hazard. Mater.* **2008**, *155*, 32–38. [CrossRef]
94. Gómez, J.; Alcántara, M.T.; Pazos, M.; Sanromán, M.A. Remediation of polluted soil by a two-stage treatment system: Desorption of phenanthrene in soil and electrochemical treatment to recover the extraction agent. *J. Hazard. Mater.* **2010**, *173*, 794–798. [CrossRef]
95. Mosmann, T. Rapid colorimetric assay for cellular growth and survival: Application to proliferation and cytotoxicity assays. *J. Immunol. Methods* **1983**, *65*, 55–63. [CrossRef] [PubMed]

MDPI AG
Grosspeteranlage 5
4052 Basel
Switzerland
Tel.: +41 61 683 77 34

Molecules Editorial Office
E-mail: molecules@mdpi.com
www.mdpi.com/journal/molecules

www.ingramcontent.com/pod-product-compliance
Lightning Source LLC
Chambersburg PA
CBHW062035220926
51850CB00010B/2552
9783725863686